AF389371

The Effect of Calorie Restriction and Intermittent Fasting on Health and Disease

The Effect of Calorie Restriction and Intermittent Fasting on Health and Disease

Editor

Hae-Young Chung

MDPI • Basel • Beijing • Wuhan • Barcelona • Belgrade • Manchester • Tokyo • Cluj • Tianjin

Editor
Hae-Young Chung
Pusan Nation University
Korea

Editorial Office
MDPI
St. Alban-Anlage 66
4052 Basel, Switzerland

This is a reprint of articles from the Special Issue published online in the open access journal *Nutrients* (ISSN 2072-6643) (available at: https://www.mdpi.com/journal/nutrients/special_issues/Calorie_Restriction).

For citation purposes, cite each article independently as indicated on the article page online and as indicated below:

LastName, A.A.; LastName, B.B.; LastName, C.C. Article Title. *Journal Name* **Year**, *Volume Number*, Page Range.

ISBN 978-3-03943-837-2 (Hbk)
ISBN 978-3-03943-838-9 (PDF)

© 2020 by the authors. Articles in this book are Open Access and distributed under the Creative Commons Attribution (CC BY) license, which allows users to download, copy and build upon published articles, as long as the author and publisher are properly credited, which ensures maximum dissemination and a wider impact of our publications.

The book as a whole is distributed by MDPI under the terms and conditions of the Creative Commons license CC BY-NC-ND.

Contents

About the Editor

Hae-Young Chung

Education and Training

1994–1995   Post-Doc. in Gerontology
                University of Texas Health Science Center, Texas, U.S.A.
1984–1988   Ph.D. in Biochemistry
                Toyama University, Toyama, Japan
1980–1982   M.S. in Pharmacology
                Seoul National University, Seoul, Korea
1976–1980   B.S. in Pharmacy
                Pusan National University, Busan, Korea

Professional Experiences

2017.9–Present   Distinguished Professor
                Department of Pharmacy,
                College of Pharmacy Pusan National University, Busan, Korea
2006.12–Present   Member of Korean Academy of Science and Technology, Korea
1989.4–Present   Professor
                Department of Pharmacy,
                College of Pharmacy Pusan National University, Busan, Korea
2009.9–2018.10   Director
                Molecular Inflammation Research Center for Aging Intervention (MRCA), Busan, Korea
2011.1–2011.12   President
                The Korean Society for Gerontology, Korea
2004.3–2006.2   Dean
                College of Pharmacy, Pusan National University, Busan, Korea

 *nutrients*

Editorial

Impacts of Calorie Restriction and Intermittent Fasting on Health and Diseases: Current Trends

Hae Young Chung [1],*, Dae Hyun Kim [1], EunJin Bang [1] and Byung Pal Yu [2]

[1] Department of Pharmacy, College of Pharmacy, Pusan National University, 2, Busandaehak-ro 63 beon-gil, Geumjeong-gu, Busan 46241, Korea; bioimmune@hanmail.net (D.H.K.); eunjin2285@gmail.com (E.B.)

[2] Department of Physiology, The University of Texas Health Science Center at San Antonio, San Antonio, TX 78229, USA; yu6936@sbcglobal.net

* Correspondence: hyjung@pusan.ac.kr; Tel.: +82-51-510-2814

Received: 4 September 2020; Accepted: 14 September 2020; Published: 25 September 2020

This special issue on the effects of calorie restriction (CR) and intermittent fasting (IF) on health and diseases includes five scholarly reviews and four original articles that provide an insight into the molecular and cellular action mechanisms of epigenetically manipulated dietary paradigms.

CR and IF are well publicized methods for health promotion and disease prevention. An interesting aspect shared by both is that their efficacy is demonstrable without modification of any dietary ingredient. CR, also often called as restriction in dietary intake, provides dietary allocation (generally 10–50% of ad libitum feeding) to the experimental group [1]. Currently, gerontologists have acknowledged CR as the only well-founded anti-aging experimental model and the gold standard in investigating aging interventions. CR intervention delays biological aging and attenuates many age-related pathological processes, thus extending both average and maximal lifespan. Notably, CR is well-known to modulate adulthood diseases, such as atherosclerosis, diabetes, obesity, cancer, and other metabolically related disorders [2].

One unique feature of CR, first proposed as growth retardation in 1935 [3], is its well-known effect on various species, such as rats, mice, hamsters, dogs, yeast, fish, flies, monkeys, and humans [4]. Although many hypotheses for CR diverse actions have been proposed previously, at present, the precise underlying mechanisms remain to be proven. One attractive and plausible mechanism is its highly effective anti-oxidative action against oxidative stress-induced aging. The anti-aging effects of CR are likely attributable to its ability to prevent ubiquitous redox stress by maintaining a homeostatic cellular oxidative status and an antioxidative defense system that contributes to appropriate cellular signaling transductions and proper gene transcription activity [5]. One of the oxidative stress-related physiological processes is the activation of inflammation. The key pro-inflammatory molecules, IL-1β, IL-6, TNFα, iNOS, and COX-2 are all alleviated by CR during aging [6]. CR implementation leads to epigenetic alterations, such as histone modifications and methylation of promoter CpG islands, resulting in epigenetically modified markers in histones and chromatin. This established a molecular mechanism for gene transcription and expression regulation [7,8]. In this regard, lymphocyte-specific protein tyrosine kinase as a new target molecule of the CR modulatory effect was found by integrative analysis using cDNA microarray and interactome [9].

Recent progress in Omics technologies has contributed to technical improvement that allows analysis at the level of the RNA, DNA, protein, and other cellular molecules and their intricate associations during aging. Therefore, it is now feasible to both resolve and coordinate complex biological processes during aging using recent methods, such as next-generation sequencing, proteomics, lipidomics, metabolomics, and epigenomics techniques. Systems biology and Omics integration can predict CR effects, cellular signaling pathways, and action mechanisms. These technical advances made in recent years provide a much-needed database on the regulation of inflammatory process. An expanded view of the inflammatory response in aging progression dubbed as "senoinflammation", which describes both chronic inflammation and metabolic dysregulation, was suggested in our previous work [10]. Based on previous studies, we observed that the senescence-associated secretory phenotype (SASP) comprises cytokines and chemokines, and these were notably upregulated during aging. This senoinflammatory response was reversed and downregulated by CR [11]. Moreover, metabolic signaling pathways at the cell level were also dysregulated in aging progression, and CR mimetics could reverse such effects. According to CR and Omics big data analysis, cytokines and chemokine pathways are upregulated during aging and modulated by CR [12]. CR is widely accepted as a positive and reasonable control for anti-aging intervention, which recovers disturbed metabolic pathways and decreases the pro-inflammatory SASP [11].

IF is a dietary pattern alternating between fasting and non-fasting periods. Specifically, the IF diet in a particular mouse strain extended both mean and maximal lifespans [13]. For example, loss of body weight and extension of maximal lifespan were observed in 2-, 6-, and 10-month-old C57BL/6J mice with IF implementation [14]. Furthermore, IF lowered the occurrence of diabetes and levels of fasting glucose and insulin [15]. These effects of IF are similar to those observed with CR. The beneficial results of IF on different cancers are also explained by many research groups [16]. The observations in animals indicate that, owing to dietary intake reduction, IF could effectively regulate the number of risk factors and thereby prevent chronic diseases. Such modulatory effects of IF are similar to those of CR. Although the effectiveness of IF has been reported primarily based on in vivo models, implementation of IF in humans has been proposed to prevent major risk factors for age-associated diseases [17]. Systems gerontology (i.e., the study on the overall mechanisms of complex aging progression as an integrated network of systemic interacting molecular pathways, organs, and components during aging) based on Omics technology and systems biology should facilitate unraveling of mechanisms underlying the anti-aging actions of CR and IF as well as their similarities and dissimilarities for better dietary strategies.

The growth hormone (GF)/insulin-like growth factor-1 (IGF-1) axis is an evolutionary well-conserved signaling axis, and the mechanisms underlying its effects on aging and longevity are modulated by CR in mammals [18]. Recently, Shimokawa's group confirmed that neuropeptide Y (Npy) and FoxO1, which are targets of CR, play various roles in aging and age-associated diseases depending on the nutritional intake [19]. The longevity effects of CR and suppression of IGF-1 signaling pathway are sexually heterogeneous. The beneficial effects of Npy during CR and the diverse roles for Npy in life stages are also described. These genes that mediate the effects of CR and control the aging process depend on nutritional states [19]. As described in a review by Higami's group, the mitochondrial biogenesis, following CR implementation, in white adipose tissue is regulated by peroxisome proliferator-activated receptor γ coactivator-1α (PGC-1α), a transcription cofactor regulated by sterol regulatory element-binding protein 1c (SREBP-1c), which is the principal regulator of fatty acid biosynthesis [20]. This research group has also suggested that CR mediates the increase in SREBP-1 and PGC-1α, which could occur because of downregulation of the leptin signaling axis and upregulation of fibroblast growth factor 21 (FGF21) expression in white adipose tissue [21].

Another cellular activity recently that was revealed as one of the major players in CR action is the involvement of autophagy. Autophagy is an important cellular management process that maintains an optimal cellular homeostatic condition under normal physiological and abnormal pathological states. Recent developments in CR research have proposed that the initiation of autophagy is associated with beneficial anti-aging response. CR-promoted autophagy plays a critical role under physiological states by sustaining proper homeostasis in the organism. In addition, autophagy induced by CR plays a protective role in different organs and tissues under many pathological and aging-associated conditions [22]. While CR is the most beneficial intervention to prolong lifespan and prevent aging-related diseases, its positive effects on skin aging and diseases are not yet fully understood. CR promotes anti-inflammatory and anti-oxidative responses, stem cell maintenance, and metabolic activities, and such effects contribute to the positive effects on skin aging [23]. Furthermore, widely known CR mimetics, for example, resveratrol, metformin, rapamycin, and peroxisome proliferator-activated receptor agonists, exhibit CR-like effects to inhibit or delay skin aging [23].

Fasting is known to account for physiological changes in the endocrine organs, particularly the pancreas. The study on every other day IF implementation for up to 3 months in a developing and matured healthy organism suggested that this regimen promotes β-cell dysfunction and muscle mass loss and increases fat reserves, especially in developing 30-day-old Wistar rats. More long-term studies are needed to explain the most effective IF regimen to minimize side effects [24]. In recent years, fasting-like intervention methods, such as IF and time-restricted feeding, have appeared as alternatives to CR. Lifespan responses in both CR and IF also differ significantly between males and females [25,26], leading to additional complications in explaining the molecular mechanisms of CR. A simple explanation of these in vivo studies is that a particular regimen of CR and IF may not be effective; however, they may be rather harmful possibly owing to genetic variations and sex [27]. Therefore, for implementation and practice of CR and IF in humans, it is suggested that personalized genomics and medicine should first be set in place to make the most of CR and IF [28].

In summary, we believe that this special issue provides useful information on the current research trends regarding the beneficial effects of CR and IF on longevity and health of humans. Further exploration of the effect of CR and IF on aging and age-related diseases will provide deeper insights into the interrelationships between health and diseases and enable beneficial interventions in aging mechanisms, thus aiding the development of new therapeutic approaches to improve health, diseases, and longevity.

Author Contributions: All authors contributed equally to the development and finalization of this Editorial. All authors have read and agreed to the published version of the manuscript.

Funding: This work was supported by the National Research Foundation of Korea (NRF) grant funded by the Korea government (MSIT) (2018R1A2A3075425).

Acknowledgments: We thank Aging Tissue Bank (Busan, Korea) for providing research information.

Conflicts of Interest: The authors declare no conflict of interest.

References

1. Simpson, S.J.; Le Couteur, D.G.; Raubenheimer, D.; Solon-Biet, S.M.; Cooney, G.J.; Cogger, V.C.; Fontana, L. Dietary protein, aging and nutritional geometry. *Ageing Res. Rev.* **2017**, *39*, 78–86. [CrossRef] [PubMed]

2. Devarajan, A.; Rajasekaran, N.S.; Valburg, C.; Ganapathy, E.; Bindra, S.; Freije, W.A. Maternal perinatal calorie restriction temporally regulates the hepatic autophagy and redox status in male rat. *Free Radic. Biol. Med.* **2019**, *130*, 592–600. [CrossRef] [PubMed]

3. McCay, C.M.; Crowell, M.F.; Maynard, L.A. The effect of retarded growth upon the length of life span and upon the ultimate body size. *J. Nutr.* **1935**, *10*, 63–79. [CrossRef]

4. Bordone, L.; Guarente, L. Calorie restriction, SIRT1 and metabolism: Understanding longevity. *Nat. Rev. Mol. Cell Biol.* **2005**, *6*, 298–305. [CrossRef]

5. Yu, B.P.; Chung, H.Y. The inflammatory process in aging. *Rev. Clin. Gerontol.* **2006**, *16*, 179–187.

6. Kim, D.H.; Kim, J.Y.; Yu, B.P.; Chung, H.Y. The activation of NF-κB through Aktinduced FoxO1 phosphorylation during aging and its modulation by calorie restriction. *Biogerontology* **2008**, *9*, 33–47. [CrossRef]

7. Kim, C.H.; Lee, E.K.; Choi, Y.J.; An, H.J.; Jeong, H.O.; Park, D.; Kim, B.C.; Yu, B.P.; Bhak, J.; Chung, H.Y. Short-term calorie restriction ameliorates genomewide, age-related alterations in DNA methylation. *Aging Cell* **2016**, *15*, 1074–1081. [CrossRef]

8. Stenvinkel, P.; Karimi, M.; Johansson, S.; Axelsson, J.; Suliman, M.; Lindholm, B.; Heimbürger, O.; Barany, P.; Alvestrand, A.; Nordfors, L.; et al. Impact of inflammation on epigenetic DNA methylation—A novel risk factor for cardiovascular disease? *J. Intern. Med.* **2007**, *261*, 488–499. [CrossRef]

9. Park, D.; Lee, E.K.; Jang, E.J.; Jeong, H.O.; Kim, B.C.; Ha, Y.M.; Hong, S.E.; Yu, B.P.; Chung, H.Y. Identification of the dichotomous role of age-related LCK in calorie restriction revealed by integrative analysis of cDNA microarray and interactome. *Age* **2013**, *35*, 1045–1060. [CrossRef]

10. Chung, H.Y.; Kim, D.H.; Lee, E.K.; Chung, K.W.; Chung, S.; Lee, B.; Seo, A.Y.; Chung, J.H.; Jung, Y.S.; Im, E.; et al. Redefining Chronic Inflammation in Aging and Age-Related Diseases: Proposal of the Senoinflammation Concept. *Aging Dis.* **2019**, *10*, 367–382. [CrossRef]

11. Kim, D.H.; Bang, E.; Arulkumar, R.; Ha, S.; Chung, K.W.; Park, M.H.; Choi, Y.J.; Yu, B.P.; Chung, H.Y. Senoinflammation: A major mediator underlying age-related metabolic dysregulation. *Exp. Gerontol.* **2020**, *134*, 110891. [CrossRef] [PubMed]

12. Bang, E.; Lee, B.; Noh, S.G.; Kim, D.H.; Jung, H.J.; Ha, S.; Yu, B.P.; Chung, H.Y. Modulation of senoinflammation by calorie restriction based on biochemical and Omics big data analysis. *BMB Rep.* **2019**, *52*, 56–63. [CrossRef] [PubMed]

13. Chung, K.W.; Kim, D.H.; Park, M.H.; Choi, Y.J.; Kim, N.D.; Lee, J.; Yu, B.P.; Chung, H.Y. Recent advances in calorie restriction research on aging. *Exp. Gerontol.* **2013**, *48*, 1049–1053. [CrossRef] [PubMed]

14. Goodrick, C.L.; Ingram, D.K.; Reynolds, M.A.; Freeman, J.R.; Cider, N. Effects of intermittent feeding upon body weight and lifespan in inbred mice: Interaction of genotype and age. *Mech. Ageing Dev.* **1990**, *55*, 69–87. [CrossRef]

15. Hsieh, E.A.; Chai, C.M.; Hellerstein, M.K. Effects of caloric restriction on cell proliferation in several tissues in mice: Role of intermittent feeding. *Am. J. Physiol. Endocrinol. Metab.* **2005**, *288*, E965–E972. [CrossRef]

16. Descamps, O.; Riondel, J.; Ducros, V.; Roussel, A.M. Mitochondrial production of reactive oxygen species and incidence of age-associated lymphoma in OF1 mice: Effect of alternate-day fasting. *Mech. Ageing Dev.* **2005**, *126*, 1185–1191. [CrossRef]

17. Varady, K.A.; Hellerstein, M.K. Alternate-day fasting and chronic disease prevention: A review of human and animal trials1'2'3. *Am. J. Clin. Nutr.* **2007**, *86*, 7–13. [CrossRef]

18. Chiba, T.; Yamaza, H.; Shimokawa, I. Role of insulin and growth hormone/insulin-like growth factor-I signaling in lifespan extension: Rodent longevity models for studying aging and calorie restriction. *Curr. Genom.* **2007**, *8*, 423–428. [CrossRef]

19. Komatsu, T.; Park, S.; Hayashi, H.; Mori, R.; Yamaza, H.; Shimokawa, I. Mechanisms of Calorie Restriction: A Review of Genes Required for the Life-Extending and Tumor-Inhibiting Effects of Calorie Restriction. *Nutrients* **2019**, *11*, 3068. [CrossRef]

20. Fujii, N.; Narita, T.; Okita, N.; Kobayashi, M.; Furuta, Y.; Chujo, Y.; Sakai, M.; Yamada, A.; Takeda, K.; Konishi, T.; et al. Sterol regulatory element-binding protein-1c orchestrates metabolic remodeling of white adipose tissue by caloric restriction. *Aging Cell* **2017**, *16*, 508–517. [CrossRef]

21. Kobayashi, M.; Uta, S.; Otsubo, M.; Deguchi, Y.; Tagawa, R.; Mizunoe, Y.; Nakagawa, Y.; Shimano, H.; Higami, Y. Srebp-1c/Fgf21/Pgc-1α Axis Regulated by Leptin Signaling in Adipocytes-Possible Mechanism of Caloric Restriction-Associated Metabolic Remodeling of White Adipose Tissue. *Nutrients* **2020**, *12*, 2054. [CrossRef] [PubMed]

22. Chung, K.W.; Chung, H.Y. The Effects of Calorie Restriction on Autophagy: Role on Aging Intervention. *Nutrients* **2019**, *11*, 2923. [CrossRef] [PubMed]

23. Choi, Y.J. Shedding Light on the Effects of Calorie Restriction and its Mimetics on Skin Biology. *Nutrients* **2020**, *12*, 1529. [CrossRef] [PubMed]

24. Munhoz, A.C.; Vilas-Boas, E.A.; Panveloski-Costa, A.C.; Leite, J.S.M.; Lucena, C.F.; Riva, P.; Emilio, H.; Carpinelli, A.R. Intermittent Fasting for Twelve Weeks Leads to Increases in Fat Mass and Hyperinsulinemia in Young Female Wistar Rats. *Nutrients* **2020**, *12*, 1029. [CrossRef] [PubMed]

25. Liao, C.Y.; Rikke, B.A.; Johnson, T.E.; Diaz, V.; Nelson, J.F. Genetic variation in the murine lifespan response to dietary restriction: From life extension to life shortening. *Aging Cell* **2010**, *9*, 92–95. [CrossRef] [PubMed]

26. Wilson, K.; Beck, J.; Nelson, C.; Hilsabeck, T.; Promislow, D.; Brem, R.; Kapahi, P. GWAS for Lifespan and Decline in Climbing Ability in Flies upon Dietary Restriction Reveal decima as a Mediator of Insulin-like Peptide Production. *Curr. Biol.* **2020**, *30*, 2749–2760. [CrossRef]

27. Di Francesco, A.; Di Germanio, C.; Bernier, M.; de Cabo, R. A time to fast. *Science* **2018**, *362*, 770–775. [CrossRef]
28. Hwangbo, D.S.; Lee, H.Y.; Abozaid, L.S.; Min, K.J. Mechanisms of Lifespan Regulation by Calorie Restriction and Intermittent Fasting in Model Organisms. *Nutrients* **2020**, *12*, 1194. [CrossRef]

© 2020 by the authors. Licensee MDPI, Basel, Switzerland. This article is an open access article distributed under the terms and conditions of the Creative Commons Attribution (CC BY) license (http://creativecommons.org/licenses/by/4.0/).

Review

Anti-Aging Effects of Calorie Restriction (CR) and CR Mimetics Based on the Senoinflammation Concept

Dae Hyun Kim [1], EunJin Bang [1], Hee Jin Jung [1], Sang Gyun Noh [1], Byung Pal Yu [2], Yeon Ja Choi [3,*] and Hae Young Chung [1,*]

[1] Department of Pharmacy, College of Pharmacy, Pusan National University, 2, Busandaehak-ro 63 beon-gil, Geumjeong-gu, Busan 46241, Korea; bioimmune@hanmail.net (D.H.K.); eunjin2285@gmail.com (E.B.); king2046@hanmail.net (H.J.J.); rskrsk92@naver.com (S.G.N.)
[2] Department of Physiology, The University of Texas Health Science Center at San Antonio, San Antonio, TX 78229, USA; yu6936@sbcglobal.net
[3] Department of Biopharmaceutical Engineering, Division of Chemistry and Biotechnology, Dongguk University, Gyeongju 38066, Korea
* Correspondence: yjchoi@dongguk.ac.kr (Y.J.C.); hyjung@pusan.ac.kr (H.Y.C.); Tel.: +82-51-510-2814 (H.Y.C.)

Received: 18 December 2019; Accepted: 3 February 2020; Published: 6 February 2020

Abstract: Chronic inflammation, a pervasive feature of the aging process, is defined by a continuous, multifarious, low-grade inflammatory response. It is a sustained and systemic phenomenon that aggravates aging and can lead to age-related chronic diseases. In recent years, our understanding of age-related chronic inflammation has advanced through a large number of investigations on aging and calorie restriction (CR). A broader view of age-related inflammation is the concept of senoinflammation, which has an outlook beyond the traditional view, as proposed in our previous work. In this review, we discuss the effects of CR on multiple phases of proinflammatory networks and inflammatory signaling pathways to elucidate the basic mechanism underlying aging. Based on studies on senoinflammation and CR, we recognized that senescence-associated secretory phenotype (SASP), which mainly comprises cytokines and chemokines, was significantly increased during aging, whereas it was suppressed during CR. Further, we recognized that cellular metabolic pathways were also dysregulated in aging; however, CR mimetics reversed these effects. These results further support and enhance our understanding of the novel concept of senoinflammation, which is related to the metabolic changes that occur in the aging process. Furthermore, a thorough elucidation of the effect of CR on senoinflammation will reveal key insights and allow possible interventions in aging mechanisms, thus contributing to the development of new therapies focused on improving health and longevity.

Keywords: aging; calorie restriction; senescence-associated secretory phenotype; senoinflammation; mimetics

---

1. Introduction

The aging process can be defined as progressive, physiological, functional deterioration throughout the lifetime of an individual by different convoluted interactions among genes and non-genetic environmental factors that eventually result in disruption of homeostasis and increased susceptibility to disease or death. The basic mechanism of the aging process is a sustained, long-term inflammatory state that is further aggravated by elevated oxidative stress due to enhanced reactive oxygen species (ROS), lipid peroxidation, and protein oxidative modifications [1]. To understand the phenomenon of aging and its significance, several concepts and terminologies have been proposed including inflammaging, molecular inflammation, micro-inflammation, pan-inflammation, and gero-inflammation. These concepts and terminologies are apposite for describing the increased chronic inflammatory events and mediators during aging [2–6].

Generally, cells are continually exposed to and damaged by exogenous and endogenous stress inducers. The cell cycle of damaged cells, which cannot be recovered from cell death, is permanently arrested and the proliferative activity of these cells becomes extinct, which is defined as cellular senescence. This cellular response largely contributes to an organism's aging. Senescent cells have been shown to release multiple inflammatory cytokines and chemokines, which is defined as senescence-associated secretory phenotype (SASP) [7]. An increase in cellular dysregulation due to the release of proinflammatory molecules such as TNF, IL-1β, IL-6, MCP-1, MIP-1α, RANTES, and IL-18 [8,9] induces age-related chronic inflammation, leading to aging and its associated diseases.

In order to understand age-related chronic inflammatory progress from a multilayered point of view, we previously proposed a novel concept of senoinflammation, which includes an expanded systemic view of chronic inflammation during the aging process.

CR is a well-known gold standard for many aging intervention strategies. A number of age-related biological changes and pathologic abnormalities can be delayed or suppressed by CR regardless of gender and species (mammalian or non-mammalian) [10]. CR has been shown to suppress oxidative damage-induced alterations and age-related diseases and to extend lifespan [11]. In this review, we focus on the diverse protective effects of CR against aging from a senoinflammatory perspective. In addition, the beneficial effects of CR mimetics and other types of dietary restrictions on anti-aging are covered.

2. Age-related Inflammation and Senoinflammation

Senescent cells produce proinflammatory senescence-associated (SA) secretome, which is referred to as the SASP. Macrophages are recruited in the secretome by chemotactic factors to clear senescent cells [12]. However, senescent macrophages with M2 polarized phenotype secrete proinflammatory cytokines and exhibit impaired phagocytosis and chemotaxis, and a downregulated rate of cellular proliferation [13–15]. It has been proposed that deficiency in the ability of aged macrophages to clear senescent cells leads to increased inflammatory response and results in chronic inflammation as SASP plays a role in the initiation of tissue inflammation [16]. Based on previous observations and evidence of the aging process at molecular and cellular levels, we coined the term "senescent inflammation (senoinflammation)" with a new framework (Figure 1) in our recent review to provide an expanded, broader view of age-related chronic inflammation and metabolic dysfunction [17].

Aging in hepatocytes is associated with various markers of cellular senescence, such as increased expression of heterochromatin protein 1β, and increased activity of senescence-associated-β-galactosidase, p21 and p16 [18]. p53 expression is an important marker of cellular senescence and DNA damage in the normal liver [19], and its regulation depends on the nutrient-sensing pathways of non-alcoholic fatty liver disease (NAFLD) [20]. Senescent hepatocytes exhibit increased lipid droplet accumulation and ROS production [21].

In understanding age-related inflammation at the molecular level, an abundance of data in our and other previous work strongly suggested that NF-κB is a key player involved in the initiation and exacerbation of tissue inflammation in the aging process and cancer [22–24]. Chronic transactivation of NF-κB has been observed in multiple tissues in various experimental models of aging and human fibroblasts and human CD4+ T cells obtained from aged individuals [25,26]. NF-κB induces proinflammatory mediators, chemokines, and adhesion molecules [27] and interacts with other transcriptional factors that are involved in the initiation and deterioration of chronic inflammatory response including signal transducers, the activator of transcription 3 (STAT3) and p53 [22]. The transcriptional activity of NF-κB occurs concurrently with crosstalk among upstream signaling components such as glycogen synthase kinase 3 (GSK3)-β, mitogen-activated protein kinase (MAPK), mammalian target of rapamycin (mTOR), and protein kinase B (PKB) [23,28].

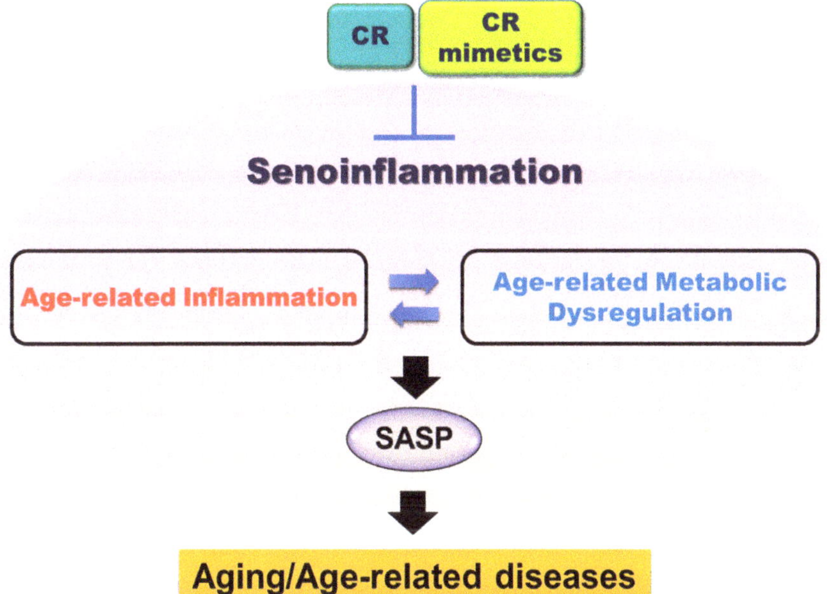

Figure 1. Possible mechanism of senoinflammation in aging. During aging, the continuous age-related inflammatory responses that cause metabolic dysregulation to form a vicious cycle. Age-related inflammation and metabolic dysregulation result in the induction and activation of many pro-inflammatory genes, metabolic and signaling pathways, and SASP. This vicious cycle of age-related chronic inflammation and metabolic dysregulation underlies the senoinflammation phenomenon, which occurs because of the interaction between senescent cells, immune cells, adipocytes, hepatocytes, and myocytes. SASP, senescence associated secretory phenotype; CR, calorie restriction.

The upregulation of systemic inflammation is associated with aging and age-related chronic diseases [29,30]. As mentioned, age-related systemic inflammation and senoinflammation are functionally distinct from acute inflammatory responses due to sustained high levels of pro-inflammatory mediators. In fact, epidemiological and experimental results suggest that persistent low-grade, chronic inflammation exists in aged animals [31,32]. A recent longitudinal, semi-supercentenarian study in Japan has demonstrated that inflammation, and not telomere length, strongly predicts successful aging at extremely old age [33]. This study concluded that chronic and systemic inflammation has a significant effect on mortality and had a correlation with the decline in cognitive function in the centenarians, thus showing that chronic inflammation is a critical risk factor in the aging process [33]. An increase in systemic inflammation is related to many aging phenotypes. For example, abruptly increased inflammation is generally associated with tissue dysfunction, metabolic syndrome, immune dysfunction, and neuronal complications [34].

CR animals live fairly longer with the right amount of nutrients, and most of the typical age-related chronic diseases are prevented or delayed in them. For example, the incidence of cancer, the main cause of death in rodents, is significantly reduced in CR animals. Similarly, reduced incidence or slower disease progression has been reported for cardiomyopathy, diabetes, chronic lung diseases, autoimmune diseases, and neurodegenerative diseases [35,36]. Preclinical and preliminary clinical studies have shown that CR or fasting can effectively prevent the development of malignant tumors through a variety of cellular responses and can improve the efficacy of therapeutics [37]. CR reverses most symptoms of immunosenescence, including decreased naïve T cell and increased memory T cell population [38], reduced T cell proliferative response to mitogens or antigens, reduced IL-2 production and NK activity [38–42], age-related increase in serum levels of TNFα and IL-6 [43], and autoimmune diseases [44,45]. More recently, CR was shown to delay immunosenescence in animals; however, this effect needs to be confirmed in humans.

Therefore, the concept of senoinflammation shows the orchestral performance of activated pro-inflammatory cytokines and transcription factors at a molecular level, immune cell senescence and SASP at a cellular level, and systemic inflammation and metabolic disorders at a systemic level during the aging process.

3. Calorie Restriction

In an initial study on CR by McCay et al. [46], the growth retardation hypothesis was investigated in a rat model by reducing food intake or CR, which slowed down the growth rate and prolonged lifespan. Various additional studies on CR have been conducted in diverse species ranging from yeast, fish, drosophila, hamsters, dogs, and non-human primates to humans [47]. Experts in the field of aging have accepted CR as an anti-aging experimental concept, which serves as an established aging intervention. As CR is a non-genetic, nutritional means to delay the aging process, it was used to identify underlying signaling mechanisms of aging, resulting in its considerable importance. Understanding elemental mechanisms underlying the effect of CR is critical as they may aid in identifying novel therapeutic molecular targets for age-associated inflammatory pathological conditions.

Previous studies conducted at the molecular level significantly support the hypothesis that CR is capable of reducing age-associated oxidative stress and suppressing systemic, chronic inflammation [48]. CR and its anti-aging effects are majorly considered due to its significant regulatory role in oxidative stress and capability to sustain appropriate cellular redox conditions [1]. CR also has beneficial effects in the inhibition of protein synthesis and the oxidization of proteins in the liver and skeletal muscle. Furthermore, it enhances immune functions and inhibits inflammatory responses during the aging process [49]. CR exerts preventive or delaying effects on age-associated diseases, such as chronic nephropathies, cardiomyopathies, diabetes, autoimmune conditions and respiratory diseases, as well as aging [48,50]. Implementation of CR in mice suppressed the degree of neurological degeneration and β-amyloid deposition in the brain tissue and subsequently promoted the generation of neurons in in vivo animal models of Alzheimer, Parkinson, and Huntington diseases [51,52]. The results of the first randomized clinical human trial on CR highlighted a reduced probability of developing age-related diseases and improved number of biomarkers showing health longevity.

4. Anti-senoinflammatory Effect of CR

Many interventions and strategies for modulating chronic inflammation and anti-aging have been scientifically demonstrated. Among many well-described anti-aging strategies, CR has been identified as one of the most powerful interventions to fight the aging process and age-related pathological conditions such as diabetes, obesity, cardiovascular diseases, rheumatoid arthritis, Alzheimer's disease and more [53]. Although the detailed molecular mechanisms and signaling pathways underlying CR still require further investigation, previous evidence for modulatory action of CR in senoinflammation suggest potential therapeutic effects of CR on aging (Table 1). For example, at a molecular level, CR exhibits powerful anti-inflammatory effects by suppressing key pro-inflammatory mediators such as NF-κB, IL-1β, IL-6, TNF, cyclooxygenase 2 (COX-2), and inducible nitric oxide synthase (iNOS) [54–56]. In addition, CR was shown to regulate the activity of pro-inflammatory upstream signaling pathway molecules such as MAPKs (ERK, JNK, p38), and NIK/IKKs. CR was also shown to regulate the DNA binding activity of NF-κB and AP-1 transcription factors and expression of their corresponding genes, *COX-1* and *iNOS* [55,57]. CR was also shown to reduce the plasma concentration of cytokines, TNF, ICAM-1 and to induce cortisol release, which suppresses the systemic inflammatory response [58,59]. In obese mice models, implementation of 30% CR for 2 months notably decreased the levels of adipose tissue cytokines and chemokines, including IL-6, IL-2, IL-1Rα, MCP-1, and CXCL16, which are considered as major components of SASP [60]. In hepatic tissue, even mild CR notably suppressed proinflammatory and lipogenic gene expression of molecules such as MCP-1, SREBPs, and peroxisome proliferator-activated receptor (PPAR)-γ [61]. These evidences suggest that CR successfully regulates the symptomatic prevalence of senoinflammation that expands to pathological conditions such as chronic inflammation, insulin resistance, and low energy metabolism [17,58,62,63].

Regarding anti-aging effects, CR is known to play an important role in suppressing oxidative stress and damage [64,65]. Cellular oxidative stress leads to formation of ROS, hydrogen peroxide, reactive nitrogen species, peroxynitrites, which then induce cellular inflammation, damage, and senescence. CR exerts its beneficial, maximal life-spanning effects by partially attenuating oxidative stress. For example, age-dependent functional decline of mitochondria in cardiac tissue, a major organelle of ROS production, has been well documented. Additionally, it was demonstrated that CR attenuates oxidative damage in an aged heart by lowering the levels of 8-oxodG, an oxidative damage DNA biomarker [66].

In addition to the anti-inflammatory effects described above, CR is also well known for regulating the expression of various genes involved in regulating energy metabolism. In regulating lipid metabolism, the PPARs could sense fatty acid molecules released from dietary lipids and their metabolites. PPARs are specialized receptors that recognize and bind lipid metabolites to transmit signals and can regulate lipid and carbohydrate metabolisms and inflammation. Among three subtypes of PPARs, PPARα and PPARγ have been well investigated and both have been suggested as regulators of inflammatory responses. In an aged rat model, the expression of *PPARα* and *PPARγ* genes was decreased and age-associated alterations were reversed by CR [63,67]. In a previous review, it has been noted that suppression of PPAR activity leads to upregulation of cytosolic IκB and NF-κB inhibitor, and suppression of NF-κB activation [68]. Such experimental evidence further strengthens the fact that PPARα agonists could alleviate age-related inflammation by suppressing NF-κB-mediated proinflammatory cytokine production [67,69].

CR modulates nutrient-signaling pathway molecules such as sirtuin proteins. One major molecule known to exert its effects in delaying aging and increasing longevity during CR is SIRT1 [70]. Sirtuins regulate protein expression in diverse cellular processes such as DNA repair, epigenetic modification of chromatin, ROS production, and metabolism. CR is well known to promote SIRT expression and activation in the liver, adipose tissue, brain and kidney by interacting with FOXOs, PGC1α, p53 and NF-κB to mediate anti-aging effects [71]. CR-mediated SIRT1 activity regulates pro-inflammatory NF-κB activation. For example, SIRT1 induces deacetylation and suppresses NF-κB activation [72,73].

Diverse research has provided an understanding of the association between aging and CR and the effects of CR on senoinflammatory and metabolic signaling pathways. The experimental evidence suggests that CR exerts beneficial effects on senoinflammation during aging by altering molecular pathways through regulation of expression and activities of core molecules such as NF-κB, PPARs, SIRT1, and others. Collective evidence on CR further supports the concept of senoinflammation during the aging process and confirms the positive role of CR against aging. In addition, the evidence strongly supports the notion that the anti-aging effects of CR are due to the alleviation of systematic physiological senoinflammatory response. However, further research is needed to clearly define the signaling mechanisms in detail.

5. Omics Big Data on Aging and CR

The immense amount of collected data in the field of biology and biomedicine research necessitates integrative data analysis to understand a complicated physiological system as a whole. Integrative dataset analysis has also provided an understanding of the underlying mechanism of aging and age-dependent changes at molecular, cellular, and physiological levels. An immense amount of data on age-related diseases enables the building of interactive networks and alterations in these networks may aid in developing aging intervention methods. Transcriptomics is the study of complete sets of RNA transcripts of a whole genome under certain conditions, which includes analysis of comparative differential gene expression in response to different conditions. As the biological aging process is complex and heterogeneous, defining specific mechanism of aging and a potential intervention method such as CR requires data integrative analysis based on age-dependent changes at the molecular level.

Among Omics Big Data analysis, gene set enrichment analysis (GSEA) allows the identification of gene types that are significantly associated with the disease. In order to analyze young and aged groups, RNA-sequencing data were collected and analyzed by the GSEA method to detect the set of differentially expressed genes or DEGs. Based on the results, it was demonstrated that proinflammatory functional genes, including cytokines, chemokines, TNF and toll-like receptors (TLRs), were notably over-represented in the aged group, whereas genes associated with metabolism, such as fatty acid metabolism and PPAR pathways, were significantly suppressed. This suggests that aging is highly associated with gene expression alterations including increased expression of inflammatory genes and decreased expression of metabolic genes. In addition, it was found that these changes were reversed by CR, a well-known aging intervention method [74].

By detecting DEGs, the Omics Big Data analysis method can identify informative epigenetic modifications along with the changes in gene expression and potential biomarkers of aging and age-related pathologic conditions. The results of a previous epigenetic study, performed using collective genomic data program called The Cancer Genome Atlas (TCGA) Program, indicated various age-related changes in the pattern of DNA hypomethylation in young and old subjects. In particular, the genes that were upregulated and hypomethylated were *AZU1*, *ELF3*, *NOX1*, *IL1B*, and *S100A12*; these genes are known to function in inflammatory responses, indicating that inflammatory genes are involved in age-related cancer onset and progression [75]. A previous briefing paper discussed the idea of big data in the field of medicine and suggested that a big network constructed by multi-omics data such as epigenomics, transcriptomics, and metabolomics can detect a significant association existing between aging and inflammation, indicating the use of a systems biology tool to identify new genes involved in inflammaging [76].

Omics Big Data is a useful tool to investigate the beneficial effects of CR. The transcriptome profile of mice liver tissue obtained after implementation of CR showed that CR improved the expression of genes associated with health, which were previously modulated by obesity. These data and other research studies collectively indicate that CR promotes beneficial outcomes leading to the expansion of life span [77]. In another study, transcriptomic data of adipose tissue indicated that CR suppresses transcription and activity of genes involved in inflammatory response, for example, the NF-κB signaling molecule gene. Additionally, in that study, diverse evidence indicated that CR has protective effects in physiological systems [78]. Furthermore, based on a global mass spectrometry-based metabolomics study, graded CR (10, 20, 30, 40% CR) modulated metabolic signaling pathways, including the carnitine synthesis and shuttle pathway and sphingosine-1-phosphate and methionine metabolism, in a graded manner. The expression of various metabolites was modulated by CR, indicating that CR could ameliorate the energy release process of hepatic fatty acids [79]. In addition, augmented gene–gene network connectivity analysis has shown that CR changed the network arrangement and biological gene centrality in a CR level-dependent manner. Therefore, the results suggested that CR-induced genes play a critical role in countering the age-associated loss of gene–gene network connectivity [80]. In order to better understand the mechanism of CR in age-associated pathological symptoms, its relation to insulin sensitivity has also been investigated. Multi-omics approaches that integrate transcriptomics, metabolomics, and microbiomics data were used to further enhance our previous knowledge that CR induced amelioration of insulin sensitivity and lifestyle, gut microbiome, and extrinsic environmental factors. Furthermore, it identified potential biomarkers that could be used for personalized weight-loss interventions [81].

Analysis of transcriptomic data from different tissues shows that aging induces the upregulation of inflammatory pathways and downregulation of metabolic pathways. In contrast, CR intervention reverses such effects, in which it downregulates the immune response and upregulates the metabolic pathways. In particular, LCK, a key signaling molecule in the development of T cells, was significantly upregulated in aged tissues and later downregulated as a consequence of CR. This identification of

LCK gene was based on integrative analysis of cDNA microarray and interactome, which showed a high degree of centrality and between centrality analyses. These results suggest that immune and inflammatory responses are increased during aging and can be modulated by CR [82]. In support of such results, Hong et al. reported that CR successfully suppressed immune response and increased lipid metabolism, thereby delaying aging and preventing age-associated diseases [83]. Kim et al. also reported that CR implementation delayed age-associated alterations of DNA methylation, which could prevent the progression of age-related diseases [84].

Other omics studies provided evidence that CR definitively modulates aging processes and prevents or delays age-associated disease progression. Analysis of hepatic transcriptome showed that CR stimulated pathways involving IGF-1, NF-κB, mTOR, and SIRTs, which collectively contribute to reduced oxidative stress and improved metabolism, further supporting that CR promotes health and could extend lifespan by interfering with age-associated signaling pathways [85]. In addition, CR suppresses adiposity and insulin resistance, consequentially suppressing a proinflammatory status. By implementing CR, DEG analysis could help identify biomarkers that are closely related to age-associated diseases, including metabolic disease [86]. In support of the beneficial effects of CR in aging, one of the proteomics analyses demonstrated that CR improved glucose and lipid energy metabolism and suppressed oxidative stress [87]. As lipid composition was one of the critical determinants of aging, using CR as an intervention method, a research group conducted LC–MS and demonstrated that CR reprogrammed the lipidome and metabolome, which lowered the protein oxidative damage, sequentially increasing lifespan and healthspan [88].

6. Preventive Effects of Other Types of Dietary Restriction in Aging

CR is usually considered for reducing overall calorie intake or food intake without malnutrition. In animal models, under CR, food intake is reduced by around 10–50% compared with ad libitum-fed controls [89]. Although CR exhibits preventive effects on age-related phenotypes, to practice and sustain CR in human life is quite challenging. In efforts to improve human health during aging, there are other types of dietary restriction and pharmacological interventions available that mimic CR. Here, we introduce other types of dietary restriction and CR mimetics, recapturing the beneficial effects of CR in the present and next sections.

Reduced intake of specific nutrients, rather than reduced intake of total calories, was considered important for health benefits of a restricted diet. The reduction of either dietary protein or sugar can reduce mortality and extend life span in Drosophila, independent of the calorie intake [90]. A study by Solon-Biet et al. in mice showed a clear correlation between the ratio of protein to carbohydrate and lifespan. Mice were fed 25 diets differing systematically in protein, carbohydrate, and fat content. The energy density and median lifespan of the mice increased by up to 30% as the protein to carbohydrate ratio decreased [91]. In addition, it is suggested that reduced intake of specific essential amino acids such as methionine, tryptophan, or branched chain amino acids can delay aging or improve health [92,93].

Intermittent (ex. alternate day fasting) and periodic (fasting that lasts three days or longer, every two or more weeks) fasting have been studied as alternative dietary interventions for long-term CR [94]. The effects of fasting on lifespan extension have been reported in various species including bacteria [95], yeast [96], worm [97], and mice [98]. Intermittent fasting has a protective effect on age-dependent diseases including diabetes, cancer, heart disease, and neurodegenerative disorders in rodents [94]. There are various regimens of fasting. Recently, the fasting mimicking diet (FMD) was developed and has shown several beneficial effects in mice including extended longevity, lowered visceral fat, reduced cancer incidence and skin lesions, rejuvenated immune system, and retarded bone mineral and hippocampal neurogenesis [99].

7. CR Mimetics in Aging

CR mimetics are compounds that mimic the benefits of CR at the molecular, cellular, and physiological levels, leading to health-promoting effects [100]. Recently, there has been an increased interest in CR mimetics due to the benefits of using these anti-aging interventions in terms of extended health and lifespan [101–103]. Although the valuable effects of CR mimetics on lifespan and health have been extensively highlighted, limitations persist because it is difficult to implement such diet regimens in humans. In this section, we summarize the current knowledge of CR mimetic compounds and highlight their typical effects.

7.1. Resveratrol

Resveratrol (3,5,4′-Trihydroxystilbene), a natural polyphenolic, phytoalexin compound found in grapes, cranberries, and peanuts, is currently the most thoroughly studied CR mimetic. Resveratrol promotes lifespan extension across a range of evolutionarily distinct sets of species, including *Saccharomyces cerevisiae*, *Caenorhabditis elegans*, and *Drosophila melanogaster*, all the way to mammals such as mice [104]. Previous studies have indicated the beneficial effects of sirtuins as the best small molecule that activate sirtuins, which extended lifespan in a yeast model [104–109]. Although only the longevity extension effect of resveratrol has been reported in *C. elegans* and *D. melanogaster*, many subsequent studies reported that resveratrol intake promotes health and plays a preventive role in age-related diseases, such as cancer [110–113], atherosclerosis [114,115], arthritis [116,117], cataract [118–120], cardiovascular disease [121], hypertension [122,123], type 2 diabetes [124–127], osteoporosis [128–130], and Alzheimer disease [131–133]. In clinical studies, resveratrol intake improved the memory capacity of elderly individuals and reduced blood lipid levels in obese and type 2 diabetic patients [134,135]. However, additional studies are required to investigate intake duration and dose-dependent metabolic effects of resveratrol supplements required to overcome metabolic irregularity in human subjects. Resveratrol suppresses SASP through SIRT1/NF-κB signaling and delays aging [136], represses cellular senescence, and improves insulin resistance in muscle [137].

7.2. Metformin

Metformin, a biguanide used as a first-line drug for treating type 2 diabetes [138], was shown to extend the lifespan of *C. elegans* [139–141], *D. melanogaster* [142], and mice [143,144]. Moreover, it was shown to delay the onset of age-related diseases, such as cancer, metabolic syndrome [145], and cognitive disorders [146]. Its mechanism of action is associated with the activation of 5′ AMP-activated protein kinase (AMPK) [147,148], inhibition of the mammalian target of rapamycin (mTOR) [149], reduction of DNA damage [150,151], and decreased insulin levels and IGF-1 signaling [152–154]. The longevity effect of metformin has not yet been identified in humans, and therefore, its mechanism of action requires further investigation. Metformin regulates mitochondrial biogenesis and cellular senescence through SIRT3 [155], and decreases oxidative stress-induced senescence by activating autophagy [156].

7.3. Rapamycin

Rapamycin, (International Nonproprietary Name: sirolimus), is an inhibitor of mTOR, which results in an extended life span and prevents age-related diseases [157–160] by mediating SIRT1 expression [161,162]. mTOR is a serine-threonine kinase that plays a role in modulating cell survival, growth, proliferation, motility, protein synthesis and transcription [163] and inducing autophagy [164–166]. In addition, mTOR promotes growth and aging in *C. elegans* [167], *D. melanogaster* [168], *S. cerevisiae* [169], as well as in mice [170,171] and rat [172,173] models. Further, it modulates glucose and lipid metabolism [174,175]. Rapamycin prevents insulin resistance in humans [176], reduces insulin resistance in hyperinsulinemia rats [177,178], and normalizes glucose metabolism in diabetic mice [179,180]. Recently, Garcia et al. [181] reported that rapamycin treatment has a mechanism similar to CR in ovarian mice, which increases *FOXO3* gene expression. Thus, the use of rapamycin as a CR mimetic needs to be investigated further for understanding significant signaling pathways that can be targeted for enhancing its therapeutic potential. Rapamycin ameliorates age-related late-life cancer by inhibiting senescence-associated inflammation [182]. It is also an effective inhibitor of cellular senescence [183].

7.4. PPAR Agonists

In addition to the anti-inflammatory effect, PPARs have diverse biological effects including the promotion of cellular proliferation, glucose and lipid metabolism, insulin sensitivity, and tissue remodeling processes [67,184]. Because of their association with multiple metabolic processes, PPARs have been suggested to play roles in pathogenic conditions such as obesity, metabolic syndrome, diabetes, NAFLD, and atherosclerosis. Therefore, PPARs have been considered as important molecular targets for the discovery and development of new drugs to treat these age-related diseases [185–188].

Fenofibrate is a PPARα agonist used for the treatment of hyperlipidemia, hyperglycemia, and hypertriglyceridemia [189,190]. PPARα activation by fenofibrate also reduces renal oxidative stress and cellular apoptosis in aging-related renal injury through AMPK-SIRT1 and AMPK-PGC1α signaling pathways [191]. The activation of PPARα, AMPK, and SIRT1 has been shown to protect aging-related renal injury. PPARβ/δ is involved in the regulation of insulin sensitivity, adipogenesis, lipid and energy metabolism, inflammation, and atherosclerosis [192–194]. A specific PPARβ/δ agonist, GW501516, attenuates inflammation, insulin resistance, and dyslipidemia, and modulates angiogenesis [192,193]. Two thiazolidinediones (TZD), rosiglitazone and pioglitazone, which are also PPARγ agonists, have been shown to be effective in the treatment of type 2 diabetes [195,196]. Further, they are also associated with human life longevity and cell senescence [197]; this has also been observed in aged rats [198]. Recently, Patel et al. [199–201] reported that a novel dual PPARα/γ agonist, saroglitazar magnesium, was used in the treatment of dyslipidemia and metabolic disorders in in vivo and healthy Indian adult subjects. These results are further supported by the results of a preclinical study conducted by Kaul et al. [201]. Notably, Xu et al. [202] reported that chiglitazar acts as a PPAR-α/β/γ pan agonist and evaluated its use in diabetic therapeutics in healthy Chinese volunteers.

Recent data support PPAR agonists as potential candidates for anti-senoinflammation therapy. Our group synthesized MHY908, new a PPARα/γ dual agonist, and showed that it has a significant inhibitory effect on age-related inflammation and insulin resistance [203]. It is reported that PPAR activation might have an effect on the prevention of cell senescence and that PPARα silencing induces cancer cell senescence. Rosiglitazone significantly suppressed olaparib (a PARP inhibitor)-induced cellular senescence and SASP in ovarian cancer [204].

Collectively, identifying the role of PPAR agonists in various metabolic or non-metabolic organs and pathological conditions will contribute to the development of new therapeutic options and promising anti-senoinflammatory chemicals for the treatment of many age-related metabolic disorders.

7.5. Ketone Bodies

Ketone bodies such as β-hydroxybutyrate (HB), acetoacetate, and acetone are water-soluble molecules that are generated from fatty acids in the mitochondrial matrix of the liver. They serve as moving energy sources for physiological systems during periods of fasting [205]. The process of ketogenesis starts within 24 h of fasting through gluconeogenesis [206]. In humans, physiological serum levels of HB are normally maintained at a low micromolar concentration, which increases to a few hundred micromoles after 12 to 16 h of the fasting period and eventually reaches 1 to 2 mM after 2 days of fasting [207]. Insulin inhibits lipolysis of adipose tissue and restricts ketogenesis, while glucagon promotes ketogenic flow by exerting its direct effect on the hepatic tissue [208]. In a study, CR or fasting interventions elevated the circulating concentration of ketone bodies, HB, compared to that in a normal feeding group [209,210]. Furthermore, it has also been reported that the implementation of a ketogenic diet exerts therapeutic effects on various age-related diseases related to insulin resistance, as well as diseases resulting from free radical damage and hypoxia [211].

HB also acts as a signaling molecule and activates cellular signaling pathways. For example, HB plays a role in endogenously inhibiting histone deacetylases (HDACs) [209]. Suppression of HDAC activity exerts beneficial metabolic and cytoprotective effects similar to those seen in HB investigations [212]. However, SIRT3 regulates diverse pathways involved in fasting metabolism, and mice without *SIRT3* genes have decreased HB concentration during fasting [188]. Ketogenic diets are also related to low levels of insulin [213,214], suppressed IGF signaling [215], induction of FOXO3 [209], and activation of AMPK [215,216] and antioxidant genes [209]. Ketone bodies exert neuroprotective and lifespan extension effects similar to CR in *C. elegans* [217]. HB upregulates transcription of antioxidant genes, including manganese superoxide dismutase (*MnSOD*) and *FOXO3*, both of which exert antioxidant effects [209]. It is thought that HB exerts its effect through signaling mechanisms comparable to that of CR by inducing co-activation of FOXO1/PGC-1α through deactivation of the PI3K/Akt pathway [218].

HB is an effector that transduces signals via G-protein coupled receptors. It represses the actions of the sympathetic nervous system and decreases energy expenditure and heart rate by blocking fatty acid signaling pathways through the G protein-coupled receptor 41 [219]. One of the most well-studied signaling effects of HB signals is via GPR109A, a member of the hydrocarboxylic acid GPCR subfamily that is expressed in adipose tissues (white and brown) [220] and immune cells [221]. Although the GPR109A receptor has protective effects, associations have been found between ketogenic dietary intervention use in stroke patients and neurodegenerative diseases [222,223]. In a TNFα or LPS-induced inflammatory setting, HB exerts anti-inflammatory effects by suppressing the release of pro-inflammatory proteins (iNOS and COX-2) and cytokines (TNF, IL-1β, IL-6 and CCL2/MCP-1), which seems to occur partially via inhibition of NF-κB translocation to the nucleus for pro-inflammatory gene activation [224,225]. However, in neurodegenerative inflammatory conditions, the effects of GPR109A-mediated HB do not appear to involve inflammatory mediator signaling via the MAPK pathway [224]. In addition to their role in providing energy fuels for various key organs and tissues, including the brain, heart or skeletal muscle, ketone bodies play critical roles as signaling mediators and modulators of inflammation and oxidation [226].

8. Conclusions

Based on the available molecular and biochemical evidence, we proposed the concept of senoinflammation in our previous review [17,227]. The concept proposes a broader perspective on age-related inflammatory response and creates a complex network among many inflammatory mediators that can lead to systemic chronic inflammation. Oxidative stress leads to improper gene regulation and genomic DNA damage during aging. Such improper gene regulation in aged senescent cells allows them to fall into a proinflammatory state, consequently changing systemic chemokine or cytokine activities. The proinflammatory SASP environment further exerts stress on the intracellular organelles, tissues, and systems, which affects the development and occurrence of metabolic disorders. It appears that a repetitive vicious cycle occurs between SASP and metabolic dysregulation as proposed in the concept of senoinflammation, and this interactive network forms the basis of the aging process and age-related diseases. However, the secretion of proinflammatory mediators, collectively termed as SASP, in response to internal and external stress leads to the chronic inflammatory condition termed as senoinflammation. Based on CR experiments and observations, cytokine, chemokine, and metabolic pathways are significantly regulated by CR and CR mimetics in the aging process. It is expected that a better understanding of senoinflammation modulatory mechanisms will provide a basis for the discovery of molecular targets that can therapeutically modulate age-related chronic inflammatory conditions and enable the development of potentially effective interventions to delay aging and prevent the occurrence of aging-associated diseases.

Table 1. Changes in parameters in senoinflammation.

	SASP Factors	Old	CR	Species	References
Cytokines	IL-1β	↑	⊢	Human, Mouse, Rat	[82,228–230]
	IL-6	↑	⊢	Human *, Mouse, Monkey	[82,231–233]
	IL-7	↑		Human, Rat	[74,82], TCGA database
	IL-13	-		Human	TCGA database
	IL-11	↑		Rat	[74,82]
	IL-6R	↑		Rat	[74,82]
	IL-2RA	↑		Rat	[74,82]
	TNF-α	↑	⊢	*C. elegans*, Mouse, Rat	[74,82,231,234]
	TNF-β	↑		Human, Rat	[74,82], TCGA database
Cheomokines	IL-8	↑	⊢	Monkey	[235]
	MCP-1 (CCL2)	↑	⊢	Mouse, Rat	[74,82,236]
	MCP-2	-	⊢	Mouse	[237]
	MIP-1α (CCL3)	-	⊢	Mouse	[236,238]
	MIP-3α	↑		Rat	[74,82]
MMPs, GFs, etc.	MMP1	↑		Human, Mouse	[239,240]
	MMP2	↑	⊢	Mouse	[241]
	MMP3	↑	⊢	Mouse, Rat	[60,74,82,242]
	MMP9	↑	⊢	Mouse, Rat	[243,244]
	MMP12	↑		Rat	[74,82]
	MMP13	↑	⊢	Rat	[245]
	MMP14	↑		Human	TCGA database
	HGF	↑		Human, Rat	[74,82], TCGA database
	EGFR	↑		Human, Rat	[74,82], TCGA database
	FAS	↑	⊢	Human, Mouse, Rat	[74,82,246–248]
	IGFBP2	↑		Human	TCGA data base
Metabolism	Insulin resistance	↑	⊢	Human, Mouse, Rat	[181,249–251]
	ER stress	↑	⊢	Human, Mouse, Rat	[252–254]
	Autophagy	↑	⊢	Human, Mouse, Rat	[255–258]
	Lipid accumulation	↑	⊢	Human, Mouse, Rat	[259–262]

* A calorie restriction (CR) diet supplemented with fish oil. SASP, senescence-associated secretory phenotype; IL-1β, Interleukin 1 beta; IL-6R, Interleukin 6 receptor; TNF-α, Tumor necrosis factor-alpha; MCP-1, Monocyte chemoattractant protein-1; MIP-1α, Macrophage inflammatory protein-1alpha; MMP, Matrix metallopeptidases; GF, Growth factor; HGF, Hepatocyte Growth Factor; EGFR, Epidermal growth factor receptor; FAS, Apoptosis Antigen 1; IGFBP2, Insulin Like Growth Factor Binding Protein 2; TCGA, The Cancer Genome Atlas.

Author Contributions: Conceptualization, design and writing of the work: D.H.K., E.B., H.J.J., S.G.N., B.P.Y., Y.J.C. and H.Y.C. All authors have read and agreed to the published version of the manuscript.

Funding: This study was supported by the Korean National Research Foundation (NRF) funded by the Korean government (NRF-2018R1A2A3075425 and NRF-2015M3A9B8029074).

Acknowledgments: We thank Aging Tissue Bank (Busan, Korea) for providing research information.

Conflicts of Interest: The authors declare no conflict of interest.

References

1. Yu, B.P. Aging and oxidative stress: Modulation by dietary restriction. *Free Radic. Biol. Med.* **1996**, *21*, 651–668. [CrossRef]
2. Bektas, A.; Schurman, S.H.; Sen, R.; Ferrucci, L. Aging, inflammation and the environment. *Exp. Gerontol.* **2018**, *105*, 10–18. [CrossRef]
3. Chhetri, J.K.; de Souto Barreto, P.; Fougere, B.; Rolland, Y.; Vellas, B.; Cesari, M. Chronic inflammation and sarcopenia: A regenerative cell therapy perspective. *Exp. Gerontol.* **2018**, *103*, 115–123. [CrossRef] [PubMed]
4. Franceschi, C.; Garagnani, P.; Vitale, G.; Capri, M.; Salvioli, S. Inflammaging and 'Garb-aging'. *Trends Endocrinol. Metab.* **2017**, *28*, 199–212. [CrossRef] [PubMed]
5. Fulop, T.; Larbi, A.; Dupuis, G.; Le Page, A.; Frost, E.H.; Cohen, A.A.; Witkowski, J.M.; Franceschi, C. Immunosenescence and Inflamm-Aging As Two Sides of the Same Coin: Friends or Foes? *Front. Immunol.* **2017**, *8*, 1960. [CrossRef] [PubMed]
6. Fougere, B.; Boulanger, E.; Nourhashemi, F.; Guyonnet, S.; Cesari, M. Chronic Inflammation: Accelerator of Biological Aging. *J. Gerontol. A Biol. Sci. Med. Sci.* **2017**, *72*, 1218–1225. [CrossRef]
7. Kuilman, T.; Peeper, D.S. Senescence-messaging secretome: SMS-ing cellular stress. *Nat. Rev. Cancer* **2009**, *9*, 81–94. [CrossRef]
8. Vasto, S.; Candore, G.; Balistreri, C.R.; Caruso, M.; Colonna-Romano, G.; Grimaldi, M.P.; Listi, F.; Nuzzo, D.; Lio, D.; Caruso, C. Inflammatory networks in ageing, age-related diseases and longevity. *Mech. Ageing Dev.* **2007**, *128*, 83–91. [CrossRef]
9. Rea, I.M.; Gibson, D.S.; McGilligan, V.; McNerlan, S.E.; Alexander, H.D.; Ross, O.A. Age and Age-Related Diseases: Role of Inflammation Triggers and Cytokines. *Front. Immunol.* **2018**, *9*, 586. [CrossRef]
10. Yu, B.P. Calorie restriction as a potent anti-aging intervention: Modulation of oxidative stress. In *Aging Interventions and Therapies*; World Scientific Publishing Co.: Singapore, 2005; pp. 193–217.
11. Stepanyan, Z.; Hughes, B.; Cliche, D.O.; Camp, D.; Hekimi, S. Genetic and molecular characterization of CLK-1/mCLK1, a conserved determinant of the rate of aging. *Exp. Gerontol.* **2006**, *41*, 940–951. [CrossRef]
12. Childs, B.G.; Gluscevic, M.; Baker, D.J.; Laberge, R.-M.; Marquess, D.; Dananberg, J.; van Deursen, J.M. Senescent cells: An emerging target for diseases of ageing. *Nat. Rev. Drug Discov.* **2017**, *16*, 718–735. [CrossRef] [PubMed]
13. Shaw, A.C.; Joshi, S.; Greenwood, H.; Panda, A.; Lord, J.M. Aging of the innate immune system. *Curr. Opin. Immunol.* **2010**, *22*, 507–513. [CrossRef] [PubMed]
14. Shaw, A.C.; Goldstein, D.R.; Montgomery, R.R. Age-dependent dysregulation of innate immunity. *Nat. Rev. Immunol.* **2013**, *13*, 875–887. [CrossRef] [PubMed]
15. Rawji, K.S.; Mishra, M.K.; Michaels, N.J.; Rivest, S.; Stys, P.K.; Yong, V.W. Immunosenescence of microglia and macrophages: Impact on the ageing central nervous system. *Brain* **2016**, *139*, 653–661. [CrossRef]
16. Oishi, Y.; Manabe, I. Macrophages in age-related chronic inflammatory diseases. *NPJ Aging Mech. Dis.* **2016**, *2*, 16018. [CrossRef]
17. Chung, H.Y.; Kim, D.H.; Lee, E.K.; Chung, K.W.; Chung, S.; Lee, B.; Seo, A.Y.; Chung, J.H.; Jung, Y.S.; Im, E.; et al. Redefining Chronic Inflammation in Aging and Age-Related Diseases: Proposal of the Senoinflammation Concept. *Aging Dis.* **2019**, *10*, 367–382. [CrossRef]
18. Irvine, K.M.; Skoien, R.; Bokil, N.J.; Melino, M.; Thomas, G.P.; Loo, D.; Gabrielli, B.; Hill, M.M.; Sweet, M.J.; Clouston, A.D.; et al. Senescent human hepatocytes express a unique secretory phenotype and promote macrophage migration. *World J. Gastroenterol.* **2014**, *20*, 17851–17862. [CrossRef]
19. Wang, C.; Jurk, D.; Maddick, M.; Nelson, G.; Martin-Ruiz, C.; von Zglinicki, T. DNA damage response and cellular senescence in tissues of aging mice. *Aging Cell* **2009**, *8*, 311–323. [CrossRef]
20. Castro, R.E.; Ferreira, D.M.; Afonso, M.B.; Borralho, P.M.; Machado, M.V.; Cortez-Pinto, H.; Rodrigues, C.M. miR-34a/SIRT1/p53 is suppressed by ursodeoxycholic acid in the rat liver and activated by disease severity in human non-alcoholic fatty liver disease. *J. Hepatol.* **2013**, *58*, 119–125. [CrossRef]
21. Ogrodnik, M.; Miwa, S.; Tchkonia, T.; Tiniakos, D.; Wilson, C.L.; Lahat, A.; Day, C.P.; Burt, A.; Palmer, A.; Anstee, Q.M.; et al. Cellular senescence drives age-dependent hepatic steatosis. *Nat. Commun.* **2017**, *8*, 15691. [CrossRef]

22. Chung, H.Y.; Lee, E.K.; Choi, Y.J.; Kim, J.M.; Kim, D.H.; Zou, Y.; Kim, C.H.; Lee, J.; Kim, H.S.; Kim, N.D.; et al. Molecular inflammation as an underlying mechanism of the aging process and age-related diseases. *J. Dent. Res.* **2011**, *90*, 830–840. [CrossRef]

23. Hoesel, B.; Schmid, J.A. The complexity of NF-kappaB signaling in inflammation and cancer. *Mol. Cancer* **2013**, *12*, 86. [CrossRef] [PubMed]

24. Patel, M.; Horgan, P.G.; McMillan, D.C.; Edwards, J. NF-kappaB pathways in the development and progression of colorectal cancer. *Transl. Res.* **2018**, *197*, 43–56. [CrossRef] [PubMed]

25. Kriete, A.; Mayo, K.L.; Yalamanchili, N.; Beggs, W.; Bender, P.; Kari, C.; Rodeck, U. Cell autonomous expression of inflammatory genes in biologically aged fibroblasts associated with elevated NF-kappaB activity. *Immun. Ageing* **2008**, *5*, 5. [CrossRef] [PubMed]

26. Bektas, A.; Zhang, Y.; Lehmann, E.; Wood, W.H., 3rd; Becker, K.G.; Madara, K.; Ferrucci, L.; Sen, R. Age-associated changes in basal NF-kappaB function in human CD4$^+$ T lymphocytes via dysregulation of PI3 kinase. *Aging (Albany NY)* **2014**, *6*, 957–974. [CrossRef]

27. Esparza-Lopez, J.; Alvarado-Munoz, J.F.; Escobar-Arriaga, E.; Ulloa-Aguirre, A.; de Jesus Ibarra-Sanchez, M. Metformin reverses mesenchymal phenotype of primary breast cancer cells through STAT3/NF-kappaB pathways. *BMC Cancer* **2019**, *19*, 728. [CrossRef]

28. Liu, L.; Liu, Y.; Liu, X.; Zhang, N.; Mao, G.; Zeng, Q.; Yin, M.; Song, D.; Deng, H. Resibufogenin suppresses transforming growth factor-beta-activated kinase 1-mediated nuclear factor-kappaB activity through protein kinase C-dependent inhibition of glycogen synthase kinase 3. *Cancer Sci.* **2018**, *109*, 3611–3622. [CrossRef]

29. Chung, H.Y.; Cesari, M.; Anton, S.; Marzetti, E.; Giovannini, S.; Seo, A.Y.; Carter, C.; Yu, B.P.; Leeuwenburgh, C. Molecular inflammation: Underpinnings of aging and age-related diseases. *Ageing Res. Rev.* **2009**, *8*, 18–30. [CrossRef]

30. Tabas, I.; Glass, C.K. Anti-inflammatory therapy in chronic disease: Challenges and opportunities. *Science* **2013**, *339*, 166–172. [CrossRef]

31. Bruunsgaard, H.; Pedersen, B.K. Age-related inflammatory cytokines and disease. *Immunol. Allergy Clin. N. Am.* **2003**, *23*, 15–39. [CrossRef]

32. Soysal, P.; Stubbs, B.; Lucato, P.; Luchini, C.; Solmi, M.; Peluso, R.; Sergi, G.; Isik, A.T.; Manzato, E.; Maggi, S.; et al. Inflammation and frailty in the elderly: A systematic review and meta-analysis. *Ageing Res. Rev.* **2016**, *31*, 1–8. [CrossRef] [PubMed]

33. Arai, Y.; Martin-Ruiz, C.M.; Takayama, M.; Abe, Y.; Takebayashi, T.; Koyasu, S.; Suematsu, M.; Hirose, N.; von Zglinicki, T. Inflammation, But Not Telomere Length, Predicts Successful Ageing at Extreme Old Age: A Longitudinal Study of Semi-supercentenarians. *EBioMedicine* **2015**, *2*, 1549–1558. [CrossRef] [PubMed]

34. Michaud, M.; Balardy, L.; Moulis, G.; Gaudin, C.; Peyrot, C.; Vellas, B.; Cesari, M.; Nourhashemi, F. Proinflammatory cytokines, aging, and age-related diseases. *J. Am. Med. Dir. Assoc.* **2013**, *14*, 877–882. [CrossRef] [PubMed]

35. Fontana, L.; Partridge, L.; Longo, V.D. Extending healthy life span—From yeast to humans. *Science* **2010**, *328*, 321–326. [CrossRef] [PubMed]

36. Mitchell, S.J.; Madrigal-Matute, J.; Scheibye-Knudsen, M.; Fang, E.; Aon, M.; Gonzalez-Reyes, J.A.; Cortassa, S.; Kaushik, S.; Gonzalez-Freire, M.; Patel, B.; et al. Effects of Sex, Strain, and Energy Intake on Hallmarks of Aging in Mice. *Cell Metab.* **2016**, *23*, 1093–1112. [CrossRef] [PubMed]

37. Brandhorst, S.; Longo, V.D. Fasting and Caloric Restriction in Cancer Prevention and Treatment. *Recent Results Cancer Res.* **2016**, *207*, 241–266. [CrossRef]

38. Chen, J.; Astle, C.M.; Harrison, D.E. Delayed immune aging in diet-restricted B6CBAT6 F1 mice is associated with preservation of naive T cells. *J. Gerontol. A Biol. Sci. Med. Sci.* **1998**, *53*, B330–B337. [CrossRef]

39. Walford, R.L.; Liu, R.K.; Gerbase-Delima, M.; Mathies, M.; Smith, G.S. Longterm dietary restriction and immune function in mice: Response to sheep red blood cells and to mitogenic agents. *Mech. Ageing Dev.* **1973**, *2*, 447–454. [CrossRef]

40. Weindruch, R.; Devens, B.H.; Raff, H.V.; Walford, R.L. Influence of dietary restriction and aging on natural killer cell activity in mice. *J. Immunol.* **1983**, *130*, 993–996.

41. Weindruch, R.; Gottesman, S.R.; Walford, R.L. Modification of age-related immune decline in mice dietarily restricted from or after midadulthood. *Proc. Natl. Acad. Sci. USA* **1982**, *79*, 898–902. [CrossRef]

42. Weindruch, R.; Walford, R.L. Dietary restriction in mice beginning at 1 year of age: Effect on life-span and spontaneous cancer incidence. *Science* **1982**, *215*, 1415–1418. [CrossRef] [PubMed]

43.	Spaulding, C.C.; Walford, R.L.; Effros, R.B. Calorie restriction inhibits the age-related dysregulation of the cytokines TNF-alpha and IL-6 in C3B10RF1 mice. *Mech. Ageing Dev.* **1997**, *93*, 87–94. [CrossRef]

44.	Fernandes, G.; Friend, P.; Yunis, E.J.; Good, R.A. Influence of dietary restriction on immunologic function and renal disease in (NZB x NZW) F1 mice. *Proc. Natl. Acad. Sci. USA* **1978**, *75*, 1500–1504. [CrossRef]

45.	Friend, P.S.; Fernandes, G.; Good, R.A.; Michael, A.F.; Yunis, E.J. Dietary restrictions early and late: Effects on the nephropathy of the NZB X NZW mouse. *Lab. Investig.* **1978**, *38*, 629–632.

46.	McCay, C.M.; Crowell, M.F.; Maynard, L.A. The effect of retarded growth upon the length of life span and upon the ultimate body size. 1935. *Nutrition* **1989**, *5*, 155–171, discussion 172.

47.	Bordone, L.; Guarente, L. Calorie restriction, SIRT1 and metabolism: Understanding longevity. *Nat. Rev. Mol. Cell Biol.* **2005**, *6*, 298–305. [CrossRef]

48.	Yu, B.P.; Chung, H.Y. Stress resistance by caloric restriction for longevity. *Ann. N. Y. Acad. Sci.* **2001**, *928*, 39–47. [CrossRef]

49.	Longo, V.D.; Finch, C.E. Evolutionary medicine: From dwarf model systems to healthy centenarians? *Science* **2003**, *299*, 1342–1346. [CrossRef]

50.	Demetrius, L. Of mice and men. When it comes to studying ageing and the means to slow it down, mice are not just small humans. *EMBO Rep.* **2005**, *6*, S39–S44. [CrossRef]

51.	Mattson, M.P.; Duan, W.; Lee, J.; Guo, Z. Suppression of brain aging and neurodegenerative disorders by dietary restriction and environmental enrichment: Molecular mechanisms. *Mech. Ageing Dev.* **2001**, *122*, 757–778. [CrossRef]

52.	Cohen, D.E.; Supinski, A.M.; Bonkowski, M.S.; Donmez, G.; Guarente, L.P. Neuronal SIRT1 regulates endocrine and behavioral responses to calorie restriction. *Genes Dev.* **2009**, *23*, 2812–2817. [CrossRef] [PubMed]

53.	Grivennikov, S.I.; Karin, M. Dangerous liaisons: STAT3 and NF-kappaB collaboration and crosstalk in cancer. *Cytokine Growth Factor Rev.* **2010**, *21*, 11–19. [CrossRef] [PubMed]

54.	Weichhart, T. mTOR as Regulator of Lifespan, Aging, and Cellular Senescence: A Mini-Review. *Gerontology* **2018**, *64*, 127–134. [CrossRef] [PubMed]

55.	Jung, K.J.; Lee, E.K.; Kim, J.Y.; Zou, Y.; Sung, B.; Heo, H.S.; Kim, M.K.; Lee, J.; Kim, N.D.; Yu, B.P.; et al. Effect of short term calorie restriction on pro-inflammatory NF-kB and AP-1 in aged rat kidney. *Inflamm. Res.* **2009**, *58*, 143–150. [CrossRef] [PubMed]

56.	Allen, B.D.; Liao, C.Y.; Shu, J.; Muglia, L.J.; Majzoub, J.A.; Diaz, V.; Nelson, J.F. Hyperadrenocorticism of calorie restriction contributes to its anti-inflammatory action in mice. *Aging Cell* **2019**, *18*, e12944. [CrossRef] [PubMed]

57.	Jin Jung, K.; Hyun Kim, D.; Kyeong Lee, E.; Woo Song, C.; Pal Yu, B.; Young Chung, H. Oxidative stress induces inactivation of protein phosphatase 2A, promoting proinflammatory NF-kappaB in aged rat kidney. *Free Radic. Biol. Med.* **2013**, *61*, 206–217. [CrossRef]

58.	Meydani, S.N.; Das, S.K.; Pieper, C.F.; Lewis, M.R.; Klein, S.; Dixit, V.D.; Gupta, A.K.; Villareal, D.T.; Bhapkar, M.; Huang, M.; et al. Long-term moderate calorie restriction inhibits inflammation without impairing cell-mediated immunity: A randomized controlled trial in non-obese humans. *Aging (Albany NY)* **2016**, *8*, 1416–1431. [CrossRef]

59.	Contreras, N.A.; Fontana, L.; Tosti, V.; Nikolich-Zugich, J. Calorie restriction induces reversible lymphopenia and lymphoid organ atrophy due to cell redistribution. *Geroscience* **2018**, *40*, 279–291. [CrossRef]

60.	Kurki, E.; Shi, J.; Martonen, E.; Finckenberg, P.; Mervaala, E. Distinct effects of calorie restriction on adipose tissue cytokine and angiogenesis profiles in obese and lean mice. *Nutr. Metab. (Lon.)* **2012**, *9*, 64. [CrossRef]

61.	Park, C.Y.; Park, S.; Kim, M.S.; Kim, H.K.; Han, S.N. Effects of mild calorie restriction on lipid metabolism and inflammation in liver and adipose tissue. *Biochem. Biophys. Res. Commun.* **2017**, *490*, 636–642. [CrossRef]

62.	Johansson, H.E.; Edholm, D.; Kullberg, J.; Rosqvist, F.; Rudling, M.; Straniero, S.; Karlsson, F.A.; Ahlstrom, H.; Sundbom, M.; Riserus, U. Energy restriction in obese women suggest linear reduction of hepatic fat content and time-dependent metabolic improvements. *Nutr. Diabetes* **2019**, *9*, 34. [CrossRef] [PubMed]

63.	Masternak, M.M.; Bartke, A. PPARs in Calorie Restricted and Genetically Long-Lived Mice. *PPAR Res.* **2007**, *2007*, 28436. [CrossRef] [PubMed]

64.	Stankovic, M.; Mladenovic, D.; Ninkovic, M.; Vucevic, D.; Tomasevic, T.; Radosavljevic, T. Effects of caloric restriction on oxidative stress parameters. *Gen. Physiol. Biophys.* **2013**, *32*, 277–283. [CrossRef] [PubMed]

65. Redman, L.M.; Smith, S.R.; Burton, J.H.; Martin, C.K.; Il'yasova, D.; Ravussin, E. Metabolic Slowing and Reduced Oxidative Damage with Sustained Caloric Restriction Support the Rate of Living and Oxidative Damage Theories of Aging. *Cell Metab.* **2018**, *27*, 805–815.e804. [CrossRef]

66. Shinmura, K. Effects of caloric restriction on cardiac oxidative stress and mitochondrial bioenergetics: Potential role of cardiac sirtuins. *Oxid. Med. Cell. Longev.* **2013**, *2013*, 528935. [CrossRef]

67. Sung, B.; Park, S.; Yu, B.P.; Chung, H.Y. Modulation of PPAR in aging, inflammation, and calorie restriction. *J. Gerontol. A Biol. Sci. Med. Sci.* **2004**, *59*, 997–1006. [CrossRef]

68. Nunn, A.V.; Bell, J.; Barter, P. The integration of lipid-sensing and anti-inflammatory effects: How the PPARs play a role in metabolic balance. *Nucl. Recept.* **2007**, *5*, 1. [CrossRef]

69. Youssef, J.; Badr, M. Role of Peroxisome Proliferator-Activated Receptors in Inflammation Control. *J. Biomed. Biotechnol.* **2004**, *2004*, 156–166. [CrossRef]

70. Canto, C.; Auwerx, J. Caloric restriction, SIRT1 and longevity. *Trends Endocrinol. Metab.* **2009**, *20*, 325–331. [CrossRef]

71. Tang, B.L. Sirt1 and the Mitochondria. *Mol. Cells* **2016**, *39*, 87–95. [CrossRef]

72. Frescas, D.; Valenti, L.; Accili, D. Nuclear trapping of the forkhead transcription factor FoxO1 via Sirt-dependent deacetylation promotes expression of glucogenetic genes. *J. Biol. Chem.* **2005**, *280*, 20589–20595. [CrossRef] [PubMed]

73. Min, S.W.; Sohn, P.D.; Cho, S.H.; Swanson, R.A.; Gan, L. Sirtuins in neurodegenerative diseases: An update on potential mechanisms. *Front. Aging Neurosci.* **2013**, *5*, 53. [CrossRef] [PubMed]

74. Park, D.; Kim, B.C.; Kim, C.H.; Choi, Y.J.; Jeong, H.O.; Kim, M.E.; Lee, J.S.; Park, M.H.; Chung, K.W.; Kim, D.H.; et al. RNA-Seq analysis reveals new evidence for inflammation-related changes in aged kidney. *Oncotarget* **2016**, *7*, 30037–30048. [CrossRef] [PubMed]

75. Kim, B.C.; Jeong, H.O.; Park, D.; Kim, C.H.; Lee, E.K.; Kim, D.H.; Im, E.; Kim, N.D.; Lee, S.; Yu, B.P.; et al. Profiling age-related epigenetic markers of stomach adenocarcinoma in young and old subjects. *Cancer Inform.* **2015**, *14*, 47–54. [CrossRef]

76. Castellani, G.C.; Menichetti, G.; Garagnani, P.; Giulia Bacalini, M.; Pirazzini, C.; Franceschi, C.; Collino, S.; Sala, C.; Remondini, D.; Giampieri, E.; et al. Systems medicine of inflammaging. *Brief. Bioinform.* **2016**, *17*, 527–540. [CrossRef]

77. Rusli, F.; Lute, C.; Boekschoten, M.V.; van Dijk, M.; van Norren, K.; Menke, A.L.; Muller, M.; Steegenga, W.T. Intermittent calorie restriction largely counteracts the adverse health effects of a moderate-fat diet in aging C57BL/6J mice. *Mol. Nutr. Food Res.* **2017**, *61*, 1600677. [CrossRef]

78. Derous, D.; Mitchell, S.E.; Green, C.L.; Wang, Y.; Han, J.D.J.; Chen, L.; Promislow, D.E.L.; Lusseau, D.; Douglas, A.; Speakman, J.R. The Effects of Graded Levels of Calorie Restriction: X. Transcriptomic Responses of Epididymal Adipose Tissue. *J. Gerontol. Ser. A Biol. Sci. Med Sci.* **2018**, *73*, 279–288. [CrossRef]

79. Green, C.L.; Mitchell, S.E.; Derous, D.; Wang, Y.; Chen, L.; Han, J.J.; Promislow, D.E.L.; Lusseau, D.; Douglas, A.; Speakman, J.R. The effects of graded levels of calorie restriction: IX. Global metabolomic screen reveals modulation of carnitines, sphingolipids and bile acids in the liver of C57BL/6 mice. *Aging Cell* **2017**, *16*, 529–540. [CrossRef]

80. Derous, D.; Mitchell, S.E.; Green, C.L.; Wang, Y.; Han, J.D.; Chen, L.; Promislow, D.E.; Lusseau, D.; Speakman, J.R.; Douglas, A. The effects of graded levels of calorie restriction: VII. Topological rearrangement of hypothalamic aging networks. *Aging* **2016**, *8*, 917–932. [CrossRef]

81. Dao, M.C.; Sokolovska, N.; Brazeilles, R.; Affeldt, S.; Pelloux, V.; Prifti, E.; Chilloux, J.; Verger, E.O.; Kayser, B.D.; Aron-Wisnewsky, J.; et al. A Data Integration Multi-Omics Approach to Study Calorie Restriction-Induced Changes in Insulin Sensitivity. *Front. Physiol.* **2018**, *9*, 1958. [CrossRef]

82. Park, D.; Lee, E.K.; Jang, E.J.; Jeong, H.O.; Kim, B.C.; Ha, Y.M.; Hong, S.E.; Yu, B.P.; Chung, H.Y. Identification of the dichotomous role of age-related LCK in calorie restriction revealed by integrative analysis of cDNA microarray and interactome. *Age* **2013**, *35*, 1045–1060. [CrossRef] [PubMed]

83. Hong, S.E.; Heo, H.S.; Kim, D.H.; Kim, M.S.; Kim, C.H.; Lee, J.; Yoo, M.A.; Yu, B.P.; Leeuwenburgh, C.; Chung, H.Y. Revealing system-level correlations between aging and calorie restriction using a mouse transcriptome. *Age* **2010**, *32*, 15–30. [CrossRef] [PubMed]

84. Kim, C.H.; Lee, E.K.; Choi, Y.J.; An, H.J.; Jeong, H.O.; Park, D.; Kim, B.C.; Yu, B.P.; Bhak, J.; Chung, H.Y. Short-term calorie restriction ameliorates genomewide, age-related alterations in DNA methylation. *Aging Cell* **2016**, *15*, 1074–1081. [CrossRef] [PubMed]

85. Derous, D.; Mitchell, S.E.; Wang, L.; Green, C.L.; Wang, Y.; Chen, L.; Han, J.J.; Promislow, D.E.L.; Lusseau, D.; Douglas, A.; et al. The effects of graded levels of calorie restriction: XI. Evaluation of the main hypotheses underpinning the life extension effects of CR using the hepatic transcriptome. *Aging* **2017**, *9*, 1770–1824. [CrossRef] [PubMed]

86. Wanders, D.; Ghosh, S.; Stone, K.P.; Van, N.T.; Gettys, T.W. Transcriptional impact of dietary methionine restriction on systemic inflammation: Relevance to biomarkers of metabolic disease during aging. *BioFactors* **2014**, *40*, 13–26. [CrossRef]

87. Valle, A.; Sastre-Serra, J.; Roca, P.; Oliver, J. Modulation of white adipose tissue proteome by aging and calorie restriction. *Aging Cell* **2010**, *9*, 882–894. [CrossRef]

88. Jove, M.; Naudi, A.; Ramirez-Nunez, O.; Portero-Otin, M.; Selman, C.; Withers, D.J.; Pamplona, R. Caloric restriction reveals a metabolomic and lipidomic signature in liver of male mice. *Aging Cell* **2014**, *13*, 828–837. [CrossRef]

89. Simpson, S.J.; Le Couteur, D.G.; Raubenheimer, D.; Solon-Biet, S.M.; Cooney, G.J.; Cogger, V.C.; Fontana, L. Dietary protein, aging and nutritional geometry. *Ageing Res. Rev.* **2017**, *39*, 78–86. [CrossRef]

90. Mair, W.; Piper, M.D.; Partridge, L. Calories do not explain extension of life span by dietary restriction in Drosophila. *PLoS Biol.* **2005**, *3*, e223. [CrossRef]

91. Solon-Biet, S.M.; McMahon, A.C.; Ballard, J.W.; Ruohonen, K.; Wu, L.E.; Cogger, V.C.; Warren, A.; Huang, X.; Pichaud, N.; Melvin, R.G.; et al. The ratio of macronutrients, not caloric intake, dictates cardiometabolic health, aging, and longevity in ad libitum-fed mice. *Cell Metab.* **2014**, *19*, 418–430. [CrossRef]

92. Grandison, R.C.; Piper, M.D.; Partridge, L. Amino-acid imbalance explains extension of lifespan by dietary restriction in Drosophila. *Nature* **2009**, *462*, 1061–1064. [CrossRef] [PubMed]

93. Fontana, L.; Cummings, N.E.; Arriola Apelo, S.I.; Neuman, J.C.; Kasza, I.; Schmidt, B.A.; Cava, E.; Spelta, F.; Tosti, V.; Syed, F.A.; et al. Decreased Consumption of Branched-Chain Amino Acids Improves Metabolic Health. *Cell Rep.* **2016**, *16*, 520–530. [CrossRef] [PubMed]

94. Longo, V.D.; Mattson, M.P. Fasting: Molecular mechanisms and clinical applications. *Cell. Metab.* **2014**, *19*, 181–192. [CrossRef] [PubMed]

95. Gonidakis, S.; Finkel, S.E.; Longo, V.D. Genome-wide screen identifies Escherichia coli TCA-cycle-related mutants with extended chronological lifespan dependent on acetate metabolism and the hypoxia-inducible transcription factor ArcA. *Aging Cell* **2010**, *9*, 868–881. [CrossRef]

96. Wei, M.; Fabrizio, P.; Hu, J.; Ge, H.; Cheng, C.; Li, L.; Longo, V.D. Life span extension by calorie restriction depends on Rim15 and transcription factors downstream of Ras/PKA, Tor, and Sch9. *PLoS. Genet.* **2008**, *4*, e13. [CrossRef]

97. Kaeberlein, T.L.; Smith, E.D.; Tsuchiya, M.; Welton, K.L.; Thomas, J.H.; Fields, S.; Kennedy, B.K.; Kaeberlein, M. Lifespan extension in Caenorhabditis elegans by complete removal of food. *Aging Cell* **2006**, *5*, 487–494. [CrossRef]

98. Goodrick, C.L.; Ingram, D.K.; Reynolds, M.A.; Freeman, J.R.; Cider, N. Effects of intermittent feeding upon body weight and lifespan in inbred mice: Interaction of genotype and age. *Mech. Ageing Dev.* **1990**, *55*, 69–87. [CrossRef]

99. Brandhorst, S.; Choi, I.Y.; Wei, M.; Cheng, C.W.; Sedrakyan, S.; Navarrete, G.; Dubeau, L.; Yap, L.P.; Park, R.; Vinciguerra, M.; et al. A Periodic Diet that Mimics Fasting Promotes Multi-System Regeneration, Enhanced Cognitive Performance, and Healthspan. *Cell Metab.* **2015**, *22*, 86–99. [CrossRef]

100. Singh, S.; Singh, A.K.; Garg, G.; Rizvi, S.I. Fisetin as a caloric restriction mimetic protects rat brain against aging induced oxidative stress, apoptosis and neurodegeneration. *Life Sci.* **2018**, *193*, 171–179. [CrossRef]

101. Choi, K.M.; Lee, H.L.; Kwon, Y.Y.; Kang, M.S.; Lee, S.K.; Lee, C.K. Enhancement of mitochondrial function correlates with the extension of lifespan by caloric restriction and caloric restriction mimetics in yeast. *Biochem. Biophys. Res. Commun.* **2013**, *441*, 236–242. [CrossRef]

102. Shintani, H.; Shintani, T.; Ashida, H.; Sato, M. Calorie Restriction Mimetics: Upstream-Type Compounds for Modulating Glucose Metabolism. *Nutrients* **2018**, *10*, 1821. [CrossRef] [PubMed]

103. Madeo, F.; Carmona-Gutierrez, D.; Hofer, S.J.; Kroemer, G. Caloric Restriction Mimetics against Age-Associated Disease: Targets, Mechanisms, and Therapeutic Potential. *Cell Metab.* **2019**, *29*, 592–610. [CrossRef] [PubMed]

104. Howitz, K.T.; Bitterman, K.J.; Cohen, H.Y.; Lamming, D.W.; Lavu, S.; Wood, J.G.; Zipkin, R.E.; Chung, P.; Kisielewski, A.; Zhang, L.L.; et al. Small molecule activators of sirtuins extend Saccharomyces cerevisiae lifespan. *Nature* **2003**, *425*, 191–196. [CrossRef] [PubMed]

105. Baur, J.A.; Pearson, K.J.; Price, N.L.; Jamieson, H.A.; Lerin, C.; Kalra, A.; Prabhu, V.V.; Allard, J.S.; Lopez-Lluch, G.; Lewis, K.; et al. Resveratrol improves health and survival of mice on a high-calorie diet. *Nature* **2006**, *444*, 337–342. [CrossRef]

106. Collins, J.J.; Evason, K.; Kornfeld, K. Pharmacology of delayed aging and extended lifespan of Caenorhabditis elegans. *Exp. Gerontol.* **2006**, *41*, 1032–1039. [CrossRef]

107. Bass, T.M.; Weinkove, D.; Houthoofd, K.; Gems, D.; Partridge, L. Effects of resveratrol on lifespan in Drosophila melanogaster and Caenorhabditis elegans. *Mech. Ageing Dev.* **2007**, *128*, 546–552. [CrossRef]

108. Rascon, B.; Hubbard, B.P.; Sinclair, D.A.; Amdam, G.V. The lifespan extension effects of resveratrol are conserved in the honey bee and may be driven by a mechanism related to caloric restriction. *Aging* **2012**, *4*, 499–508. [CrossRef]

109. Wang, C.; Wheeler, C.T.; Alberico, T.; Sun, X.; Seeberger, J.; Laslo, M.; Spangler, E.; Kern, B.; de Cabo, R.; Zou, S. The effect of resveratrol on lifespan depends on both gender and dietary nutrient composition in Drosophila melanogaster. *Age* **2013**, *35*, 69–81. [CrossRef]

110. Baur, J.A.; Sinclair, D.A. Therapeutic potential of resveratrol: The in vivo evidence. *Nat. Rev. Drug Discov.* **2006**, *5*, 493–506. [CrossRef]

111. Tsai, Y.F.; Chen, C.Y.; Chang, W.Y.; Syu, Y.T.; Hwang, T.L. Resveratrol suppresses neutrophil activation via inhibition of Src family kinases to attenuate lung injury. *Free Radic. Biol. Med.* **2019**, *145*, 67–77. [CrossRef]

112. Wang, L.L.; Shi, D.L.; Gu, H.Y.; Zheng, M.Z.; Hu, J.; Song, X.H.; Shen, Y.L.; Chen, Y.Y. Resveratrol attenuates inflammatory hyperalgesia by inhibiting glial activation in mice spinal cords. *Mol. Med. Rep.* **2016**, *13*, 4051–4057. [CrossRef] [PubMed]

113. Aghamiri, S.; Jafarpour, A.; Zandsalimi, F.; Aghemiri, M.; Shoja, M. Effect of resveratrol on the radiosensitivity of 5-FU in human breast cancer MCF-7 cells. *J. Cell. Biochem.* **2019**, *120*, 15671–15677. [CrossRef] [PubMed]

114. Chassot, L.N.; Scolaro, B.; Roschel, G.G.; Cogliati, B.; Cavalcanti, M.F.; Abdalla, D.S.P.; Castro, I.A. Comparison between red wine and isolated trans-resveratrol on the prevention and regression of atherosclerosis in LDLr ((-/-)) mice. *J. Nutr. Biochem.* **2018**, *61*, 48–55. [CrossRef] [PubMed]

115. Figueira, L.; Gonzalez, J.C. Effect of resveratrol on seric vascular endothelial growth factor concentrations during atherosclerosis. *Clin. Investig. Arterioscler.* **2018**, *30*, 209–216. [CrossRef]

116. Lomholt, S.; Mellemkjaer, A.; Iversen, M.B.; Pedersen, S.B.; Kragstrup, T.W. Resveratrol displays anti-inflammatory properties in an ex vivo model of immune mediated inflammatory arthritis. *BMC Rheumatol.* **2018**, *2*, 27. [CrossRef]

117. Oz, B.; Yildirim, A.; Yolbas, S.; Celik, Z.B.; Etem, E.O.; Deniz, G.; Akin, M.; Akar, Z.A.; Karatas, A.; Koca, S.S. Resveratrol inhibits Src tyrosine kinase, STAT3, and Wnt signaling pathway in collagen induced arthritis model. *BioFactors* **2019**, *45*, 69–74. [CrossRef]

118. Wang, H.M.; Li, G.X.; Zheng, H.S.; Wu, X.Z. Protective effect of resveratrol on lens epithelial cell apoptosis in diabetic cataract rat. *Asian Pac. J. Trop. Med.* **2015**, *8*, 153–156. [CrossRef]

119. Zheng, Y.; Liu, Y.; Ge, J.; Wang, X.; Liu, L.; Bu, Z.; Liu, P. Resveratrol protects human lens epithelial cells against H2O2-induced oxidative stress by increasing catalase, SOD-1, and HO-1 expression. *Mol. Vis.* **2010**, *16*, 1467–1474.

120. Higashi, Y.; Higashi, K.; Mori, A.; Sakamoto, K.; Ishii, K.; Nakahara, T. Anti-cataract Effect of Resveratrol in High-Glucose-Treated Streptozotocin-Induced Diabetic Rats. *Biol. Pharm. Bull.* **2018**, *41*, 1586–1592. [CrossRef]

121. Huo, X.; Zhang, T.; Meng, Q.; Li, C.; You, B. Resveratrol Effects on a Diabetic Rat Model with Coronary Heart Disease. *Med Sci. Monit. Int. Med J. Exp. Clin. Res.* **2019**, *25*, 540–546. [CrossRef]

122. Sarkar, O.; Li, Y.; Anand-Srivastava, M.B. Resveratrol prevents the development of high blood pressure in spontaneously hypertensive rats through the inhibition of enhanced expression of Gialpha proteins (1). *Can. J. Physiol. Pharmacol.* **2019**, *97*, 872–879. [CrossRef] [PubMed]

123. Yousefian, M.; Shakour, N.; Hosseinzadeh, H.; Hayes, A.W.; Hadizadeh, F.; Karimi, G. The natural phenolic compounds as modulators of NADPH oxidases in hypertension. *Phytomed. Int. J. Phytother. Phytopharm.* **2019**, *55*, 200–213. [CrossRef] [PubMed]

124. Szkudelski, T.; Szkudelska, K. Resveratrol and diabetes: From animal to human studies. *Biochim. Biophys. Acta* **2015**, *1852*, 1145–1154. [CrossRef] [PubMed]

125. Ozturk, E.; Arslan, A.K.K.; Yerer, M.B.; Bishayee, A. Resveratrol and diabetes: A critical review of clinical studies. *Biomed. Pharmacother. Biomed. Pharmacother.* **2017**, *95*, 230–234. [CrossRef] [PubMed]

126. Cheang, W.S.; Wong, W.T.; Wang, L.; Cheng, C.K.; Lau, C.W.; Ma, R.C.W.; Xu, A.; Wang, N.; Huang, Y.; Tian, X.Y. Resveratrol ameliorates endothelial dysfunction in diabetic and obese mice through sirtuin 1 and peroxisome proliferator-activated receptor delta. *Pharmacol. Res.* **2019**, *139*, 384–394. [CrossRef] [PubMed]

127. Javid, A.Z.; Hormoznejad, R.; Yousefimanesh, H.A.; Haghighi-Zadeh, M.H.; Zakerkish, M. Impact of resveratrol supplementation on inflammatory, antioxidant, and periodontal markers in type 2 diabetic patients with chronic periodontitis. *Diabetes Metab. Syndr.* **2019**, *13*, 2769–2774. [CrossRef] [PubMed]

128. Wang, X.; Chen, L.; Peng, W. Protective effects of resveratrol on osteoporosis via activation of the SIRT1-NF-kappaB signaling pathway in rats. *Exp. Ther. Med.* **2017**, *14*, 5032–5038. [CrossRef]

129. Feng, Y.L.; Jiang, X.T.; Ma, F.F.; Han, J.; Tang, X.L. Resveratrol prevents osteoporosis by upregulating FoxO1 transcriptional activity. *Int. J. Mol. Med.* **2018**, *41*, 202–212. [CrossRef]

130. Li, J.; Xin, Z.; Cai, M. The role of resveratrol in bone marrow-derived mesenchymal stem cells from patients with osteoporosis. *J. Cell. Biochem.* **2019**, *120*, 16634–16642. [CrossRef]

131. Kim, D.; Nguyen, M.D.; Dobbin, M.M.; Fischer, A.; Sananbenesi, F.; Rodgers, J.T.; Delalle, I.; Baur, J.A.; Sui, G.; Armour, S.M.; et al. SIRT1 deacetylase protects against neurodegeneration in models for Alzheimer's disease and amyotrophic lateral sclerosis. *EMBO J.* **2007**, *26*, 3169–3179. [CrossRef]

132. Alkhouli, M.F.; Hung, J.; Squire, M.; Anderson, M.; Castro, M.; Babu, J.R.; Al-Nakkash, L.; Broderick, T.L.; Plochocki, J.H. Exercise and resveratrol increase fracture resistance in the 3xTg-AD mouse model of Alzheimer's disease. *BMC Complement. Altern. Med.* **2019**, *19*, 39. [CrossRef] [PubMed]

133. Cosin-Tomas, M.; Senserrich, J.; Arumi-Planas, M.; Alquezar, C.; Pallas, M.; Martin-Requero, A.; Sunol, C.; Kaliman, P.; Sanfeliu, C. Role of Resveratrol and Selenium on Oxidative Stress and Expression of Antioxidant and Anti-Aging Genes in Immortalized Lymphocytes from Alzheimer's Disease Patients. *Nutrients* **2019**, *11*, 1764. [CrossRef] [PubMed]

134. Timmers, S.; Konings, E.; Bilet, L.; Houtkooper, R.H.; van de Weijer, T.; Goossens, G.H.; Hoeks, J.; van der Krieken, S.; Ryu, D.; Kersten, S.; et al. Calorie restriction-like effects of 30 days of resveratrol supplementation on energy metabolism and metabolic profile in obese humans. *Cell Metab.* **2011**, *14*, 612–622. [CrossRef]

135. Witte, A.V.; Kerti, L.; Margulies, D.S.; Floel, A. Effects of resveratrol on memory performance, hippocampal functional connectivity, and glucose metabolism in healthy older adults. *J. Neurosci. Off. J. Soc. Neurosci.* **2014**, *34*, 7862–7870. [CrossRef]

136. Liu, S.; Zheng, Z.; Ji, S.; Liu, T.; Hou, Y.; Li, S.; Li, G. Resveratrol reduces senescence-associated secretory phenotype by SIRT1/NF-kappaB pathway in gut of the annual fish Nothobranchius guentheri. *Fish Shellfish Immunol.* **2018**, *80*, 473–479. [CrossRef]

137. Chang, Y.C.; Liu, H.W.; Chen, Y.T.; Chen, Y.A.; Chen, Y.J.; Chang, S.J. Resveratrol protects muscle cells against palmitate-induced cellular senescence and insulin resistance through ameliorating autophagic flux. *J. Food Drug Anal.* **2018**, *26*, 1066–1074. [CrossRef]

138. Kirpichnikov, D.; McFarlane, S.I.; Sowers, J.R. Metformin: An update. *Ann. Intern. Med.* **2002**, *137*, 25–33. [CrossRef]

139. Cabreiro, F.; Au, C.; Leung, K.Y.; Vergara-Irigaray, N.; Cocheme, H.M.; Noori, T.; Weinkove, D.; Schuster, E.; Greene, N.D.; Gems, D. Metformin retards aging in C. elegans by altering microbial folate and methionine metabolism. *Cell* **2013**, *153*, 228–239. [CrossRef]

140. De Haes, W.; Frooninckx, L.; Van Assche, R.; Smolders, A.; Depuydt, G.; Billen, J.; Braeckman, B.P.; Schoofs, L.; Temmerman, L. Metformin promotes lifespan through mitohormesis via the peroxiredoxin PRDX-2. *Proc. Natl. Acad. Sci. USA* **2014**, *111*, E2501–E2509. [CrossRef]

141. Wu, L.; Zhou, B.; Oshiro-Rapley, N.; Li, M.; Paulo, J.A.; Webster, C.M.; Mou, F.; Kacergis, M.C.; Talkowski, M.E.; Carr, C.E.; et al. An Ancient, Unified Mechanism for Metformin Growth Inhibition in C. elegans and Cancer. *Cell* **2016**, *167*, 1705–1718.e1713. [CrossRef]

142. Slack, C.; Foley, A.; Partridge, L. Activation of AMPK by the putative dietary restriction mimetic metformin is insufficient to extend lifespan in Drosophila. *PLoS ONE* **2012**, *7*, e47699. [CrossRef] [PubMed]

143. Onken, B.; Driscoll, M. Metformin induces a dietary restriction-like state and the oxidative stress response to extend C. elegans Healthspan via AMPK, LKB1, and SKN-1. *PLoS ONE* **2010**, *5*, e8758. [CrossRef] [PubMed]

144. Martin-Montalvo, A.; Mercken, E.M.; Mitchell, S.J.; Palacios, H.H.; Mote, P.L.; Scheibye-Knudsen, M.; Gomes, A.P.; Ward, T.M.; Minor, R.K.; Blouin, M.J.; et al. Metformin improves healthspan and lifespan in mice. *Nat. Commun.* **2013**, *4*, 2192. [CrossRef] [PubMed]

145. Barzilai, N.; Crandall, J.P.; Kritchevsky, S.B.; Espeland, M.A. Metformin as a Tool to Target Aging. *Cell Metab.* **2016**, *23*, 1060–1065. [CrossRef] [PubMed]

146. Chen, J.; Ou, Y.; Li, Y.; Hu, S.; Shao, L.W.; Liu, Y. Metformin extends C. elegans lifespan through lysosomal pathway. *Elife* **2017**, *6*, e31268. [CrossRef] [PubMed]

147. Collier, C.A.; Bruce, C.R.; Smith, A.C.; Lopaschuk, G.; Dyck, D.J. Metformin counters the insulin-induced suppression of fatty acid oxidation and stimulation of triacylglycerol storage in rodent skeletal muscle. *Am. J. Physiol. Endocrinol. Metab.* **2006**, *291*, E182–E189. [CrossRef] [PubMed]

148. Kim, Y.D.; Park, K.G.; Lee, Y.S.; Park, Y.Y.; Kim, D.K.; Nedumaran, B.; Jang, W.G.; Cho, W.J.; Ha, J.; Lee, I.K.; et al. Metformin inhibits hepatic gluconeogenesis through AMP-activated protein kinase-dependent regulation of the orphan nuclear receptor SHP. *Diabetes* **2008**, *57*, 306–314. [CrossRef]

149. Wu, K.; Tian, R.; Huang, J.; Yang, Y.; Dai, J.; Jiang, R.; Zhang, L. Metformin alleviated endotoxemia-induced acute lung injury via restoring AMPK-dependent suppression of mTOR. *Chem. Biol. Interact.* **2018**, *291*, 1–6. [CrossRef]

150. Algire, C.; Moiseeva, O.; Deschenes-Simard, X.; Amrein, L.; Petruccelli, L.; Birman, E.; Viollet, B.; Ferbeyre, G.; Pollak, M.N. Metformin reduces endogenous reactive oxygen species and associated DNA damage. *Cancer Prev. Res.* **2012**, *5*, 536–543. [CrossRef]

151. Marinello, P.C.; da Silva, T.N.; Panis, C.; Neves, A.F.; Machado, K.L.; Borges, F.H.; Guarnier, F.A.; Bernardes, S.S.; de-Freitas-Junior, J.C.; Morgado-Diaz, J.A.; et al. Mechanism of metformin action in MCF-7 and MDA-MB-231 human breast cancer cells involves oxidative stress generation, DNA damage, and transforming growth factor beta1 induction. *Tumour Biol. J. Int. Soc. Oncodev. Biol. Med.* **2016**, *37*, 5337–5346. [CrossRef]

152. Sarfstein, R.; Friedman, Y.; Attias-Geva, Z.; Fishman, A.; Bruchim, I.; Werner, H. Metformin downregulates the insulin/IGF-I signaling pathway and inhibits different uterine serous carcinoma (USC) cells proliferation and migration in p53-dependent or -independent manners. *PLoS ONE* **2013**, *8*, e61537. [CrossRef] [PubMed]

153. Zhang, Y.; Li, M.X.; Wang, H.; Zeng, Z.; Li, X.M. Metformin down-regulates endometrial carcinoma cell secretion of IGF-1 and expression of IGF-1R. *Asian Pac. J. Cancer Prev.* **2015**, *16*, 221–225. [CrossRef] [PubMed]

154. Abo-Elmatty, D.M.; Ahmed, E.A.; Tawfik, M.K.; Helmy, S.A. Metformin enhancing the antitumor efficacy of carboplatin against Ehrlich solid carcinoma grown in diabetic mice: Effect on IGF-1 and tumoral expression of IGF-1 receptors. *Int. Immunopharmacol.* **2017**, *44*, 72–86. [CrossRef] [PubMed]

155. Karnewar, S.; Neeli, P.K.; Panuganti, D.; Kotagiri, S.; Mallappa, S.; Jain, N.; Jerald, M.K.; Kotamraju, S. Metformin regulates mitochondrial biogenesis and senescence through AMPK mediated H3K79 methylation: Relevance in age-associated vascular dysfunction. *Biochim. Biophys. Acta Mol. Basis Dis.* **2018**, *1864*, 1115–1128. [CrossRef] [PubMed]

156. Kuang, Y.; Hu, B.; Feng, G.; Xiang, M.; Deng, Y.; Tan, M.; Li, J.; Song, J. Metformin prevents against oxidative stress-induced senescence in human periodontal ligament cells. *Biogerontology* **2020**, *21*, 13–27. [CrossRef]

157. Dodds, S.G.; Livi, C.B.; Parihar, M.; Hsu, H.K.; Benavides, A.D.; Morris, J.; Javors, M.; Strong, R.; Christy, B.; Hasty, P.; et al. Adaptations to chronic rapamycin in mice. *Pathobiol. Aging Age Relat. Dis.* **2016**, *6*, 31688. [CrossRef]

158. Swindell, W.R. Rapamycin in mice. *Aging* **2017**, *9*, 1941–1942. [CrossRef]

159. Blagosklonny, M.V. Fasting and rapamycin: Diabetes versus benevolent glucose intolerance. *Cell Death Dis.* **2019**, *10*, 607. [CrossRef]

160. Papadopoli, D.; Boulay, K.; Kazak, L.; Pollak, M.; Mallette, F.; Topisirovic, I.; Hulea, L. mTOR as a central regulator of lifespan and aging. *F1000Research* **2019**, *8*. [CrossRef]

161. Liu, Y.C.; Gao, X.X.; Zhang, Z.G.; Lin, Z.H.; Zou, Q.L. PPAR Gamma Coactivator 1 Beta (PGC-1beta) Reduces Mammalian Target of Rapamycin (mTOR) Expression via a SIRT1-Dependent Mechanism in Neurons. *Cell. Mol. Neurobiol.* **2017**, *37*, 879–887. [CrossRef]

162. Zhu, X.; Chu, H.; Jiang, S.; Liu, Q.; Liu, L.; Xue, Y.; Zheng, S.; Wan, W.; Qiu, J.; Wang, J.; et al. Sirt1 ameliorates systemic sclerosis by targeting the mTOR pathway. *J. Dermatol. Sci.* **2017**, *87*, 149–158. [CrossRef] [PubMed]

163. Tokunaga, C.; Yoshino, K.; Yonezawa, K. mTOR integrates amino acid- and energy-sensing pathways. *Biochem. Biophys. Res. Commun.* **2004**, *313*, 443–446. [CrossRef] [PubMed]

164. Salminen, A.; Kaarniranta, K. Regulation of the aging process by autophagy. *Trends Mol. Med.* **2009**, *15*, 217–224. [CrossRef] [PubMed]

165. Kim, Y.C.; Guan, K.L. mTOR: A pharmacologic target for autophagy regulation. *J. Clin. Investig.* **2015**, *125*, 25–32. [CrossRef]

166. Munson, M.J.; Ganley, I.G. MTOR, PIK3C3, and autophagy: Signaling the beginning from the end. *Autophagy* **2015**, *11*, 2375–2376. [CrossRef]

167. Dall, K.B.; Faergeman, N.J. Metabolic regulation of lifespan from a C. elegans perspective. *Genes Nutr.* **2019**, *14*, 25. [CrossRef]

168. Kapahi, P.; Zid, B.M.; Harper, T.; Koslover, D.; Sapin, V.; Benzer, S. Regulation of lifespan in Drosophila by modulation of genes in the TOR signaling pathway. *Curr. Biol.* **2004**, *14*, 885–890. [CrossRef]

169. Kaeberlein, M.; Powers, R.W., 3rd; Steffen, K.K.; Westman, E.A.; Hu, D.; Dang, N.; Kerr, E.O.; Kirkland, K.T.; Fields, S.; Kennedy, B.K. Regulation of yeast replicative life span by TOR and Sch9 in response to nutrients. *Science* **2005**, *310*, 1193–1196. [CrossRef]

170. Lamming, D.W.; Ye, L.; Katajisto, P.; Goncalves, M.D.; Saitoh, M.; Stevens, D.M.; Davis, J.G.; Salmon, A.B.; Richardson, A.; Ahima, R.S.; et al. Rapamycin-induced insulin resistance is mediated by mTORC2 loss and uncoupled from longevity. *Science* **2012**, *335*, 1638–1643. [CrossRef]

171. Wu, J.J.; Liu, J.; Chen, E.B.; Wang, J.J.; Cao, L.; Narayan, N.; Fergusson, M.M.; Rovira, I.I.; Allen, M.; Springer, D.A.; et al. Increased mammalian lifespan and a segmental and tissue-specific slowing of aging after genetic reduction of mTOR expression. *Cell Rep.* **2013**, *4*, 913–920. [CrossRef]

172. Li, J.; Zhao, R.; Zhao, H.; Chen, G.; Jiang, Y.; Lyu, X.; Wu, T. Reduction of Aging-Induced Oxidative Stress and Activation of Autophagy by Bilberry Anthocyanin Supplementation via the AMPK-mTOR Signaling Pathway in Aged Female Rats. *J. Agric. Food Chem.* **2019**, *67*, 7832–7843. [CrossRef] [PubMed]

173. Di Francesco, A.; Diaz-Ruiz, A.; de Cabo, R.; Bernier, M. Intermittent mTOR Inhibition Reverses Kidney Aging in Old Rats. *J. Gerontol. Ser. A Biol. Sci. Med Sci.* **2018**, *73*, 843–844. [CrossRef] [PubMed]

174. Wang, Y.; Li, X.; He, Z.; Chen, W.; Lu, J. Rapamycin attenuates palmitate-induced lipid aggregation by up-regulating sirt-1 signaling in AML12 hepatocytes. *Die Pharm.* **2016**, *71*, 733–737. [CrossRef]

175. Fang, Y.; Hill, C.M.; Darcy, J.; Reyes-Ordonez, A.; Arauz, E.; McFadden, S.; Zhang, C.; Osland, J.; Gao, J.; Zhang, T.; et al. Effects of rapamycin on growth hormone receptor knockout mice. *Proc. Natl. Acad. Sci. USA* **2018**, *115*, E1495–E1503. [CrossRef] [PubMed]

176. Krebs, M.; Brunmair, B.; Brehm, A.; Artwohl, M.; Szendroedi, J.; Nowotny, P.; Roth, E.; Furnsinn, C.; Promintzer, M.; Anderwald, C.; et al. The Mammalian target of rapamycin pathway regulates nutrient-sensitive glucose uptake in man. *Diabetes* **2007**, *56*, 1600–1607. [CrossRef] [PubMed]

177. Ueno, M.; Carvalheira, J.B.; Tambascia, R.C.; Bezerra, R.M.; Amaral, M.E.; Carneiro, E.M.; Folli, F.; Franchini, K.G.; Saad, M.J. Regulation of insulin signalling by hyperinsulinaemia: Role of IRS-1/2 serine phosphorylation and the mTOR/p70 S6K pathway. *Diabetologia* **2005**, *48*, 506–518. [CrossRef]

178. Zhou, W.; Ye, S. Rapamycin improves insulin resistance and hepatic steatosis in type 2 diabetes rats through activation of autophagy. *Cell Biol. Int.* **2018**, *42*, 1282–1291. [CrossRef]

179. He, S.; Zhang, Y.; Wang, D.; Tao, K.; Zhang, S.; Wei, L.; Chen, Q. Rapamycin/GABA combination treatment ameliorates diabetes in NOD mice. *Mol. Immunol.* **2016**, *73*, 130–137. [CrossRef]

180. Reifsnyder, P.C.; Flurkey, K.; Te, A.; Harrison, D.E. Rapamycin treatment benefits glucose metabolism in mouse models of type 2 diabetes. *Aging* **2016**, *8*, 3120–3130. [CrossRef]

181. Garcia, D.N.; Saccon, T.D.; Pradiee, J.; Rincon, J.A.A.; Andrade, K.R.S.; Rovani, M.T.; Mondadori, R.G.; Cruz, L.A.X.; Barros, C.C.; Masternak, M.M.; et al. Effect of caloric restriction and rapamycin on ovarian aging in mice. *Geroscience* **2019**, *41*, 395–408. [CrossRef]

182. Laberge, R.M.; Sun, Y.; Orjalo, A.V.; Patil, C.K.; Freund, A.; Zhou, L.; Curran, S.C.; Davalos, A.R.; Wilson-Edell, K.A.; Liu, S.; et al. MTOR regulates the pro-tumorigenic senescence-associated secretory phenotype by promoting IL1A translation. *Nat. Cell Biol.* **2015**, *17*, 1049–1061. [CrossRef] [PubMed]

183. Horvath, S.; Lu, A.T.; Cohen, H.; Raj, K. Rapamycin retards epigenetic ageing of keratinocytes independently of its effects on replicative senescence, proliferation and differentiation. *Aging (Albany NY)* **2019**, *11*, 3238–3249. [CrossRef] [PubMed]

184. Bishop-Bailey, D. Peroxisome proliferator-activated receptors in the cardiovascular system. *Br. J. Pharmacol.* **2000**, *129*, 823–834. [CrossRef] [PubMed]

185. Gross, B.; Pawlak, M.; Lefebvre, P.; Staels, B. PPARs in obesity-induced T2DM, dyslipidaemia and NAFLD. *Nat. Rev. Endocrinol.* **2017**, *13*, 36–49. [CrossRef] [PubMed]

186. Jay, M.A.; Ren, J. Peroxisome proliferator-activated receptor (PPAR) in metabolic syndrome and type 2 diabetes mellitus. *Curr. Diabetes Rev.* **2007**, *3*, 33–39. [CrossRef] [PubMed]

187. Kersten, S.; Desvergne, B.; Wahli, W. Roles of PPARs in health and disease. *Nature* **2000**, *405*, 421–424. [CrossRef] [PubMed]

188. Mansour, M. The roles of peroxisome proliferator-activated receptors in the metabolic syndrome. *Prog. Mol. Biol. Transl. Sci.* **2014**, *121*, 217–266. [CrossRef]

189. Hamilton-Craig, I.; Kostner, K.M.; Woodhouse, S.; Colquhoun, D. Use of fibrates in clinical practice: Queensland Lipid Group consensus recommendations. *Int. J. Evid. Based Healthc.* **2012**, *10*, 181–190. [CrossRef]

190. Hong, Y.A.; Lim, J.H.; Kim, M.Y.; Kim, T.W.; Kim, Y.; Yang, K.S.; Park, H.S.; Choi, S.R.; Chung, S.; Kim, H.W.; et al. Fenofibrate improves renal lipotoxicity through activation of AMPK-PGC-1alpha in db/db mice. *PLoS ONE* **2014**, *9*, e96147. [CrossRef]

191. Kim, E.N.; Lim, J.H.; Kim, M.Y.; Kim, H.W.; Park, C.W.; Chang, Y.S.; Choi, B.S. PPARalpha agonist, fenofibrate, ameliorates age-related renal injury. *Exp. Gerontol.* **2016**, *81*, 42–50. [CrossRef]

192. Bojic, L.A.; Huff, M.W. Peroxisome proliferator-activated receptor delta: A multifaceted metabolic player. *Curr. Opin. Lipidol.* **2013**, *24*, 171–177. [CrossRef] [PubMed]

193. Ding, Y.; Yang, K.D.; Yang, Q. The role of PPARdelta signaling in the cardiovascular system. *Prog. Mol. Biol. Transl. Sci.* **2014**, *121*, 451–473. [CrossRef] [PubMed]

194. Vazquez-Carrera, M. Unraveling the Effects of PPARbeta/delta on Insulin Resistance and Cardiovascular Disease. *Trends Endocrinol. Metab.* **2016**, *27*, 319–334. [CrossRef]

195. Rizos, C.V.; Kei, A.; Elisaf, M.S. The current role of thiazolidinediones in diabetes management. *Arch. Toxicol.* **2016**, *90*, 1861–1881. [CrossRef]

196. Yki-Jarvinen, H. Thiazolidinediones. *N. Engl. J. Med.* **2004**, *351*, 1106–1118. [CrossRef]

197. Barbieri, M.; Bonafe, M.; Rizzo, M.R.; Ragno, E.; Olivieri, F.; Marchegiani, F.; Franceschi, C.; Paolisso, G. Gender specific association of genetic variation in peroxisome proliferator-activated receptor (PPAR)gamma-2 with longevity. *Exp. Gerontol.* **2004**, *39*, 1095–1100. [CrossRef]

198. Yang, H.C.; Deleuze, S.; Zuo, Y.; Potthoff, S.A.; Ma, L.J.; Fogo, A.B. The PPARgamma agonist pioglitazone ameliorates aging-related progressive renal injury. *J. Am. Soc. Nephrol.* **2009**, *20*, 2380–2388. [CrossRef]

199. Patel, H.; Giri, P.; Patel, P.; Singh, S.; Gupta, L.; Patel, U.; Modi, N.; Shah, K.; Jain, M.R.; Srinivas, N.R.; et al. Preclinical evaluation of saroglitazar magnesium, a dual PPAR-alpha/gamma agonist for treatment of dyslipidemia and metabolic disorders. *Xenobiotica* **2018**, *48*, 1268–1277. [CrossRef]

200. Patel, M.R.; Kansagra, K.A.; Parikh, D.P.; Parmar, D.V.; Patel, H.B.; Soni, M.M.; Patil, U.S.; Patel, H.V.; Patel, J.A.; Gujarathi, S.S.; et al. Effect of Food on the Pharmacokinetics of Saroglitazar Magnesium, a Novel Dual PPARalphagamma Agonist, in Healthy Adult Subjects. *Clin. Drug Investig.* **2018**, *38*, 57–65. [CrossRef]

201. Kaul, U.; Parmar, D.; Manjunath, K.; Shah, M.; Parmar, K.; Patil, K.P.; Jaiswal, A. New dual peroxisome proliferator activated receptor agonist-Saroglitazar in diabetic dyslipidemia and non-alcoholic fatty liver disease: Integrated analysis of the real world evidence. *Cardiovasc. Diabetol.* **2019**, *18*, 80. [CrossRef]

202. Xu, H.R.; Zhang, J.W.; Chen, W.L.; Ning, Z.Q.; Li, X.N. Pharmacokinetics, Safety and Tolerability of Chiglitazar, A Novel Peroxisome Proliferator-Activated Receptor (PPAR) Pan-Agonist, in Healthy Chinese Volunteers: A Phase I Study. *Clin. Drug Investig.* **2019**, *39*, 553–563. [CrossRef] [PubMed]

203. Park, M.H.; Kim, D.H.; Kim, M.J.; Lee, E.K.; An, H.J.; Jeong, J.W.; Kim, H.R.; Kim, S.J.; Yu, B.P.; Moon, H.R.; et al. Effects of MHY908, a New Synthetic PPARalpha/gamma Dual Agonist, on Inflammatory Responses and Insulin Resistance in Aged Rats. *J. Gerontol. A Biol. Sci. Med. Sci.* **2016**, *71*, 300–309. [CrossRef] [PubMed]

204. Wang, Z.; Gao, J.; Ohno, Y.; Liu, H.; Xu, C. Rosiglitazone ameliorates senescence and promotes apoptosis in ovarian cancer induced by olaparib. *Cancer Chemother. Pharmacol.* **2020**, 1–12. [CrossRef] [PubMed]

205. Veech, R.L.; Chance, B.; Kashiwaya, Y.; Lardy, H.A.; Cahill, G.F., Jr. Ketone bodies, potential therapeutic uses. *IUBMB Life* **2001**, *51*, 241–247. [CrossRef] [PubMed]

206. Heitmann, R.N.; Dawes, D.J.; Sensenig, S.C. Hepatic ketogenesis and peripheral ketone body utilization in the ruminant. *J. Nutr.* **1987**, *117*, 1174–1180. [CrossRef] [PubMed]

207. Cahill, G.F., Jr.; Herrera, M.G.; Morgan, A.P.; Soeldner, J.S.; Steinke, J.; Levy, P.L.; Reichard, G.A., Jr.; Kipnis, D.M. Hormone-fuel interrelationships during fasting. *J. Clin. Investig.* **1966**, *45*, 1751–1769. [CrossRef] [PubMed]

208. Hegardt, F.G. Mitochondrial 3-hydroxy-3-methylglutaryl-CoA synthase: A control enzyme in ketogenesis. *Biochem. J.* **1999**, *338*, 569–582. [CrossRef]

209. Shimazu, T.; Hirschey, M.D.; Newman, J.; He, W.; Shirakawa, K.; Le Moan, N.; Grueter, C.A.; Lim, H.; Saunders, L.R.; Stevens, R.D.; et al. Suppression of oxidative stress by beta-hydroxybutyrate, an endogenous histone deacetylase inhibitor. *Science* **2013**, *339*, 211–214. [CrossRef]

210. Bae, H.R.; Kim, D.H.; Park, M.H.; Lee, B.; Kim, M.J.; Lee, E.K.; Chung, K.W.; Kim, S.M.; Im, D.S.; Chung, H.Y. beta-Hydroxybutyrate suppresses inflammasome formation by ameliorating endoplasmic reticulum stress via AMPK activation. *Oncotarget* **2016**, *7*, 66444–66454. [CrossRef]

211. Maalouf, M.; Sullivan, P.G.; Davis, L.; Kim, D.Y.; Rho, J.M. Ketones inhibit mitochondrial production of reactive oxygen species production following glutamate excitotoxicity by increasing NADH oxidation. *Neuroscience* **2007**, *145*, 256–264. [CrossRef]

212. Newman, J.C.; Verdin, E. Ketone bodies as signaling metabolites. *Trends Endocrinol. Metab.* **2014**, *25*, 42–52. [CrossRef] [PubMed]

213. Kinzig, K.P.; Honors, M.A.; Hargrave, S.L. Insulin sensitivity and glucose tolerance are altered by maintenance on a ketogenic diet. *Endocrinology* **2010**, *151*, 3105–3114. [CrossRef] [PubMed]

214. Garbow, J.R.; Doherty, J.M.; Schugar, R.C.; Travers, S.; Weber, M.L.; Wentz, A.E.; Ezenwajiaku, N.; Cotter, D.G.; Brunt, E.M.; Crawford, P.A. Hepatic steatosis, inflammation, and ER stress in mice maintained long term on a very low-carbohydrate ketogenic diet. *Am. J. Physiol. Gastrointest. Liver Physiol.* **2011**, *300*, G956–G967. [CrossRef] [PubMed]

215. McDaniel, S.S.; Rensing, N.R.; Thio, L.L.; Yamada, K.A.; Wong, M. The ketogenic diet inhibits the mammalian target of rapamycin (mTOR) pathway. *Epilepsia* **2011**, *52*, e7–e11. [CrossRef]

216. Kennedy, A.R.; Pissios, P.; Otu, H.; Roberson, R.; Xue, B.; Asakura, K.; Furukawa, N.; Marino, F.E.; Liu, F.F.; Kahn, B.B.; et al. A high-fat, ketogenic diet induces a unique metabolic state in mice. *Am. J. Physiol. Endocrinol. Metab.* **2007**, *292*, E1724–E1739. [CrossRef]

217. Veech, R.L.; Bradshaw, P.C.; Clarke, K.; Curtis, W.; Pawlosky, R.; King, M.T. Ketone bodies mimic the life span extending properties of caloric restriction. *IUBMB Life* **2017**, *69*, 305–314. [CrossRef]

218. Kim, D.H.; Park, M.H.; Ha, S.; Bang, E.J.; Lee, Y.; Lee, A.K.; Lee, J.; Yu, B.P.; Chung, H.Y. Anti-inflammatory action of beta-hydroxybutyrate via modulation of PGC-1alpha and FoxO1, mimicking calorie restriction. *Aging (Albany NY)* **2019**, *11*, 1283–1304. [CrossRef]

219. Kimura, I.; Inoue, D.; Maeda, T.; Hara, T.; Ichimura, A.; Miyauchi, S.; Kobayashi, M.; Hirasawa, A.; Tsujimoto, G. Short-chain fatty acids and ketones directly regulate sympathetic nervous system via G protein-coupled receptor 41 (GPR41). *Proc. Natl. Acad. Sci. USA* **2011**, *108*, 8030–8035. [CrossRef]

220. Tunaru, S.; Kero, J.; Schaub, A.; Wufka, C.; Blaukat, A.; Pfeffer, K.; Offermanns, S. PUMA-G and HM74 are receptors for nicotinic acid and mediate its anti-lipolytic effect. *Nat. Med.* **2003**, *9*, 352–355. [CrossRef]

221. Ahmed, K.; Tunaru, S.; Offermanns, S. GPR109A, GPR109B and GPR81, a family of hydroxy-carboxylic acid receptors. *Trends Pharmacol. Sci.* **2009**, *30*, 557–562. [CrossRef]

222. Fu, S.P.; Liu, B.R.; Wang, J.F.; Xue, W.J.; Liu, H.M.; Zeng, Y.L.; Huang, B.X.; Li, S.N.; Lv, Q.K.; Wang, W.; et al. beta-Hydroxybutyric acid inhibits growth hormone-releasing hormone synthesis and secretion through the GPR109A/extracellular signal-regulated 1/2 signalling pathway in the hypothalamus. *J. Neuroendocrinol.* **2015**, *27*, 212–222. [CrossRef] [PubMed]

223. Rahman, M.; Muhammad, S.; Khan, M.A.; Chen, H.; Ridder, D.A.; Muller-Fielitz, H.; Pokorna, B.; Vollbrandt, T.; Stolting, I.; Nadrowitz, R.; et al. The beta-hydroxybutyrate receptor HCA2 activates a neuroprotective subset of macrophages. *Nat. Commun.* **2014**, *5*, 3944. [CrossRef]

224. Fu, S.P.; Li, S.N.; Wang, J.F.; Li, Y.; Xie, S.S.; Xue, W.J.; Liu, H.M.; Huang, B.X.; Lv, Q.K.; Lei, L.C.; et al. BHBA suppresses LPS-induced inflammation in BV-2 cells by inhibiting NF-kappaB activation. *Mediat. Inflamm.* **2014**, *2014*, 983401. [CrossRef] [PubMed]

225. Gambhir, D.; Ananth, S.; Veeranan-Karmegam, R.; Elangovan, S.; Hester, S.; Jennings, E.; Offermanns, S.; Nussbaum, J.J.; Smith, S.B.; Thangaraju, M.; et al. GPR109A as an anti-inflammatory receptor in retinal pigment epithelial cells and its relevance to diabetic retinopathy. *Investig. Ophthalmol. Vis. Sci.* **2012**, *53*, 2208–2217. [CrossRef] [PubMed]

226. Puchalska, P.; Crawford, P.A. Multi-dimensional Roles of Ketone Bodies in Fuel Metabolism, Signaling, and Therapeutics. *Cell Metab.* **2017**, *25*, 262–284. [CrossRef]

227. Bang, E.; Lee, B.; Noh, S.G.; Kim, D.H.; Jung, H.J.; Ha, S.; Yu, B.P.; Chung, H.Y. Modulation of senoinflammation by calorie restriction based on biochemical and Omics big data analysis. *BMB Rep.* **2019**, *52*, 56–63. [CrossRef]

228. Park, H.S.; Nam, H.S.; Seo, H.S.; Hwang, S.J. Change of periodontal inflammatory indicators through a 4-week weight control intervention including caloric restriction and exercise training in young Koreans: A pilot study. *BMC Oral Health* **2015**, *15*, 109. [CrossRef]

229. Youm, Y.H.; Nguyen, K.Y.; Grant, R.W.; Goldberg, E.L.; Bodogai, M.; Kim, D.; D'Agostino, D.; Planavsky, N.; Lupfer, C.; Kanneganti, T.D.; et al. The ketone metabolite beta-hydroxybutyrate blocks NLRP3 inflammasome-mediated inflammatory disease. *Nat. Med.* **2015**, *21*, 263–269. [CrossRef]

230. Wang, J.; Vanegas, S.M.; Du, X.; Noble, T.; Zingg, J.M.; Meydani, M.; Meydani, S.N.; Wu, D. Caloric restriction favorably impacts metabolic and immune/inflammatory profiles in obese mice but curcumin/piperine consumption adds no further benefit. *Nutr. Metab.* **2013**, *10*, 29. [CrossRef]

231. Yamada, K.; Takizawa, S.; Ohgaku, Y.; Asami, T.; Furuya, K.; Yamamoto, K.; Takahashi, F.; Hamajima, C.; Inaba, C.; Endo, K.; et al. MicroRNA 16-5p is upregulated in calorie-restricted mice and modulates inflammatory cytokines of macrophages. *Gene* **2020**, *725*, 144191. [CrossRef]

232. Su, H.Y.; Lee, H.C.; Cheng, W.Y.; Huang, S.Y. A calorie-restriction diet supplemented with fish oil and high-protein powder is associated with reduced severity of metabolic syndrome in obese women. *Eur. J. Clin. Nutr.* **2015**, *69*, 322–328. [CrossRef] [PubMed]

233. Willette, A.A.; Bendlin, B.B.; McLaren, D.G.; Canu, E.; Kastman, E.K.; Kosmatka, K.J.; Xu, G.; Field, A.S.; Alexander, A.L.; Colman, R.J.; et al. Age-related changes in neural volume and microstructure associated with interleukin-6 are ameliorated by a calorie-restricted diet in old rhesus monkeys. *Neuroimage* **2010**, *51*, 987–994. [CrossRef] [PubMed]

234. Das, U.N. A defect in the activity of Delta6 and Delta5 desaturases may be a factor predisposing to the development of insulin resistance syndrome. *Prostaglandins Leukot Essent Fat. Acids* **2005**, *72*, 343–350. [CrossRef] [PubMed]

235. Willette, A.A.; Coe, C.L.; Birdsill, A.C.; Bendlin, B.B.; Colman, R.J.; Alexander, A.L.; Allison, D.B.; Weindruch, R.H.; Johnson, S.C. Interleukin-8 and interleukin-10, brain volume and microstructure, and the influence of calorie restriction in old rhesus macaques. *Age* **2013**, *35*, 2215–2227. [CrossRef] [PubMed]

236. Liu, B.; Page, A.J.; Hatzinikolas, G.; Chen, M.; Wittert, G.A.; Heilbronn, L.K. Intermittent Fasting Improves Glucose Tolerance and Promotes Adipose Tissue Remodeling in Male Mice Fed a High-Fat Diet. *Endocrinology* **2019**, *160*, 169–180. [CrossRef]

237. Sharov, A.A.; Falco, G.; Piao, Y.; Poosala, S.; Becker, K.G.; Zonderman, A.B.; Longo, D.L.; Schlessinger, D.; Ko, M. Effects of aging and calorie restriction on the global gene expression profiles of mouse testis and ovary. *BMC Biol.* **2008**, *6*, 24. [CrossRef]

238. Chen, J.; Mo, R.; Lescure, P.A.; Misek, D.E.; Hanash, S.; Rochford, R.; Hobbs, M.; Yung, R.L. Aging is associated with increased T-cell chemokine expression in C57BL/6 mice. *J. Gerontol. A Biol. Sci. Med. Sci.* **2003**, *58*, 975–983. [CrossRef]

239. Babu, H.; Ambikan, A.T.; Gabriel, E.E.; Svensson Akusjarvi, S.; Palaniappan, A.N.; Sundaraj, V.; Mupanni, N.R.; Sperk, M.; Cheedarla, N.; Sridhar, R.; et al. Systemic Inflammation and the Increased Risk of Inflamm-Aging and Age-Associated Diseases in People Living With HIV on Long Term Suppressive Antiretroviral Therapy. *Front. Immunol.* **2019**, *10*, 1965. [CrossRef]

240. Lan, C.E.; Hung, Y.T.; Fang, A.H.; Ching-Shuang, W. Effects of irradiance on UVA-induced skin aging. *J. Dermatol. Sci.* **2019**, *94*, 220–228. [CrossRef]

241. Wang, M.; Zhang, L.; Zhu, W.; Zhang, J.; Kim, S.H.; Wang, Y.; Ni, L.; Telljohann, R.; Monticone, R.E.; McGraw, K.; et al. Calorie Restriction Curbs Proinflammation That Accompanies Arterial Aging, Preserving a Youthful Phenotype. *J. Am. Heart Assoc.* **2018**, *7*, e009112. [CrossRef]

242. Caria, C.; Gotardo, E.M.F.; Santos, P.S.; Acedo, S.C.; de Morais, T.R.; Ribeiro, M.L.; Gambero, A. Extracellular matrix remodeling and matrix metalloproteinase inhibition in visceral adipose during weight cycling in mice. *Exp. Cell Res.* **2017**, *359*, 431–440. [CrossRef] [PubMed]

243. Varani, J.; Bhagavathula, N.; Aslam, M.N.; Fay, K.; Warner, R.L.; Hanosh, A.; Barron, A.G.; Miller, R.A. Inhibition of retinoic acid-induced skin irritation in calorie-restricted mice. *Arch. Dermatol. Res.* **2008**, *300*, 27–35. [CrossRef] [PubMed]

244. Wiggins, J.E.; Patel, S.R.; Shedden, K.A.; Goyal, M.; Wharram, B.L.; Martini, S.; Kretzler, M.; Wiggins, R.C. NFkappaB promotes inflammation, coagulation, and fibrosis in the aging glomerulus. *J. Am. Soc. Nephrol.* **2010**, *21*, 587–597. [CrossRef] [PubMed]

245. Kim, H.J.; Jung, K.J.; Seo, A.Y.; Choi, J.S.; Yu, B.P.; Chung, H.Y. Calorie restriction modulates redox-sensitive AP-1 during the aging process. *J. Am. Aging Assoc.* **2002**, *25*, 123–130. [CrossRef] [PubMed]

246. Reddy Avula, C.P.; Lawrence, R.A.; Zaman, K.; Fernandes, G. Inhibition of intracellular peroxides and apoptosis of lymphocytes in lupus-prone B/W mice by dietary n-6 and n-3 lipids with calorie restriction. *J. Clin. Immunol.* **2002**, *22*, 206–219. [CrossRef] [PubMed]

247. Lin, Y.Y.; Hsieh, P.S.; Cheng, Y.J.; Cheng, S.M.; Chen, C.J.; Huang, C.Y.; Kuo, C.H.; Kao, C.L.; Shyu, W.C.; Lee, S.D. Anti-apoptotic and Pro-survival Effects of Food Restriction on High-Fat Diet-Induced Obese Hearts. *Cardiovasc. Toxicol.* **2017**, *17*, 163–174. [CrossRef]

248. Ando, K.; Higami, Y.; Tsuchiya, T.; Kanematsu, T.; Shimokawa, I. Impact of aging and life-long calorie restriction on expression of apoptosis-related genes in male F344 rat liver. *Microsc. Res. Tech.* **2002**, *59*, 293–300. [CrossRef]

249. Nylen, C.; Lundell, L.S.; Massart, J.; Zierath, J.R.; Naslund, E. Short-term low-calorie diet remodels skeletal muscle lipid profile and metabolic gene expression in obese adults. *Am. J. Physiol. Endocrinol. Metab.* **2019**, *316*, E178–E185. [CrossRef]

250. Palee, S.; Minta, W.; Mantor, D.; Sutham, W.; Jaiwongkam, T.; Kerdphoo, S.; Pratchayasakul, W.; Chattipakorn, S.C.; Chattipakorn, N. Combination of exercise and calorie restriction exerts greater efficacy on cardioprotection than monotherapy in obese-insulin resistant rats through the improvement of cardiac calcium regulation. *Metabolism* **2019**, *94*, 77–87. [CrossRef]

251. Kim, D.H.; Park, M.H.; Lee, E.K.; Choi, Y.J.; Chung, K.W.; Moon, K.M.; Kim, M.J.; An, H.J.; Park, J.W.; Kim, N.D.; et al. The roles of FoxOs in modulation of aging by calorie restriction. *Biogerontology* **2015**, *16*, 1–14. [CrossRef]

252. Lopez-Domenech, S.; Abad-Jimenez, Z.; Iannantuoni, F.; de Maranon, A.M.; Rovira-Llopis, S.; Morillas, C.; Banuls, C.; Victor, V.M.; Rocha, M. Moderate weight loss attenuates chronic endoplasmic reticulum stress and mitochondrial dysfunction in human obesity. *Mol. Metab.* **2019**, *19*, 24–33. [CrossRef]

253. Kim, G.W.; Lin, J.E.; Snook, A.E.; Aing, A.S.; Merlino, D.J.; Li, P.; Waldman, S.A. Calorie-induced ER stress suppresses uroguanylin satiety signaling in diet-induced obesity. *Nutr. Diabetes* **2016**, *6*, e211. [CrossRef] [PubMed]

254. Ding, S.; Jiang, J.; Zhang, G.; Bu, Y.; Zhang, G.; Zhao, X. Resveratrol and caloric restriction prevent hepatic steatosis by regulating SIRT1-autophagy pathway and alleviating endoplasmic reticulum stress in high-fat diet-fed rats. *PLoS ONE* **2017**, *12*, e0183541. [CrossRef]

255. Walters, R.O.; Arias, E.; Diaz, A.; Burgos, E.S.; Guan, F.; Tiano, S.; Mao, K.; Green, C.L.; Qiu, Y.; Shah, H.; et al. Sarcosine Is Uniquely Modulated by Aging and Dietary Restriction in Rodents and Humans. *Cell Rep.* **2018**, *25*, 663–676.e666. [CrossRef] [PubMed]

256. Gregosa, A.; Vinuesa, A.; Todero, M.F.; Pomilio, C.; Rossi, S.P.; Bentivegna, M.; Presa, J.; Wenker, S.; Saravia, F.; Beauquis, J. Periodic dietary restriction ameliorates amyloid pathology and cognitive impairment in PDAPP-J20 mice: Potential implication of glial autophagy. *Neurobiol. Dis.* **2019**, *132*, 104542. [CrossRef]

257. Devarajan, A.; Rajasekaran, N.S.; Valburg, C.; Ganapathy, E.; Bindra, S.; Freije, W.A. Maternal perinatal calorie restriction temporally regulates the hepatic autophagy and redox status in male rat. *Free Radic. Biol. Med.* **2019**, *130*, 592–600. [CrossRef]

258. Marino, G.; Pietrocola, F.; Madeo, F.; Kroemer, G. Caloric restriction mimetics: Natural/physiological pharmacological autophagy inducers. *Autophagy* **2014**, *10*, 1879–1882. [CrossRef]

259. van Eyk, H.J.; van Schinkel, L.D.; Kantae, V.; Dronkers, C.E.A.; Westenberg, J.J.M.; de Roos, A.; Lamb, H.J.; Jukema, J.W.; Harms, A.C.; Hankemeier, T.; et al. Caloric restriction lowers endocannabinoid tonus and improves cardiac function in type 2 diabetes. *Nutr. Diabetes* **2018**, *8*, 6. [CrossRef]

260. Wang, S.; Huang, M.; You, X.; Zhao, J.; Chen, L.; Wang, L.; Luo, Y.; Chen, Y. Gut microbiota mediates the anti-obesity effect of calorie restriction in mice. *Sci. Rep.* **2018**, *8*, 13037. [CrossRef] [PubMed]

261. Chung, K.W.; Lee, E.K.; Lee, M.K.; Oh, G.T.; Yu, B.P.; Chung, H.Y. Impairment of PPARalpha and the Fatty Acid Oxidation Pathway Aggravates Renal Fibrosis during Aging. *J. Am. Soc. Nephrol.* **2018**, *29*, 1223–1237. [CrossRef]
262. Kim, J.Y.; Kim, D.H.; Choi, J.; Park, J.K.; Jeong, K.S.; Leeuwenburgh, C.; Yu, B.P.; Chung, H.Y. Changes in lipid distribution during aging and its modulation by calorie restriction. *Age* **2009**, *31*, 127–142. [CrossRef] [PubMed]

 © 2020 by the authors. Licensee MDPI, Basel, Switzerland. This article is an open access article distributed under the terms and conditions of the Creative Commons Attribution (CC BY) license (http://creativecommons.org/licenses/by/4.0/).

Review

Mechanisms of Calorie Restriction: A Review of Genes Required for the Life-Extending and Tumor-Inhibiting Effects of Calorie Restriction

Toshimitsu Komatsu [1], Seongjoon Park [1], Hiroko Hayashi [1], Ryoichi Mori [1], Haruyoshi Yamaza [2] and Isao Shimokawa [1,*]

[1] Department of Pathology, Nagasaki University School of Medicine and Graduate School of Biomedical Sciences, Nagasaki 852-8521, Japan; komatsut@nagasaki-u.ac.jp (T.K.); psj1026@nagasaki-u.ac.jp (S.P.); hayashih@nagasaki-u.ac.jp (H.H.); ryoichi@nagasaki-u.ac.jp (R.M.)

[2] Section of Pediatric Dentistry, Division of Oral Health, Growth and Development, Faculty of Dental Science, Kyushu University, 3-1-1 Maidashi, Higashi-ku, Fukuoka 812-8582, Japan; hyamaza@dent.kyushu-u.ac.jp

* Correspondence: shimo@nagasaki-u.ac.jp; Tel.: +81-95-819-7051

Received: 8 November 2019; Accepted: 13 December 2019; Published: 16 December 2019

Abstract: This review focuses on mechanisms of calorie restriction (CR), particularly the growth hormone (GH)/insulin-like growth factor-1 (IGF-1) axis as an evolutionary conserved signal that regulates aging and lifespan, underlying the effects of CR in mammals. Topics include (1) the relation of the GH-IGF-1 signal with chronic low-level inflammation as one of the possible causative factors of aging, that is, inflammaging, (2) the isoform specificity of the forkhead box protein O (FoxO) transcription factors in CR-mediated regulation of cancer and lifespan, (3) the role for FoxO1 in the tumor-inhibiting effect of CR, (4) pleiotropic roles for FoxO1 in the regulation of disorders, and (5) sirtuin (Sirt) as a molecule upstream of FoxO. From the evolutionary view, the necessity of neuropeptide Y (Npy) for the effects of CR and the pleiotropic roles for Npy in life stages are also emphasized. Genes for mediating the effects of CR and regulating aging are context-dependent, particularly depending on nutritional states.

Keywords: calorie restriction; FoxO transcription factor; sirtuin; neuropeptide Y; pleiotropy of CR genes

1. Introduction

Laboratory rodents often continue to gain body weight after puberty until certain age points under standard husbandry conditions in which animals can freely access food, often called ad libitum (AL) feeding. In contrast, restriction of food intake in mice by 30% to 40%, when initiated at young age, e.g., 12 weeks of age, limits the gain in body weight [1]. Dietary regimens involving food restriction, here referred to as calorie restriction (CR), reduce morbidity and mortality in experimental animals compared with AL feeding animals [1]. Since the original report on the effects of CR in 1935 [2], many laboratories have confirmed the effects of CR and investigated the underlying molecular mechanisms. Notably, a recent meta-analysis of lifespan studies in laboratory rodents described various responses to CR, that is, no extension or even shortened lifespans in CR rodents [3]. The analysis also suggested potentially substantial effects of the genotypes of animals as well as husbandry on experimental outcomes in CR studies. A genome-scale metabolic model and transcriptome data in the male Sprague Dawley rat liver also revealed varied metabolic responses of the liver with different levels and durations of CR [4]. Recent studies have also identified sexual dimorphism in physiological responses to CR [5,6]. Kane et al. [6] found that female rodents may have a greater response to CR than male rodents, with modulated sensitivities to mechanistic target of rapamycin (Mtor), growth hormone

(GH), insulin-like growth factor-1 (IGF-1) or fibroblastic growth factor 21 and greater inflammatory and mitochondrial health responses [6]. Therefore, the effects of CR may not be universal as originally expected. Nonetheless, CR models in a range of organisms have contributed to our understanding of the aging process.

Epistasis analyses using mutant strains in lower organisms such as *Caenorhabditis elegans* (*C. elegans*) have revealed genes required for the effects of CR (referred to here as CR genes) and the signal pathways mediating the effects of CR. In *C. elegans*, a number of genes such as *aak-2*, *daf-16*, *skin-1*, *clk-1*, and *pha-4* have been reported to be associated with the life-prolonging effect of CR [7]. Some of these genes also mediate the effects of CR in mice. Previous studies also reported that mutations of single genes (referred to here as longevity genes) can extend lifespan even in AL feeding animals. Many of these genes can be functionally categorized into genes associated with nutrient sensing or metabolic responses [8]. Among these gene mutations, reduction- or loss-of-function mutations of genes in the growth hormone (GH)-insulin-like growth factor-1 (IGF-1) signaling consistently extend lifespan in a range of organisms [8]. Since CR is known to decrease the plasma concentration of GH and IGF-1, the GH-IGF-1 pathway is considered an evolutionary conserved pathway for longevity and a main aspect of the mechanism of CR [9].

Thus far, a total of 112 CR genes in yeast, 62 in nematode, 27 in drosophila, and seven in mice have been identified and are listed in the database [10]. Among these genes, forkhead box protein O 3 (*Foxo3*) and sirtuin 1 (*Sirt1*) genes are common in mice, nematodes, and flies. CR and longevity gene models have elucidated signal pathways for the extension of lifespan, although the signal pathways are context dependent. This review will focus on FoxO transcription factors and the sirtuin deacetylases that are upstream of FoxO. We also discuss roles for neuropeptide Y (*Npy*), which is essential for the effects of CR in mice and thus mammals.

2. A Central Role for GH and IGF-1 in the Regulation of Lifespan

Previous studies reported that a single gene mutation can prolong the lifespan of experimental animals under AL conditions of standard diets. Genetic analyses of long-lived strains of nematodes identified a mutation in a single gene, *age-1* [11]. This gene was later found to encode a component of phosphoinositide 3-kinase (PI3K) that is important for growth factor signals such as those modulated by IGF-1 [12]. Kenyon et al. showed that mutation of *daf-2*, which is associated with resistance larval development, prolongs the lifespan of *C. elegans*, and *daf-16* is required [13]. Daf-2 and Daf-16 correspond to the receptor for IGF-1 and the FoxO transcription factor in mammals, respectively. Thus, in *C. elegans*, attenuated IGF-1 signaling promotes activation of Daf-16, leading to transcriptional modulation of its target genes and extended lifespan. Even in *Drosophila*, mutations in genes in insulin-like signaling such as *Inr* (insulin-like receptor) and *chico* (insulin-receptor substrate) can extend the lifespan [14]. In these conditions, *dFoxo* is also required.

The lifespan of *C. elegans* can also be extended by suppression of target of rapamycin (Tor in mammalians, mechanistic target of rapamycin, Mtor), which is downstream of IGF-1 signaling and promotes cell proliferation and division when nutrients are abundant [15]. Mtor forms complexes (mTORC1 and mTORC2) with other molecules in nutrition and energy rich conditions. These complexes activate transcription and translation when insulin and growth factors concomitantly rise. Mtor complexes promote protein synthesis and cell division while inhibiting autophagy. All genetic manipulations that suppress Mtor identified thus far can extend the lifespan of *Drosophila* [16].

In mammals, GH is upstream of the IGF-1 signal. GH is competitively controlled by Somatostatin and GH-releasing hormone (Ghrh), which are secreted from hypothalamic neurons. In mice, reduction-of-function gene mutations in molecules involved in the signal between Ghrh and IGF-1 consistently prolong lifespan [17]. Furthermore, longevity is achieved by inhibition of mTORC1 by rapamycin [18], deletion of the *S6K* gene [19], and suppression of Mtor [15,20].

Together these results in a range of experimental animals indicate that signal attenuation of the IGF-1 signal, activation of FoxO transcription factor, and suppression of Mtor are key mechanisms for slowing aging and prolonging lifespan (Figure 1). However, it should be noted that the life-extending effect of the reduced IGF-1 signaling could be sexually dimorphic. In Igf1 receptor (*Igf1r*) gene heterozygotic knockout (*Igf1*$^{+/-}$) mouse models of different genetic backgrounds (129/SvPas and C57BL/6J) [21,22], only female *Igf1r*$^{+/-}$ mice significantly outlived controls, whereas this was not observed for males. Genetic studies in humans also indicate that genetic variations causing reduced IGF-1 signaling are beneficial for survival in old age only in females but not in males [23–25]. The mechanisms underlying the sexual dimorphism for longevity in the context of reduction of IGF-1 signaling remain to be explored.

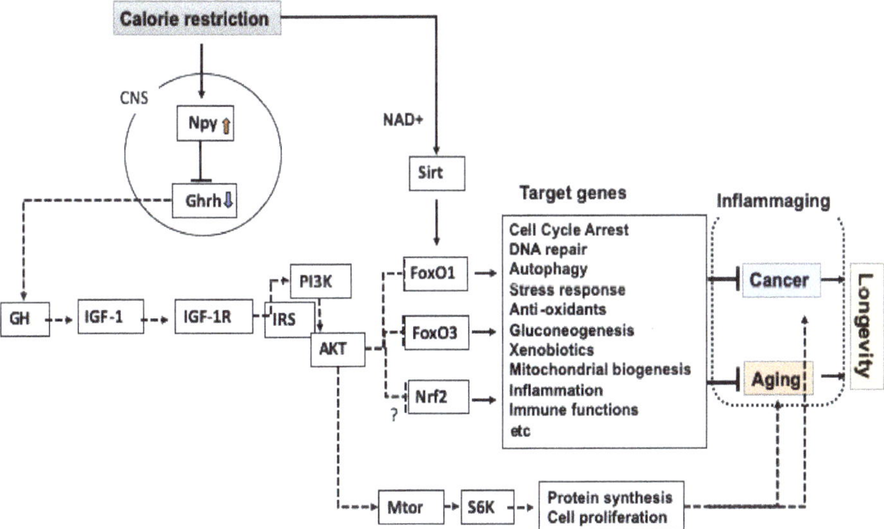

Figure 1. Mechanisms underlying the life-extending and tumor-inhibiting effects of calorie restriction in mammals. Inhibition of the growth hormone (GH)-insulin-like growth factor-1 (IGF-1) signal promotes activation of FoxO1, FoxO3, and Nrf2 transcription factors, resulting in activated target genes and inhibition of cancer and aging, and thus counteracting "inflammaging." By contrast, attenuation of the GH-IGF-1 signal reduces activity of mechanistic target of rapamycin (Mtor). In *C. elegans*, the Daf-2 pathway regulates skn-1 (mammalian Nrf2), although no evidence for this regulation is shown in mammals. Sirtuins (Sirts) modulate activities of key molecules mediating calorie restriction (CR) effects by epigenetic mechanisms, mostly deacetylation. Dotted and straight lines represent attenuated and strengthened signals, respectively. Arrows and bars represent activation and inhibition of target molecules, respectively.

3. Possible Relation of the GH-IGF-1 Axis with Inflammaging

Generalized chronic low-level inflammation may be a causative factor for aging and related diseases. This low-level inflammation associated with the aging process is called inflammaging [26]. It is also called sterile inflammation because it is not caused by infection.

Senescent cells that increase in tissues with age not only stop undergoing cell division but also express inflammatory cytokines such as IL-1α, IFNβ, chemokines, matrix metalloproteases, and growth factors, inducing local inflammation in tissues. This phenomenon is called the senescence-accelerated secretory phenotype (SASP) [27]. A recent report showed that the average lifespan of mice can be extended by selectively removing senescent cells expressing p16 [Ink4a] from tissues [28]. The removal of senescent cells also suppresses age-related lipodystrophy, tumorigenesis, glomerulosclerosis of the kidney, and so on. This report indicates that senescent cells are involved in the progression of aging in animals.

SASP was reported to be caused by induction of type I interferon (IFN-I) signal by activating transcription of the retrotransposable element long-interspersed element-1 (L1) [29]. In normally proliferating cells, L1 is suppressed by a cell monitoring mechanism, e.g., three prime repair exonuclease 1 (Trex1) and the RB transcriptional corepressor 1 (Rb1). In senescent cells, these repression mechanisms are disrupted, leading to activation of L1 via activation of FoxA1 [29].

During exposure of cells to harmful stimuli and conditions, such as chemicals and ischemia, the injured cells excrete or secrete various molecules such as ATP, uric acid, oxidatively modified DNA, and aggregated proteins, which are called damage-associated molecular pattern molecules (DAMPs) [30]. DAMPs bind to receptors, such as Toll-like receptors, in the cell membrane and cytoplasm of resident macrophages and elicit innate immune and inflammatory responses. The sensor of innate immunity, called the nucleotide-binding domain, leucine-rich-containing family, pyrin domain-containing-3 (NLRP3) inflammasome, is activated, and macrophages secrete inflammatory cytokines and exacerbate inflammatory responses. These inflammatory signals have been reported to damage surrounding parenchymal cells. As with senescent cells, this minor but chronic inflammatory response at the cellular level is thought to be the main factor driving aging and disease, including cancer.

GH is not only secreted from anterior pituitary cells but is also expressed in various cells including immune cells. The GH receptor (Ghr) is also expressed in a wide range of cells, and the GH signaling system is activated by endocrine action as well as by autocrine and paracrine actions. Recent studies suggested that the sensitivity of the GH-IGF-1 signal increases with age and that DAMP inflammasome is likely to be activated [31]. This activation of the inflammasome is suppressed in long-lived Ghr-deficient mice [31]. Since CR also attenuates GH-IGF-1 signals, suppression of the inflammasome could be a main component of the anti-aging mechanism of CR.

4. Differential Regulation of Cancer and Lifespan by CR via FoxO Transcription Factors

Although multiple studies have used a variety of CR regimens in invertebrates and vertebrates, most of the CR regimens extend lifespans in these organisms. To analyze the signal pathways mediating the life-prolonging effects of CR, epistasis experiments in which mutation of a single gene causes attenuation or abrogation of the life-prolonging effect of CR have been conducted, mostly in *C. elegans*. These experiments have identified CR genes such as *daf-16*, *skn-1*, *pha-4*, *clk-1*, *aak-2*, and *hsf-1*, although the genes required for the effect of CR may differ depending on methods of CR [7]. These CR genes suggest the importance of regulation of insulin-like signaling, stress response, and mitochondrial bioenergetics.

CR is known to lower plasma concentrations of insulin and IGF-1 in mammals [9]. Therefore, IGF-1 signaling and thus FoxO transcription factors, the mammalian orthologs of Daf-16, were considered to play roles in the effects of CR. As mentioned, loss of Daf-16 abrogates the life-extending effects of CR in *C. elegans* [7]. In mammals, the FoxO transcription factor family includes four isoforms, FoxO1, FoxO3, FoxO4, and FoxO6 [32]. We tested the hypothesis that FoxOs are involved in the effects of CR by two lifespan studies using *Foxo1* and *Foxo3* knockout mice.

For the FoxO1 study, we used *Foxo1* $^{+/-}$ mice, because *Foxo1*$^{-/-}$ mice died around embryonic day 11 due to defects in the branchial arches and remarkably impaired vascular development of embryos and yolk sacs, indicating the necessity of *Foxo1* in development of the cardiovascular system [33]. In comparison, *Foxo1*$^{+/-}$ mice grow normally through pre- and post-natal stages. In *Foxo1*$^{+/-}$ mice, expression levels of *Foxo1* mRNA in the examined tissues were reduced by 50% compared with levels in wild-type (WT) mice [34]. Results of the lifespan study indicated that CR extended lifespan in *Foxo1*$^{+/-}$ mice to the same extent as in WT mice; however, unlike CR in WT mice, CR did not significantly reduce the proportion of *Foxo1*$^{+/-}$ mice bearing spontaneously occurring tumors. By contrast, in *Foxo3*$^{+/-}$ and *Foxo3*$^{-/-}$ mice, CR did not extend lifespan [35], indicating the requirement of the *Foxo3* gene in the life-extending effect of CR in mice. With AL feeding, *Foxo3*$^{+/-}$ mice displayed no reduction of lifespan compared with WT mice, while *Foxo3*$^{-/-}$ mice showed a slight reduction of lifespan [35]. Although the effect of CR in *Foxo1*-null mice should be investigated, our lifespan studies suggest a differential regulation of cancer and lifespan by CR via *Foxo1* and *Foxo3* (Figure 1).

In early studies using mice with germline-deleted *Foxo1*, -3, and -4 genes, deletion of a single gene had little effect on longevity or cancer development in postnatal life under AL conditions [36]. One exception was female *Foxo3*$^{-/-}$ mice, in which occurrence of pituitary adenoma was accelerated compared with WT and *Foxo3*$^{+/-}$ mice. Conditional, widespread somatic *Foxo* deletion in adult tissues was also achieved using an interferon-inducible Mx-Cre transgene to circumvent embryonic lethality of FoxO1 deficiency [36]. When the three *Foxo* genes were deleted in mice at the same time, tumor incidences were elevated, while lifespan was shortened. In particular, the incidences of thymic lymphoma and hemangioma were increased, compared with groups of mice retaining at least one *Foxo* allele [36]. *Foxo1*, -3, and -4 gene-deleted mice showed premature death due to the tumors. These experiments indicated that FoxO transcription factor genes are complementary as for development of cancers [36]. In other words, even if one *Foxo* gene is deleted, the other *Foxo* genes can substitute in AL conditions. However, our lifespan studies [34,35] indicate differential roles for FoxO1 and FoxO3 in the effects of CR.

The Daf-16 isoform-specific extension of lifespan in the context of reduction of IGF signaling has been also reported in *C. elegans*. The *daf-16* genomic locus encodes three groups of transcripts (a, b, d/f/h) that are transcribed from distinct promoters [37–39]. Depending on the experimental setting, Daf-16a and/or Daf-16d/f/h, but not Daf-16b, play a major role in the extension of lifespan [39,40].

Numerous human genetic studies have indicated a correlation of the minor alleles of the *FOXO3* gene with longevity [41]. However, there is no significant correlation of *FOXO1* genotypes with longevity. A genome-wide meta-analysis of 30844 adults of European ancestry from 21 studies confirmed that the known longevity-associated *FOXO3* variant rs2153960 is a genome-wide significant SNP for lowering IGF-1 concentrations [42].

The isoform specificity of FoxO and Daf-16 in the life-extending effect of CR as well as the human genomic studies for longevity suggest that the mechanism of lifespan control has been evolutionarily conserved. A recent study showed that FoxO6 functions in age-related insulin resistance and inflammation in rats with AL feeding, although CR was not investigated [43]. Thus, potential roles for FoxO4 and FoxO6 in the effects of CR remain to be elucidated.

How FoxO3 regulates lifespan remains unclear. Our microarray analysis in *Foxo1*$^{+/-}$ and *Foxo3*$^{+/-}$ CR mouse liver show that many genes are differentially regulated in these mice under conditions of CR (Tables S1 and S2). The pathway analysis of the differentially expressed genes between WT-CR and *Foxo1*$^{+/-}$ CR mice as well as WT-CR and *Foxo3*$^{+/-}$ CR mice also suggests that a number of inflammation and immune response pathways are activated in *Foxo3*$^{+/-}$ CR mice. By contrast, in *Foxo1*$^{+/-}$ CR mice, some T-cell functions may be down-regulated. FoxO3 is reported to prevent excess activation of interferone (IFN)-I in response to viral infection [44]. A ternary complex consisting of FoxO3, nuclear co-repressor 2 (Ncor2) and histone deacetylase 3 (Hdac3) exists on the promoter of interferon regulatory factor 7 (*Irf7*); a loss of FoxO3 enhances histone acetylation in the promoter of *Irs7* gene, resulting in increased levels of *Irf7* mRNA in macrophages [44]. An experimental stimulation of IFN-I activated the PI3K/Akt pathway, which in turn led to FoxO3 degradation. FoxO3 could play a role in optimizing host defense mechanisms against harmful stimuli as well as under unstimulated conditions by limiting the transcription of Irf7 and then Irf7-induced target inflammatory genes. This process could minimize activation of the inflammasome. Reduction of FoxO3 may disrupt the FoxO3-Irf7 regulatory circuit under CR conditions.

5. FoxO1 Mediates the Tumor-Inhibiting Effect of CR

A tumor-inhibiting but not life-extending effect of CR has been reported in *Nrf2*$^{-/-}$ mice [45]. The Nrf2 transcription factor belongs to the cap'n'collar basic-leucine zipper (CNC-bZIP) family. Under unstressed cellular conditions, Nrf2 binds to Keap1 in the cytoplasm and is degraded by the ubiquitin-proteasome system [46]. When the redox environment changes due to oxidative stress, Nrf2 and Keap1 dissociate and Nrf2 moves into the nucleus. Nrf2 forms heterodimers with other transcription factor cofactors such as Maf, binds to the antioxidant response element (ARE) in the promoter of target genes, and promotes the gene expressions of antioxidants and phase II detoxification enzymes.

The role of Nrf2 in the tumor-inhibiting effect of CR was tested in a two-stage skin carcinogenesis model using 7,12-dimethyl-benz (a) anthracene (DMBA) and 12-O-tetradecanoyl phorbol 13-acetate (TPA) [45]. In WT mice, CR significantly suppressed the development of skin tumors; in $Nrf2^{-/-}$ mice, the CR-mediated inhibitory effect on tumor growth was absent. In $Nrf2^{+/-}$ mice, the inhibitory effect was attenuated, suggesting a gene-dose dependency of $Nrf2$ in the inhibition of tumor formation by CR. However, even in $Nrf2^{-/-}$ mice, the life-extending effect of CR was preserved [45]. These results suggest that Nrf2 plays a major role in the tumor-inhibiting effect of CR but not for lifespan extension.

We also conducted a two-stage skin carcinogenesis model experiment to confirm the potential role for FoxO1 in the effect of CR, following the protocol of Pearson et al. [46] Appendix A. Briefly, DMBA was applied in the dorsal skin of mice at 31 weeks of age; after two weeks, TPA treatment was initiated. Skin papilloma started occurring six weeks after DMBA application, and thus four weeks later starting TPA treatment (Figure 2A). Our findings showed that haploinsufficiency of $Foxo1$ accelerated the occurrence of skin papilloma, particularly in CR conditions (WT-CR vs. $Foxo1^{+/-}$ CR, $p = 0.0103$ by log-rank test; Figure 2A), confirming the role for FoxO1 in the tumor-inhibiting effect of CR, as in Nrf2.

TPA promotes epidermal hyperplasia in the pre-neoplastic phase via reactive oxygen species (ROS) production and inflammation [47]. We assessed the epidermal thickness during TPA treatment and the proliferative activity of epidermal cells after TPA treatment. Epidermal thickness was greater in $Foxo1^{+/-}$ mice than in WT mice (genotype, $p = 0.0331$ by 3-f ANOVA: Figure 2B), although diet did not affect the thickness (diet, $p = 0.2993$ by 3-f ANOVA). Epidermal cell proliferation was greater in $Foxo1^{+/-}$ mice (genotype, $p = 0.0193$ by 3-f ANOVA; Figure 2C); the rate of apoptosis did not differ between WT and $Foxo1^{+/-}$ mice (data not shown). Thus, the reduced FoxO1 promotes epidermal cell proliferation in response to TPA.

We also evaluated macrophage infiltration in the dermis as an index of inflammation. Although the number of macrophages was less in the CR groups compared with the AL groups, indicating an effect of CR (diet, $p = 0.0001$ by 3-f ANOVA: Figure 2D), the number of macrophages did not significantly differ between WT and $Foxo1^{+/-}$ mice (genotype, $p = 0.1374$ by 3-f ANOVA).

We further analyzed the expression of genes associated with inflammatory stimuli. TPA induces cyclooxygenase-2 (COX-2), an enzyme that catalyzes arachidonic acid to prostaglandin G2, i.e., produces chemical mediators of inflammation, in the mouse skin [48]. The mRNA expression levels of COX-2 were elevated after TPA treatment (Figure 2E). The expression levels were significantly lower in WT-DR mice compared than other groups of mice; the effect is diminished in $Foxo1^{+/-}$ CR mice (diet; $p = 0.0123$; genotype x diet; $p = 0.0399$ by 3f-ANOVA). Although other inflammatory responses induced by TPA were mostly reduced in response to CR, even in $Foxo1^{+/-}$ CR mouse skin (data not shown), reduction of FoxO1 in the CR condition might affect the regulation of prostaglandins formation and then impair the tumor-inhibiting effect of CR.

The above results indicate that both FoxO1 and Nrf2 are necessary for the tumor-inhibiting effect of CR. Some Nrf2 target genes are also regulated by FoxO1 [49], such as heme oxygenasae (decycling) 1 (Hmox-1), glutamate-cysteine ligase, and catalytic subunit (Gclc) genes. Gclc is the rate-limiting enzyme in the synthesis of Glutathione (GSH). We also analyzed the expression of Nrf2-target genes and found that *Gclc* and *Hmox-1* mRNA expression levels were lower in $Foxo1^{+/-}$ mice than in WT mice (only the data of *Gclc* mRNA are shown in Figure 2F; genotype; $p = 0.0207$: diet; $p = 0.0175$: genotype x diet; $p = 0.5809$ by 3f-ANOVA). In a study in *C. elegans*, Daf-16 and Skn-1 target genes were also reported to be preferentially regulated to induce conditions of mitohormesis [49], which could be a mechanism of CR.

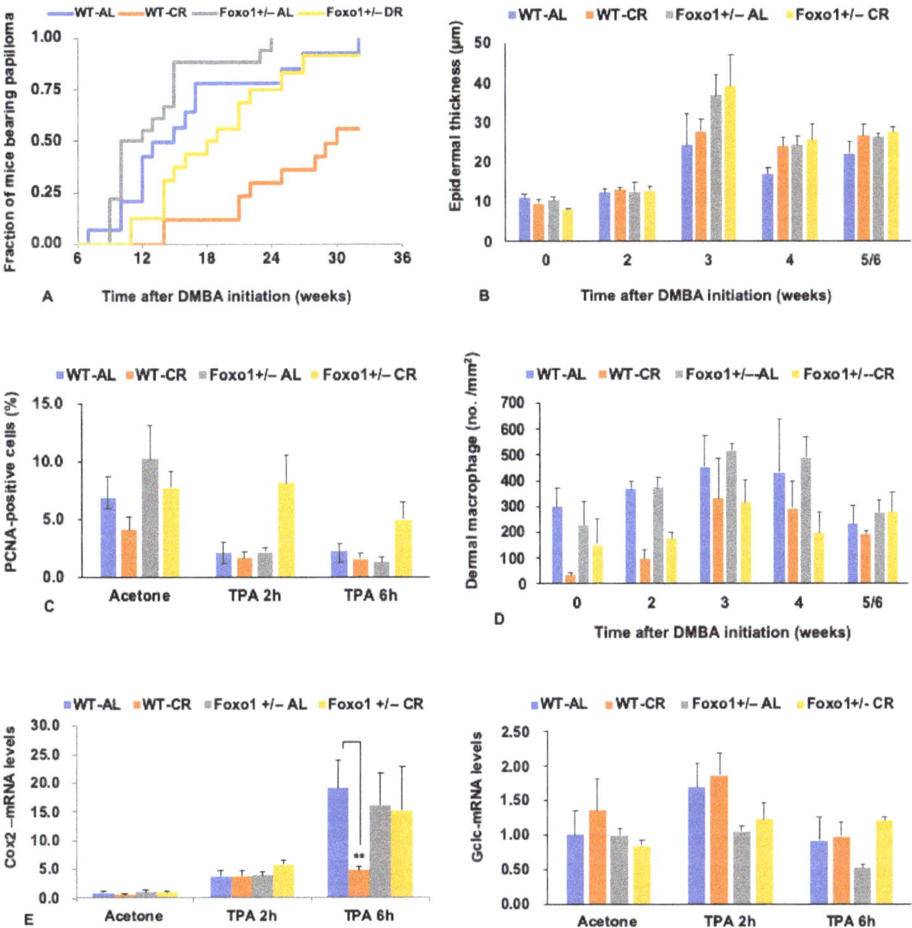

Figure 2. The inhibiting effects of CR on skin tumorigenesis are diminished in *Foxo1*[+/−] mice. (**A**) The fraction of mice bearing papilloma (diameter ≤ 1 mm) increased over time more rapidly in *Foxo1*[+/−] CR mice compared with wild-type (WT) CR mice (p = 0.0103 by Log-rank test); WT- ad libitum (AL) mice versus vs. *Foxo1*[+/−] AL mice, p = 0.0663. DMBA, 7,12-dimethyl-benz (a) anthracene. (**B**) Epidermal thickness during TPA (12-O-tetradecanoyl phorbol 13-acetate) treatment. Statistics: genotype, p = 0.0331; diet, p = 0.2993; time; p < 0.0001 by 3-f ANOVA. (**C**) Epidermal cell proliferation in response to TPA treatment. PCNA, proliferating cell nuclear antigen. Statistics: genotype, p = 0.0193; genotype x diet, p = 0.0401 by 3-f ANOVA. (**D**) The number of dermal macrophages immunohistochemically stained with F4/80 antibody. Statistics: diet; p = 0.0001; genotype; p = 0.1374; time, p = 0.0152 by 3-f ANOVA. (**E**) *Cox2* mRNA expression levels in response to TPA treatment (normalized by *Atp5f1* mRNA levels). Statistics: diet; p = 0.0123; genotype x diet; p = 0.0399 by 3f-ANOVA). **p < 0.01 by Tukey's honestly significant difference (HSD) test. (**F**) *Gclc* mRNA expression levels in response to TPA treatment (normalized by *Atp5f1* mRNA levels). Statistics: genotype; p = 0.0207; diet; p = 0.0175.

6. Pleiotropic Roles for Foxo1 in Disorders

As indicated above, reduction of FoxO1 diminishes the anti-tumor effect of CR. However, reduction of FoxO1 does not affect lifespans in AL and CR conditions [34]. Multiple studies have reported beneficial effects of reduced FoxO1 on lesions or disorders, although these results depended on nutritional states. In AL conditions, *Foxo1*[+/−] mice showed accelerated skin wound healing with enhanced keratinocyte migration, reduced granulation tissue formation, and decreased collagen density, accompanied by an attenuated inflammatory response, compared with WT mice [50]. *Foxo3*[−/−] mice did not show any significant phenotypes for skin wound healing and inflammation. Reduction of FoxO1 also ameliorates glucose intolerance and insulin insensitivity in diabetic models [51], probably due to diminution of gluconeogenesis. Indeed, FoxO1 is reported to increase hepatic gluconeogenesis through upregulation of glucose 6-phosphatase (G6pc) and phosphoenolpyruvate carboxykinase (Pck1) in CR and fasting conditions [42].

Stress resistance is one of the hallmarks in CR animals, although whether this trait directly causes the extension of lifespan remains unclear. We tested stress resistance in *Foxo1*[+/−] mice using the endotoxin shock model following the procedures of Kamohara [52]. The survival rates and endoplasmic reticulum (ER) stress were monitored after lipopolysaccharide (LPS) injection. ER stress-responsive alkaline phosphatase (ESTRAP) mice [53] were used to monitor ER stress in vivo (details are described elsewhere [52]). In ESTRAP mice, secreted alkaline phosphatase (SEAP) is constitutively expressed, and SEAP activity in the blood is reduced by ER stress [53]. The experiment was performed using mice at six months of age. The results showed that the survival rate was extended by CR, particularly in *Foxo1*[+/−] mice (diet, $p = 0.0263$; genotype × diet, $p = 0.0498$ by likelihood ratio test: Figure 3A). After LPS injection, SEAP activity decreased until 8 h in each group and then stayed constant or recovered between 8 and 24 h (Figure 3B). Between 2 and 8 h after LPS injection, that is, in the acute phase of ER stress, CR attenuated ER stress (diet, $p < 0.0001$ by 3-f ANOVA), and the attenuating effect was greater in *Foxo1*[+/−] mice compared with *Ctrl* mice (genotype, $p = 0.0093$ by 3-f ANOVA: Figure 3B). These data indicate that reduction of FoxO1 enhances stress resistance in both AL and CR conditions. Therefore, in the present experimental setting, reduction of FoxO1 could enhance stress resistance, particularly in the CR condition.

In summary, these findings demonstrate that FoxO1 exerts pleiotropic effects on disorders depending on nutritional conditions.

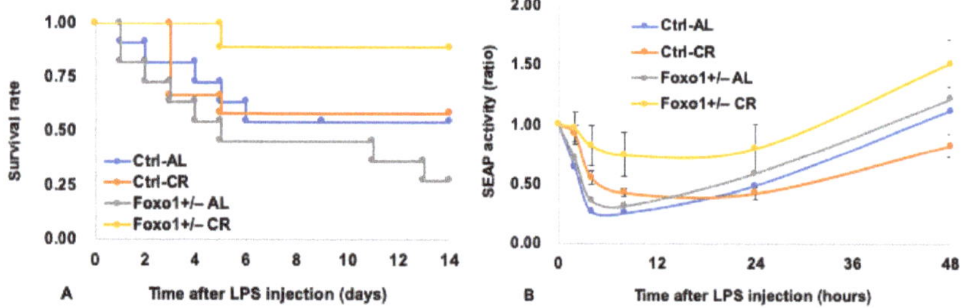

Figure 3. *Foxo1*[+/−] mice become tolerant to endotoxin, particularly in CR conditions. (**A**) Survival rates after LPS injection. Initial numbers of mice in each group, $n = 9$–12. Statistics: diet, $p = 0.0263$; genotype x diet, $p = 0.0498$ by likelihood ratio test. (**B**) Reduction of the rate of secreted alkaline phosphatase (SEAP) activity (ratios relative to the values at 0 h in respective groups) in the blood after intraperitoneal injection of lipopolysaccharide (LPS). The data represent means ± SE ($n = 9$–12). No error bars were added in *Foxo1*[+/−] AL and *Ctrl* AL groups to avoid overlapping. Statistics (when data between 2 and 8 h were analyzed): genotype, $p = 0.0093$; diet, $p < 0.0001$ by 3-f ANOVA).

7. Sirtuin as a Molecule Upstream of FoxO

Sirtuins (Sirts) are evolutionarily conserved proteins that catalyze the deacetylation and adenosine diphosphate (ADP) ribosylation of target proteins using the oxidized from of nicotinamide adenine dinucleotide (NAD[+]) as a coenzyme [54]. Since overexpression of *Sir2p*, one of the sirtuin genes in budding yeast, was reported to extend replicative lifespan of yeast, many studies have been conducted to demonstrate the importance of Sirts in the regulation of aging in various experimental models. Although findings in the early phase of investigation were controversial, re-examinations have confirmed Sirts as key molecules in the regulation of aging [55].

Although results in epistasis analyses using mutant strains in lower organisms are sometime affected by experimental settings such as CR or CR-like regimens [7,55], a number of studies have indicated the necessity of *Sirt1* in the life-extending effect of CR. In *Drosophila*, deletion of *Sirt1* (*dSir2*) in the fat body abolishes the life-extending effect of CR [56].

There are seven mammalian orthologs of nematode Sir2 (Sirt1–7) [54]. Sirt1, Sirt6, and Sirt7 exist mainly in the nucleus; Sirt2 is localized in the cytoplasm and Sirt3, Sirt4, and Sirt5 are in mitochondria [54]. Sirt targets in the nucleus include histones, transcription factors, and transcriptional regulators, which control gene expression by deacetylation. One example is the deacetylation of FoxO1 and PGC-1α by Sirt1 in the liver and increased expression of target genes related to gluconeogenesis and fatty acid oxidation [54].

Sirt in mitochondria regulates mitochondrial function, that is, energy metabolism, by deacetylating mitochondrial constituent proteins, energy metabolic pathway proteins, and mitochondrial transcription factors. Previous studies have shown that Sirt3 is deeply involved in the deacetylation of mitochondrial proteins [57].

SIRT4 and SIRT6 mainly function as ADP-ribosylating enzymes [44]. Accumulated evidence has indicated the importance of Sirts in the regulation of energy metabolism and genome instability.

Several studies have indicated the necessity of Sirt1 for the effects of CR in mice. Aged mouse kidney displays pathology such as glomerular and interstitial fibrosis, which is improved by CR [58]. However, this effect of CR is abrogated in *Sirt1$^{+/-}$* mice. From this model, Kume et al. reported Sirt1-dependent deacetylation of FoxO3 as a key mechanism that promotes mitochondrial autophagy and then reduces oxidative insult in kidneys. Finally, CR extends lifespan in *Sirt1$^{+/-}$* mice but not in *Sirt1$^{-/-}$* mice [59]. These findings indicate that one of the mechanisms underlying the effect of CR involves epigenetic regulation of FoxO3 by Sirt1 (Figure 1).

8. Roles for Npy in the Effect of CR

An evolutionary perspective predicted that the effect of CR is derived from adaptive responses in animals to harsh conditions such as a famine [60]. Animals should have acquired a system(s) to survive in famine, which frequently occurs in the wild. If not, a species could undergo extinction. When food resources are abundant, animals consume as much food to promote growth and reproduction, while simultaneously storing the remaining energy as fat in the body. In starvation, animals use the stored fat as an energy source by activation of the lipolytic pathway. During food shortage, physiological functions such as reproduction, growth, and heat are suppressed to inhibit excessive consumption of energy. The transition to Dauer states in nematodes and hibernation in some mammals could be extreme examples. The effects of CR may derive from these adaptive processes.

A previous study in *C. elegans* showed that two ciliated neurons that are part of the amphid sensilla (ASI neurons) in the head, in which chemosensitive receptors are present, are necessary for the lifespan extension by CR [61]. These neurons express *skn-1* as described above and are activated when dietary energy is reduced to promote the development of Dauer via the endocrine system [62]. The evolutionary view of CR and the finding in *C. elegans* suggest a possible implication of hypothalamic neurons in the effect of CR in vertebrates.

We focused on Npy, an orexigenic neuropeptide expressed in neurons in the arcuate nuclei of the hypothalamus and the sympathetic nervous system (SNS). Published data indicate that Npy plays a role in neuroendocrine adaptation to negative energy balance in mammals by suppressing activities of growth, reproduction, and heat, whereas it stimulates the glucocorticoid system [63,64]. Npy-overexpressing rats show improvements of cardiac function by regulation of the SNS and marginal extension of lifespan [65]. Npy is also known as a neuroprotective peptide [66]. These traits induced by Npy are consistent with the hallmarks of CR animals. Our previous study also confirmed that CR elevated mRNA levels of Npy in the arcuate nuclei in rats [67].

We tested a potential role for Npy in the effects of CR using $Npy^{-/-}$ mice. The life-extending effect in $Npy^{-/-}$ mice was attenuated to about one-third compared with WT mice [68]. Oxidative stress tolerance and tumor suppressive effects were also diminished. Initially, Npy deficiency was predicted to compromise neuroendocrine adaptative processes to CR, especially the inhibition of the GH-IGF-1 system, because experiments have shown that Npy is required to suppress Ghrh in fasting [69]. However, hypothalamic *Ghrh* mRNA, plasma IGF-1, insulin, and downstream Mtor in tissues, which could be related to the anti-aging effect of CR, were inhibited as in the WT mice [68]. In addition, plasma concentrations of leptin, adiponectin, and corticosterone in $Npy^{-/-}$ mice were also altered similarly as in WT mice [68]. These results indicate the complementary action of the neuroendocrine system in response to CR even without Npy. These findings also suggest that the Npy-related anti-aging mechanism of CR does not overlap with previously noted IGF-1, Mtor, adiponectin, and corticosteroids [68].

$Npy^{-/-}$ mice showed a slight increase of food intake normalized body weight in both AL and CR conditions during the lifespan study, indicating a reduction of food efficiency (a unit amount of food or calorie to gain or maintain body weight) compared with WT mice [59], suggesting an increment of energy expenditure in $Npy^{-/-}$ mice. In male mice, Npy deficiency increased mortality in CR condition until middle age, probably due to an excess loss of body fat [68]. Indeed, in male $Npy^{-/-}$ CR mice, lipolysis and/or thermogenesis was elevated through significant activation of the adrenergic receptor $\beta3$ and hormone-sensitive lipase pathway [70]. The premature death in CR mice was inhibited by administration of acipimox, a lipolysis inhibitor [71]. These findings suggest that the life-extending effect of CR requires inhibition of an excess loss of body energy through antagonizing the SNS. The study also emphasized the substantial role for Npy in the SNS in maintaining fat tissue under CR conditions in male mice. The extent of lifespan extension by CR negatively correlates with rates of white adipose tissue reduction by CR [72], i.e., mouse strains with less reduction of body fat by CR live longer than other strains in which fat is reduced greatly in response to CR. Therefore, the ability of energy preservation by Npy is essential for the life-extending effect of CR.

By contrast, deficiency of Npy may act beneficially in the life-stage under AL conditions. In female $Npy^{-/-}$ mice with AL feeding, the aging-related increase in body fat was minimized, leading to an improvement of insulin sensitivity via inhibition of inflammation in the fat [70]. Antagonism of Npy may be a promising target for drug discovery to prevent obesity in middle age, although Npy deficiency may cause unintentional weight loss in aged people.

Together these studies indicate that Npy exerts pleiotropic effects in health conditions depending on nutritional conditions.

9. Role for Npy in the Tumor-Inhibiting Effect of CR

Npy is expressed abundantly in the brain, adrenal gland, and adipose tissue. Npy circulates in the blood and is secreted locally from the SNS. Minor et al. reported that the tumor-inhibiting effect of CR was diminished in $Npy^{-/-}$ mice using a chemically induced skin tumor model [73]. The authors also suggested a role for hypothalamic arcuate nuclei (ARC) in the tumor-inhibiting effect of CR in a model of monosodium glutamate (MSG)-injected mice. MSG induces neuronal cell death in the ARC when injected subcutaneously on postnatal day five. In MSG-injected mice, no Npy-mRNA expression was observed in the ARC but serum Npy was detectable, suggesting a major role for hypothalamic Npy in the tumor-inhibiting effect of CR [73].

Our lifespan study also demonstrated a diminution of the tumor-inhibiting effect of CR in $Npy^{-/-}$ mice, suggesting a role for Npy in the tumor-inhibiting effect of CR [68]. The SNS not only releases catecholamines such as norepinephrine (NE) but also secretes Npy [74]. Recent studies have indicated that activation of the SNS exacerbates steatohepatitis, which is now recognized as one of the predisposing conditions of hepatocellular carcinoma (HCC) in obese people [75]. An experimental study also suggested that the SNS promotes HCC via activation of hepatic stellate cells and Kupffer cells [76,77]. Therefore, we investigated a potential role for Npy in a carcinogen-induced HCC model in different nutritional settings including CR as well as high calorie-intake regimens [78].

The complete details of the experiment are described elsewhere [78]. Briefly, HCC was induced by an intraperitoneal injection of diethylnitrosamine (DEN) in male C57BL6/J mice null for *Npy* ($Npy^{-/-}$) and their haplotype ($Npy^{+/-}$) as a control (Ctrl) on postnatal day 15 [78]. $Npy^{+/-}$ mice showed increased lifespan and reduced incidence of spontaneously occurring tumors in response to CR to the same extent as WT mice, indicating that one allele of the *Npy* gene is enough for the effects of CR [69]. Experimental mice, fed a standard diet AL after weaning, were subjected to three dietary regimens at 12 weeks of age and thereafter, that is, standard diet fed AL, 30% CR with the standard diet, and a high fat diet (HFD) fed AL. Mice were killed at 28 and 48 weeks to analyze the occurrence of HCC and growth of HCC in the liver.

CR inhibited the occurrence of microscopic HCC at 28 weeks in $Npy^{-/-}$ mice as well as *Ctrl* mice, compared to the AL and HFD groups (diet, $p = 0.0195$ by logistic regression; CR vs. AL, $p = 0.0192$; CR vs. HFD, $p = 0.0310$: Figure 4A), although the baseline proportion of mice bearing microscopic HCC at 28 weeks was slightly but significantly greater in $Npy^{-/-}$ mice (genotype, $p = 0.0426$ by logistic regression). Since DEN was administered at 15 days of age before initiating CR, Npy might exert an inhibitory effect on initiation of the carcinogen. At 48 weeks, many mice in the AL and the HFD groups showed macroscopic (≤ 1 mm in diameter) HCC [78], and thus the liver weight alone or the liver weight normalized by the body weight (Lw/Bw) represents an index of growth of HCC [78]. In the CR groups, the Lw/Bw was not increased between 28 and 48 weeks in $Npy^{-/-}$ and *Ctrl* mice, indicating that CR almost completely inhibits the growth of HCC even in the absence of Npy (Figure 4B). The AL and HFD feedings promoted the growth of HCC between 28 and 48 weeks more in $Npy^{-/-}$ mice than in *Ctrl* mice (genotype x age, $p < 0.0001$ by 3-f ANOVA; Npy−/− (48) vs. Ctrl (48), $p < 0.0001$ by Tukey's honestly significant difference (HSD) test), although there was no significant difference in the tumor growth between the AL and HFD feedings (Figure 4B). These findings suggest a role for Npy in tumor growth in conditions of overnutrition, but not for CR. This HCC model indicates that Npy is not necessary for the tumor-inhibiting effect of CR.

Steatohepatitis, which was quantitated by the number of inflammatory foci/unit area, was exacerbated between 28 and 48 weeks (age, $p = 0.0117$ by 3-f ANOVA; Figure 4C). CR consistently inhibited steatohepatitis, compared with the AL and HFD groups, in both *Ctrl* and $Npy^{-/-}$ mice (CR vs. AL, $p < 0.0001$; CR vs. HFD, $p < 0.0001$ by Tukey's HSD test). Therefore, CR might inhibit the occurrence and growth of HCC via suppressing steatohepatitis, one of the predisposing conditions for HCC even in the absence of Npy. Overall, steatohepatitis was less progressed in $Npy^{-/-}$ mice compared with *Ctrl* mice (genotype, $p = 0.0069$ by 3-f ANOVA), although the degree of steatohepatitis was differently altered by the diets between 28 and 48 weeks (Figure 4C). These findings suggest that Npy is involved in inhibition of growth of HCC but not simply through preventing steatohepatitis in AL and HFD conditions.

Npy is reported to counterbalance the actions of NE in stressed conditions [74]. In a Npy-overexpressing rat model, Npy attenuated sympathetic signals at baseline and diminished the immediate sympathetic response to stress [65]. Npy was also reported to minimize stress-induced bone loss through a suppression of NE circuits [79]. These findings indicated a counteracting role for Npy over NE in the activation of the SNS, a potentially exacerbating factor of many types of disorders in well-fed conditions (Figure 4D).

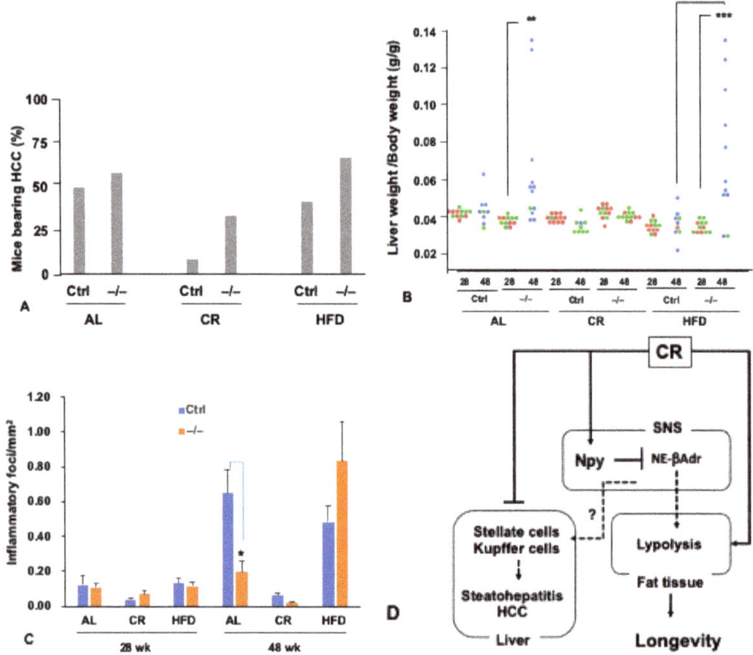

Figure 4. Effects of Npy deficiency in the occurrence and growth of hepatocellular carcinoma (HCC) and steatohepatitis. Ctrl and −/− represent male control (*Npy*$^{+/-}$) mice and *Npy*$^{-/-}$ mice, respectively. Figure 4A–C were modified from original data published by Kinoshita, A., et al. [78]. (**A**) The proportion of mice bearing microscopic HCC (diameter < 1 mm) at 28 weeks. $n = 12$ in each group. Statistics: diet, $p = 0.0195$ by logistic regression; CR vs. AL, $p = 0.0192$, CR vs. HFD, $p = 0.0310$. (**B**) Growth of HCC. Liver weights were normalized to body weights (g/g) in mice sacrificed at 28 and 48 weeks of age. $n = 12$ in each group at 28 weeks; at 48 weeks, $n = 9$–12. Closed circles: blue, liver bearing macroscopic HCC (diameter ≥ 1 mm); red, liver bearing microscopic (diameter < 1 mm) HCC; green, liver bearing no HCC. Statistics: genotype x age, $p < 0.0001$ by 3-f ANOVA; Npy−/− (48) vs. Ctrl (48), $p < 0.0001$ by Tukey's HSD test); **$p < 0.001$. ***$p < 0.0001$ by Tukey's HSD test. (**C**) The degree of steatohepatitis (the number of inflammatory foci/mm^2 in the liver). Bars represent means ± SE ($n = 8$~12 in each group). Statistics: age, $p = 0.0117$; diet, $p < 0.0001$; genotype, $p = 0.0069$ by 3-f ANOVA; *$p < 0.05$ by Tukey's HSD test. (**D**) A schematic model for the antagonizing role of Npy over norepinephrine (NE)-adrenergic receptor (Adr) signaling in the sympathetic nervous system (SNS) under CR conditions. CR promotes lipolysis, whereas CR inhibits an excess loss of fat via the action of Npy in the SNS. This trait is necessary for the life-extending effect of CR. CR inhibits steatohepatitis and HCC without Npy.

10. Conclusions

We reviewed and discussed underlying mechanisms of CR from an aspect of CR genes. It should be stressed that the isoform specificity of FoxO transcription factors for longevity becomes apparent under CR conditions but not AL conditions. Npy and FoxO1 both play pleiotropic roles in aging and related disorders, depending on the nutritional state. As briefly described in Sections 1 and 2, the life-extending effects of CR and reduced IGF-1 signaling are also sexually dimorphic. Genes associated with regulation of the aging process should be investigated carefully in a context-dependent manner, i.e., abilities of physiological adaptation for individuals against environmental challenges, particularly food shortage.

Supplementary Materials: The following are available online at http://www.mdpi.com/2072-6643/11/12/3068/s1, Table S1: Pathways signified by differentially regulated genes between WT-CR and *Foxo3*$^{+/-}$ CR livers in the fasting phase, Table S2: Pathways signified by differentially regulated genes between WT-CR and *Foxo1*[+/−] CR livers in the fasting phase.

Author Contributions: I.S. organized the review as a corresponding author and wrote most of the manuscript. T.K. and R.M. provided the data of skin carcinogenesis and critically reviewed the corresponding sections. H.Y. provided the data of ER stress and reviewed the sections related to stress response. S.P. and H.H. discussed and reviewed the sections on Npy.

Funding: This research was funded by JSPS KAKENHI Grant Numbers JP15H04682, JP19H04033.

Acknowledgments: We thank Edanz Group (www.edanzediting.com/ac) for editing a draft of this manuscript.

Conflicts of Interest: The authors declare no conflicts of interest.

Appendix A

Appendix A.1. Microarray Analysis

The microarray analysis using mouse liver tissues at six months of age was conducted, following the procedure published elsewhere [50]. Total RNA was extracted from liver tissues harvested from wild type (WT)-calorie restricted (CR), *Foxo1*[+/−] CR, and *Foxo3*[+/−] CR mice in the fasting phase, i.e., 3 to 4 h prior to the feeding. The details for isolation of RNA are also described in the Appendix A.2.3. The animal husbandry was described elsewhere [34,35]. Cyanine 3-labeled complementary RNA was generated from 200 ng of total RNA using the Low Input Quick AmpLabeling Kit, one color (Agilent Technologies, Santa Clara, CA), and then those was purified by the RNeasy mini kit (Qiagen), according to the manufacturer's instructions. Fragmented cyanine 3-labeled complementary RNA (600 ng) was hybridized to SurePrint G3 mouse GE microarray, 8×60 K (Agilent Technologies), at 65 °C for 17 h. Subsequently, the microarray was washed and scanned using the DNA microarray scanner (Agilent Technologies). Data analysis was entrusted to Ingenuity Systems, Inc (Redwood City, CA) with Ingenuity iReport. We compared the gene expression in the liver, between WT and *Foxo1*[+/−] CR groups and between WT-CR and *Foxo3*[+/−] CR groups. The filtering standard was a fold change cutoff of 1.5, with statistical significance of $p < 0.05$. Bio-data mining processes were executed by iReport based ingenuity knowledge base to identify enriched canonical pathways for the differentially expressed genes with $p < 0.05$ by Fisher's exact tests.

Appendix A.2. The Multistage Skin Tumorigenesis Model

Appendix A.2.1. Experimental Protocol

We followed the experimental procedure of Pearson et al. [36] for the multistage skin tumorigenesis model. Briefly, the number of mice in each group was 18 at the start of the experiment. At 31 weeks of age, a single dose of 25 µg of 7,12-dimethybenz(a)anthracene (DMBA, Sigma, St. Louis, MO) dissolved in 100 µl of acetone was applied topically in a dorsal area shaved in each mouse. Two weeks later, a topical treatment of 12-O-tetradecanoylphorbol-13-acetate (TPA, Sigma, St. Louis, MO, 4 µg dissolved in 100 µl of acetone) began and continued twice a week to the dorsal skin area until one papilloma with a diameter ≥ 1 mm was identified. Skin biopsy was done at the following time points: 0, 2, 3, 4, 5, and 6 weeks after DMBA treatment. Collected skin samples were cut in two specimens. One was immediately frozen in liquid nitrogen and stored at −80 °C. The other one was fixed in 4% formaldehyde for histological and immunohistochemical analyses. Once a skin tumor was identified, TPA treatment was stopped and left mice without any treatment. The experiment was terminated at 32 weeks after DMBA application. At the termination, mice were euthanized at 2 and 6 h after TPA or acetone treatment. The skin tissues without tumors were collected from each mouse, frozen in liquid nitrogen, and stored at −80 °C in order to analyze mRNA and protein expression.

Appendix A.2.2. Histology and Immunohistochemistry

Skin biopsy samples, fixed in 4% formaldehyde, were routinely processed and embedded in paraffin. Epidermal thickness was measured in hematoxylin and eosin (H.E.) stained sections under microscope. Details of immunohistochemical staining for proliferating cell nuclear antigen (PCNA) and F4/80 are described in elsewhere [41]. The primary antibodies for PCNA and for F4/80 were purchased from Abcam (Cambridge, U.K.). PCNA-positive cells in the epidermis and F4/80-positive cells in the dermis were counted for 10 high power views in each skin sample with microscope.

Appendix A.2.3. RNA Isolation and Quantitative Polymerase Chain Reaction

Total RNA was purified from skin using a RNeasy lipid tissue Mini kit (Qiagen). The quality of extracted RNA was evaluated as the densitometric ratio of 28S and 18S ribosomal RNA. The extracted RNA was converted to cDNA using RevaTra Ace® qPCR RT kit (TOYOBO, Tokyo, JAPAN) according to the manufacturer's instructions and amplified using THUNDERBIRD SYBR qPCR Mix (TOYOBO) with Light Cycler 480 (Roche Diagnostics Japan, Tokyo, Japan). All of the primers used in the present analysis were purchased from TAKARA with reference to the Perfect Real time Support System (TAKARA, Siga, Japan). All samples and standard curves were tested in duplicate. We confirmed that expression levels of a housekeeping gene, *Atp5f1*-mRNA, were not significantly affected by the diet or the genotype or TPA treatment, and then used for normalization of specific mRNA expression levels.

Appendix A.2.4. Statistical Analysis

Difference in the fraction of mice bearing skin papilloma between groups was tested with the log-rank test. Epidermal thickness, proliferating cell nuclear antigen (PCNA)-positive cells, the number of dermal macrophages, and the expression levels of *Cox2* and Gclc-mRNAs were analyzed using 3-factor ANOVA with Tukey's honestly significant difference (HSD) test.

References

1. Weindruch, R.; Walford, R.L. *The Retardation of Aging and Disease by Dietary Restriction*; Charles C Thomas Publisher: Springfield, IL, USA, 1988; pp. 31–115.
2. McCay, C.M.; Crowell, M.F.; Maynard, L.A. The effect of retarded growth upon the length of life span and upon the ultimate body size. *J. Nutr.* **1935**, *10*, 63–79. [CrossRef]
3. Swindle, W.R. Dietary restriction in rats and mice: A meta-analysis and review of the evidence fro genotype-dependent effects on lifespan. *Ageing Res. Rev.* **2012**, *11*, 254–270. [CrossRef] [PubMed]
4. Baloni, P.; Sangar, V.; Yurkovich, J.; Robinson, M.; Taylor, S.; Karbowski, C.; Hamadeh, H.; He, Y.; Price, N. Genome-scale metabolic model of the rat liver predicts effects of diet restriction. *Sci. Rep.* **2019**, *9*, 9807.
5. Mitchell, S.; Madrigal-Matute, J.; Scheibye-Knudsen, M.; Fang, E.; Aon, M.; González-Reyes, J.; Cortassa, S.; Kaushik, S.; Gonzalez-Freire, M.; Patel, B.; et al. Effects of Sex, Strain, and Energy Intake on Hallmarks of Aging in Mice. *Cell Metab.* **2016**, *23*, 1093–1112. [CrossRef] [PubMed]
6. Kane, A.; Sinclair, D.; Mitchell, J.; Mitchell, S. Sex differences in the response to dietary restriction in rodents. *Curr. Opin. Physiol.* **2018**, *6*, 28–34. [CrossRef]
7. Greer, E.L.; Brunet, A. Different dietary restriction regimens extend lifespan by both independent and overlapping genetic pathways in C. elegans. *Aging Cell* **2009**, *8*, 113–127. [CrossRef]
8. Shimokawa, I.; Chiba, T.; Yamaza, H.; Komatsu, T. Longevity genes: Insights from calorie restriction and genetic longevity models. *Mol. Cells* **2008**, *26*, 427–435.
9. Longo, V.D.; Finch, C.E. Evolutionary medicine: From dwarf model systems to healthy centenarians? *Science* **2003**, *299*, 1342–1346. [CrossRef]
10. Available online: http://genomics.senescence.info/diet/ (accessed on 6 November 2019).
11. Johnson, T.E. Increased life-span of age-1 mutants in Caenorhabditis elegans and lower Gompertz rate of aging. *Science* **1990**, *249*, 908–912. [CrossRef]
12. Morris, J.Z.; Tissenbaum, H.A.; Ruvkun, G. A phosphatidylinositol-3-OH kinase family member regulating longevity and diapause in Caenorhabditis elegans. *Nature* **1996**, *382*, 536–539. [CrossRef]
13. Kenyon, C.; Chang, J.; Gensch, E.; Rudner, A.; Tabtiang, R. A C. elegans mutant that lives twice as long as wild type. *Nature* **1993**, *366*, 461–464. [CrossRef] [PubMed]
14. Slack, C.; Foley, A.; Partridge, L. Activation of AMPK by the Putative Dietary Restriction Mimetic Metformin Is Insufficient to Extend Lifespan in Drosophila. *PLoS ONE* **2012**, *7*, e47699. [CrossRef] [PubMed]

15. Vellai, T.; Takacs-Vellai, K.; Zhang, Y.; Kovacs, A.L.; Orosz, L. Influence of TOR kinase on lifespan in C. elegans. *Nature* **2003**, *426*, 620. [CrossRef] [PubMed]
16. Kapahi, P.; Zid, B.M.; Harper, T.; Koslover, D.; Sapin, V.; Benzer, S. Regulation of lifespan in Drosophila by modulation of genes in the TOR signaling pathway. *Curr. Biol.* **2004**, *14*, 885–890. [CrossRef] [PubMed]
17. Shimokawa, I. Growth hormone and IGF-1 axis in aging and longevity. In *Hormones in Ageing and Longevity*, 1st ed.; Rattan, S., Sharma, R., Eds.; Springer International Publishing AG: Cham, Switzerland, 2017; pp. 91–106.
18. Harrison, D.E.; Strong, R.; Sharp, Z.D.; Nelson, J.F.; Astle, C.M.; Flurkey, K.; Nadon, N.L.; Wilkinson, J.E.; Frenkel, K.; Carter, C.S.; et al. Rapamycin fed late in life extends lifespan in genetically heterogeneous mice. *Nature* **2009**, *460*, 392–395. [CrossRef]
19. Selman, C.; Tullet, J.M.A.; Wieser, D.; Irvine, E.; Lingard, S.J.; Choudhury, A.I.; Claret, M.; Al-Qassab, H.; Carmignac, D.; Ramadani, F.; et al. Ribosomal Protein S6 Kinase 1 Signaling Regulates Mammalian Life Span. *Science* **2009**, *326*, 140–144. [CrossRef]
20. Wu, J.J.; Liu, J.; Chen, E.B.; Wang, J.J.; Cao, L.; Narayan, N.; Fergusson, M.M.; Rovira, I.I.; Allen, M.; Springer, D.A.; et al. Increased mammalian lifespan and a segmental and tissue-specific slowing of aging after genetic reduction of mTOR expression. *Cell Rep.* **2013**, *4*, 913–920. [CrossRef]
21. Holzenberger, M.; Dupont, J.; Ducos, B.; Leneuve, P.; Géloën, A.; Even, P.; Cervera, P.; Bouc, Y. IGF-1 receptor regulates lifespan and resistance to oxidative stress in mice. *Nature* **2003**, *421*, 182–187. [CrossRef]
22. Xu, J.; Gontier, G.; Chaker, Z.; Lacube, P.; Dupont, J.; Holzenberger, M. Longevity effect of IGF-1R(+/-) mutation depends on genetic background-specific receptor activation. *Aging Cell* **2014**, *13*, 19–28. [CrossRef]
23. Suh, Y.; Atzmon, G.; Cho, M.; Hwang, D.; Liu, B.; Leahy, D.; Barzilai, N.; Cohen, P. Functionally significant insulin-like growth factor I receptor mutations in centenarians. *Proc. Natl. Acad. Sci. USA* **2008**, *105*, 3438–3442. [CrossRef]
24. Milman, S.; Atzmon, G.; Huffman, D.; Wan, J.; Crandall, J.; Cohen, P.; Barzilai, N. Low insulin-like growth factor-1 level predicts survival in humans with exceptional longevity. *Aging Cell* **2014**, *13*, 769–771. [CrossRef] [PubMed]
25. van Heemst, D.; Beekman, M.; Mooijaart, S.; Heijmans, B.; Brandt, B.; Zwaan, B.; Slagboom, P.; Westendorp, R. Reduced insulin/IGF-1 signalling and human longevity. *Aging Cell* **2005**, *4*, 79–85. [CrossRef] [PubMed]
26. Franceschi, C.; Zaikin, A.; Gordleeva, S.; Ivanchenko, M.; Bonifazi, F.; Storci, G.; Bonafè, M. Inflammaging 2018: An update and a model. *Semin. Immunol.* **2018**, *40*, 1–5. [CrossRef] [PubMed]
27. Coppé, J.-P.; Patil, C.K.; Rodier, F.; Sun, Y.; Muñoz, D.P.; Goldstein, J.; Nelson, P.S.; Desprez, P.-Y.; Campisi, J. Senescence-associated secretory phenotypes reveal cell-nonautonomous functions of oncogenic RAS and the p53 tumor suppressor. *PLoS Biol.* **2008**, *6*, 2853–2868. [CrossRef] [PubMed]
28. Baker, D.J.; Childs, B.G.; Durik, M.; Wijers, M.E.; Sieben, C.J.; Zhong, J.; Saltness, R.A.; Jeganathan, K.B.; Verzosa, G.C.; Pezeshki, A.; et al. Naturally occurring p16(Ink4a)-positive cells shorten healthy lifespan. *Nature* **2016**, *530*, 184–189. [CrossRef] [PubMed]
29. De Cecco, M.; Ito, T.; Petrashen, A.P.; Elias, A.E.; Skvir, N.J.; Criscione, S.W.; Caligiana, A.; Brocculi, G.; Adney, E.M.; Boeke, J.D.; et al. L1 drives IFN in senescent cells and promotes age-associated inflammation. *Nature* **2019**, *566*, 73–78. [CrossRef]
30. Huang, J.; Xie, Y.; Sun, X.; Zeh, H.J., 3rd; Kang, R.; Lotze, M.T.; Tang, D. DAMPs, ageing, and cancer: The 'DAMP Hypothesis'. *Ageing Res. Rev.* **2015**, *24*, 3–16. [CrossRef]
31. Spadaro, O.; Goldberg, E.L.; Camell, C.D.; Youm, Y.-H.; Kopchick, J.J.; Nguyen, K.Y.; Bartke, A.; Sun, L.Y.; Dixit, V.D. Growth Hormone Receptor Deficiency Protects against Age-Related NLRP3 Inflammasome Activation and Immune Senescence. *Cell Rep.* **2016**, *14*, 1571–1580. [CrossRef]
32. Greer, E.L.; Brunet, A. FOXO transcription factors at the interface between longevity and tumor suppression. *Oncogene* **2005**, *24*, 7410–7425. [CrossRef]
33. Furuyama, T.; Kitayama, K.; Shimoda, Y.; Ogawa, M.; Sone, K.; Yoshida-Araki, K.; Hisatsune, H.; Nishikawa, S.-I.; Nakayama, K.; Nakayama, K.; et al. Abnormal angiogenesis in Foxo1 (Fkhr)-deficient mice. *J. Biol. Chem.* **2004**, *279*, 34741–34749. [CrossRef]
34. Yamaza, H.; Komatsu, T.; Wakita, S.; Kijogi, C.; Park, S.; Hayashi, H.; Chiba, T.; Mori, R.; Furuyama, T.; Mori, N.; et al. FoxO1 is involved in the antineoplastic effect of calorie restriction. *Aging Cell* **2010**, *9*, 372–382. [CrossRef] [PubMed]

35. Shimokawa, I.; Komatsu, T.; Hayashi, N.; Kim, S.E.; Kawata, T.; Park, S.; Hayashi, H.; Yamaza, H.; Chiba, T.; Mori, R. The life-extending effect of dietary restriction requires Foxo3 in mice. *Aging Cell* **2015**, *14*, 707–709. [CrossRef] [PubMed]

36. Paik, J.-H.; Kollipara, R.; Chu, G.; Ji, H.; Xiao, Y.; Ding, Z.; Miao, L.; Tothova, Z.; Horner, J.W.; Carrasco, D.R.; et al. FoxOs are lineage-restricted redundant tumor suppressors and regulate endothelial cell homeostasis. *Cell* **2007**, *128*, 309–323. [CrossRef] [PubMed]

37. Lin, K.; Dorman, J.B.; Rodan, A.; Kenyon, C. daf-16: An HNF-3/forkhead family member that can function to double the life-span of Caenorhabditis elegans. *Science* **1997**, *278*, 1319–1322. [CrossRef]

38. Ogg, S.; Paradis, S.; Gottlieb, S.; Patterson, G.I.; Lee, L.; Tissenbaum, H.A.; Ruvkun, G. The Fork head transcription factor DAF-16 transduces insulin-like metabolic and longevity signals in C. elegans. *Nature* **1997**, *389*, 994–999. [CrossRef]

39. Kwon, E.-S.; Narasimhan, S.D.; Yen, K.; Tissenbaum, H.A. A new DAF-16 isoform regulates longevity. *Nature* **2010**, *466*, 498–502. [CrossRef]

40. Lin, K.; Hsin, H.; Libina, N.; Kenyon, C. Regulation of the Caenorhabditis elegans longevity protein DAF-16 by insulin/IGF-1 and germline signaling. *Nat. Genet.* **2001**, *28*, 139–145. [CrossRef]

41. Willcox, B.J.; Donlon, T.A.; He, Q.; Chen, R.; Grove, J.S.; Yano, K.; Masaki, K.H.; Willcox, D.C.; Rodriguez, B.; Curb, J.D. FOXO3A genotype is strongly associated with human longevity. *Proc. Natl. Acad. Sci. USA* **2008**, *105*, 13987–13992. [CrossRef]

42. Teumer, A.; Qi, Q.; Nethander, M.; Aschard, H.; Bandinelli, S.; Beekman, M.; Berndt, S.I.; Bidlingmaier, M.; Broer, L. CHARGE Longevity Working Group. Genomewide meta-analysis identifies loci associated with IGF-I and IGFBP-3 levels with impact on age-related traits. *Aging Cell* **2016**, *15*, 811–824. [CrossRef]

43. Kim, D.H.; Lee, B.G.; Lee, J.W.; Kim, M.E.; Lee, J.S.; Chung, J.H.; Yu, B.P.; Dong, H.H.; Chung, H.Y. FoxO6-mediated IL-1β induces hepatic insulin resistance and age-related inflammation via the TF/PAR2 pathway in aging and diabetic mice. *Redox Biol.* **2019**, *24*, 101184. [CrossRef]

44. Litvak, V.; Ratushny, A.V.; Lampano, A.E.; Schmitz, F.; Huang, A.C.; Raman, A.; Rust, A.G.; Bergthaler, A.; Aitchison, J.D.; Aderem, A. A FOXO3–IRF7 gene regulatory circuit limits inflammatory sequelae of antiviral responses. *Nature* **2012**, *490*, 421–425. [CrossRef] [PubMed]

45. Pearson, K.J.; Lewis, K.N.; Price, N.L.; Chang, J.W.; Perez, E.; Cascajo, M.V.; Tamashiro, K.L.; Poosala, S.; Csiszar, A.; Ungvari, Z.; et al. Nrf2 mediates cancer protection but not prolongevity induced by caloric restriction. *Proc. Natl. Acad. Sci. USA* **2008**, *105*, 2325–2330. [CrossRef] [PubMed]

46. Sykiotis, G.P.; Bohmann, D. Stress-activated cap"n"collar transcription factors in aging and human disease. *Sci. Signal.* **2010**, *3*, re3. [CrossRef] [PubMed]

47. Schwarz, M.; Münzel, P.A.; Braeuning, A. Non-melanoma skin cancer in mouse and man. *Arch. Toxicol* **2013**, *87*, 783–798. [CrossRef] [PubMed]

48. Passos, G.F.; Medeiros, R.; Marcon, R.; Nascimento, A.F.G.; Calixto, J.B.; Pianowski, L.F. The role of PKC/ERK1/2 signaling in the anti-inflammatory effect of tetracyclic triterpene euphol on TPA-induced skin inflammation in mice. *Eur. J. Pharm.* **2013**, *698*, 413–420. [CrossRef]

49. Przybysz, A.J.; Choe, K.P.; Roberts, L.J.; Strange, K. Increased age reduces DAF-16 and SKN-1 signaling and the hormetic response of Caenorhabditis elegans to the xenobiotic juglone. *Mech. Ageing Dev.* **2009**, *130*, 357–369. [CrossRef]

50. Mori, R.; Tanaka, K.; de Kerckhove, M.; Okamoto, M.; Kashiyama, K.; Tanaka, K.; Kim, S.; Kawata, T.; Komatsu, T.; Park, S.; et al. Reduced FOXO1 Accelerates Skin Wound Healing and Attenuates Scarring. *Am. J. Pathol.* **2014**, *184*, 2465–2467. [CrossRef]

51. Nakae, J.; Oki, M.; Cao, Y. The FoxO transcription factors and metabolic regulation. *FEBS Lett.* **2008**, *582*, 54–67. [CrossRef]

52. Kamohara, R.; Yamaza, H.; Tsuchiya, T.; Komatsu, T.; Park, S.; Hayashi, H.; Chiba, T.; Mori, R.; Otabe, S.; Yamada, K.; et al. Overexpression of the adiponectin gene mimics the metabolic and stress resistance effects of calorie restriction, but not the anti-tumor effect. *Exp. Gerontol.* **2015**, *64*, 46–54. [CrossRef]

53. Kitamura, M.; Hiramatsu, N. Real-time monitoring of ER stress in living cells and animals using ESTRAP assay. *Method. Enzymol.* **2011**, *490*, 93–106.

54. Chalkiadaki, A.; Guarente, L. Sirtuins mediate mammalian metabolic responses to nutrient availability. *Nat. Rev. Endocrinol.* **2012**, *8*, 287–296. [CrossRef] [PubMed]

55. Guarente, L. SIRT1 and other sirtuins in metabolism. *Genes. Dev.* **2014**, *25*, 138–145.

56. Banerjee, K.K.; Ayyub, C.; Zeeshan, A.S.; Mandot, V.; Prasad, G. dSir2 in the adult fat body, but not in muscles, regulates life span in a diet-dependent manner. *Cell Rep.* **2012**, *2*, 1485–1491. [CrossRef] [PubMed]

57. Hirschey, M.D.; Shimazu, T.; Jing, E.; Grueter, C.A.; Collins, A.M.; Aouizerat, B.; Stančáková, A.; Goetzman, E.; Lam, M.M.; Schwer, B.; et al. SIRT3 deficiency and mitochondrial protein hyperacetylation accelerate the development of the metabolic syndrome. *Mol. Cell* **2011**, *44*, 177–190. [CrossRef] [PubMed]

58. Kume, S.; Uzu, T.; Horiike, K.; Chin-Kanasaki, M.; Isshiki, K.; Araki, S.-I.; Sugimoto, T.; Haneda, M.; Kashiwagi, A.; Koya, D. Calorie restriction enhances cell adaptation to hypoxia through Sirt1-dependent mitochondrial autophagy in mouse aged kidney. *J. Clin. Investig.* **2010**, *120*, 1043–1055. [CrossRef] [PubMed]

59. Mercken, E.M.; Hu, J.; Krzysik-Walker, S.; Wei, M.; Li, Y.; McBurney, M.W.; de Cabo, R.; Longo, V.D. SIRT1 but not its increased expression is essential for lifespan extension in caloric-restricted mice. *Aging Cell* **2014**, *13*, 193–196. [CrossRef]

60. Holliday, R. Food, reproduction and longevity: Is the extended lifespan of calorie-restricted animals an evolutionary adaptation? *Bioessays* **1989**, *10*, 125–127. [CrossRef]

61. Bishop, N.A.; Guarente, L. Two neurons mediate diet-restriction-induced longevity in C. elegans. *Nature* **2007**, *447*, 545–549. [CrossRef]

62. Bargmann, C.I.; Horvitz, H.R. Chemosensory neurons with overlapping functions direct chemotaxis to multiple chemicals in C. elegans. *Neuron* **1991**, *7*, 729–742. [CrossRef]

63. Shimokawa, I.; Higami, Y. Leptin signaling and aging: Insight from caloric restriction. *Mech. Ageing Dev.* **2001**, *122*, 1511–1519. [CrossRef]

64. Sainsbury, A.; Zhang, L. Role of the hypothalamus in the neuroendocrine regulation of body weight and composition during energy deficit. *Obes. Rev.* **2011**, *13*, 234–257. [CrossRef] [PubMed]

65. Michalkiewicz, M.; Knestaut, K.M.; Bytchkova, E.Y.; Michalkiewicz, T. Hypotension and reduced catecholamines in neuropeptide Y transgenic rats. *Hypertension* **2003**, *41*, 1056–1062. [CrossRef] [PubMed]

66. Smiałowska, M.; Domin, H.; Zieba, B.; Koźniewska, E.; Michalik, R.; Piotrowski, P.; Kajta, M. Neuroprotective effects of neuropeptide Y-Y2 and Y5 receptor agonists in vitro and in vivo. *Neuropeptides* **2009**, *43*, 235–249. [CrossRef] [PubMed]

67. Shimokawa, I.; Fukuyama, T.; Yanagihara-Outa, K.; Tomita, M.; Komatsu, T.; Higami, Y.; Tuchiya, T.; Chiba, T.; Yamaza, Y. Effects of caloric restriction on gene expression in the arcuate nucleus. *Neurobiol. Aging* **2003**, *24*, 117–123. [CrossRef]

68. Chiba, T.; Tamashiro, Y.; Park, D.; Kusudo, T.; Fujie, R.; Komatsu, T.; Kim, S.E.; Park, S.; Hayashi, H.; Mori, R.; et al. A key role for neuropeptide Y in lifespan extension and cancer suppression via dietary restriction. *Sci. Rep.* **2014**, *4*, 4517. [CrossRef]

69. Luque, R.M.; Park, S.; Kineman, R.D. Severity of the catabolic condition differentially modulates hypothalamic expression of growth hormone-releasing hormone in the fasted mouse: Potential role of neuropeptide Y and corticotropin-releasing hormone. *Endocrinology* **2007**, *148*, 300–309. [CrossRef]

70. Park, S.; Fujishita, C.; Komatsu, T.; Kim, S.E.; Chiba, T.; Mori, R.; Shimokawa, I. NPY antagonism reduces adiposity and attenuates age-related imbalance of adipose tissue metabolism. *FASEB J.* **2014**, *28*, 5337–5348. [CrossRef]

71. Park, S.; Komatsu, T.; Kim, S.E.; Tanaka, K.; Hayashi, H.; Mori, R.; Shimokawa, I. Neuropeptide Y resists excess loss of fat by lipolysis in calorie-restricted mice: A trait potential for the life-extending effect of calorie restriction. *Aging Cell* **2017**, *16*, 339–348. [CrossRef]

72. Liao, C.-Y.; Rikke, B.A.; Johnson, T.E.; Gelfond, J.A.L.; Diaz, V.; Nelson, J.F. Fat maintenance is a predictor of the murine lifespan response to dietary restriction. *Aging Cell* **2011**, *10*, 629–639. [CrossRef]

73. Minor, R.K.; López, M.; Younts, C.M.; Jones, B.; Pearson, K.J.; Anson, M.R.; Dieguez, C.; de Cabo, R. The arcuate nucleus and neuropeptide Y contribute to the antitumorigenic effect of calorie restriction. *Aging Cell* **2011**, *10*, 483–492. [CrossRef]

74. Farzi, A.; Reichmann, F.; Holzer, P. The homeostatic role of neuropeptide Y in immune function and its impact on mood and behaviour. *Acta Physiol. (Oxf.)* **2015**, *213*, 603–627. [CrossRef] [PubMed]

75. Chida, Y.; Sudo, N.; Kubo, C. Does stress exacerbate liver diseases? *J. Gastroenterol. Hepatol.* **2006**, *21*, 202–208. [CrossRef] [PubMed]

76. Sigala, B.; McKee, C.; Soeda, J.; Pazienza, V.; Morgan, M.; Lin, C.I.; Selden, C.; Vander Borght, S.; Mazzoccoli, G.; Roskams, T.; et al. Sympathetic nervous system catecholamines and neuropeptide Y neurotransmitters are upregulated in human NAFLD and modulate the fibrogenic function of hepatic stellate cells. *PLoS ONE* **2013**, *8*, e72928. [CrossRef] [PubMed]

77. Huan, H.-B.; Wen, X.-D.; Chen, X.-J.; Wu, L.; Wu, L.L.; Zhang, L.; Yang, D.P.; Zhang, X.; Bie, P.; Qian, C.; et al. Sympathetic nervous system promotes hepatocarcinogenesis by modulating inflammation through activation of alpha1-adrenergic receptors of Kupffer cells. *Brain Behav. Immun.* **2016**, *59*, 118–134. [CrossRef] [PubMed]

78. Kinoshita, A.; Hayashi, H.; Kusano, N.; Inoue, S.; Komatsu, T.; Mori, R.; Park, S.; Takatsuki, M.; Eguchi, S.; Shimokawa, I. Neuropeptide Y inhibits hepatocarcinogenesis in overnutrition in mice. *Acta Medica Nagasakiensia* **2019**, *41*, 711–722.

79. Baldock, P.A.; Lin, S.; Zhang, L.; Karl, T.; Shi, Y.; Driessler, F.; Zengin, A.; Hörmer, B.; Lee, N.J.; Wong, I.P.; et al. Neuropeptide y attenuates stress-induced bone loss through suppression of noradrenaline circuits. *J. Bone Miner. Res.* **2014**, *29*, 2238–2249. [CrossRef]

© 2019 by the authors. Licensee MDPI, Basel, Switzerland. This article is an open access article distributed under the terms and conditions of the Creative Commons Attribution (CC BY) license (http://creativecommons.org/licenses/by/4.0/).

Article

Srebp-1c/Fgf21/Pgc-1α Axis Regulated by Leptin Signaling in Adipocytes—Possible Mechanism of Caloric Restriction-Associated Metabolic Remodeling of White Adipose Tissue

Masaki Kobayashi [1], Seira Uta [1], Minami Otsubo [1], Yusuke Deguchi [1], Ryoma Tagawa [1], Yuhei Mizunoe [2], Yoshimi Nakagawa [3], Hitoshi Shimano [2,4,5] and Yoshikazu Higami [1,6,*]

[1] Laboratory of Molecular Pathology and Metabolic Disease, Faculty of Pharmaceutical Sciences, Tokyo University of Science, Chiba 278-8510, Japan; kobayashim@rs.tus.ac.jp (M.K.); 3B18508@ed.tus.ac.jp (S.U.); 3a15020@ed.tus.ac.jp (M.O.); 3B16060@ed.tus.ac.jp (Y.D.); tagawar@rs.tus.ac.jp (R.T).

[2] Department of Internal Medicine (Endocrinology and Metabolism), Faculty of Medicine, University of Tsukuba, Ibaraki 305-8575, Japan; ymizunoe@md.tsukuba.ac.jp (Y.M.); hshimano@md.tsukuba.ac.jp (H.S.)

[3] Division of Complex Biosystem Research, Department of Research and Development, Institute of Natural Medicine, University of Toyama, Toyama 930-0194, Japan; ynaka@inm.u-toyama.ac.jp

[4] Life Science Center for Survival Dynamics, Tsukuba Advanced Research Alliance (TARA), University of Tsukuba, Ibaraki 305-8575, Japan

[5] AMED-CREST, Japan Agency for Medical Research and Development (AMED), Tokyo 100-1004, Japan

[6] Research Institute for Biomedical Sciences, Tokyo University of Science, Chiba 278-8510, Japan

[*] Correspondence: higami@rs.tus.ac.jp; Tel./Fax: +81-4-7121-3676

Received: 6 June 2020; Accepted: 8 July 2020; Published: 10 July 2020

Abstract: Caloric restriction (CR) improves whole body metabolism, suppresses age-related pathophysiology, and extends lifespan in rodents. Metabolic remodeling, including fatty acid (FA) biosynthesis and mitochondrial biogenesis, in white adipose tissue (WAT) plays an important role in the beneficial effects of CR. We have proposed that CR-induced mitochondrial biogenesis in WAT is mediated by peroxisome proliferator-activated receptor γ coactivator-1α (PGC-1α), which is transcriptionally regulated by sterol regulatory element-binding protein 1c (SREBP-1c), a master regulator of FA biosynthesis. We have also proposed that the CR-associated upregulation of SREBP-1 and PGC-1α might result from the attenuation of leptin signaling and the upregulation of fibroblast growth factor 21 (FGF21) in WAT. However, the detailed molecular mechanisms remain unclear. Here, we interrogate the regulatory mechanisms involving leptin signaling, SREBP-1c, FGF21, and PGC-1α using *Srebp-1c* knockout (KO) mice, mouse embryonic fibroblasts, and 3T3-L1 adipocytes, by altering the expression of SREBP-1c or FGF21. We show that a reduction in leptin signaling induces the expression of proteins involved in FA biosynthesis and mitochondrial biogenesis via SREBP-1c in adipocytes. The upregulation of SREBP-1c activates PGC-1α transcription via FGF21, but it is unlikely that the FGF21-associated upregulation of PGC-1α expression is a predominant contributor to mitochondrial biogenesis in adipocytes.

Keywords: caloric restriction; fatty acid biosynthesis; mitochondrial biogenesis; adipocyte

---

1. Introduction

It is well known that white adipose tissue (WAT) is involved in the pathogenesis of age-related diseases including type 2 diabetes, atherosclerosis, and other cardiovascular and cerebrovascular diseases [1]. It has recently been shown that WAT quality, including adipocyte size, mitochondrial biogenesis, and adipokine expression profile, is a key player in lifespan regulation [2–6].

Caloric restriction (CR) is the most robust, reproducible, and simple experimental manipulation that is capable of improving whole body metabolism, delaying the onset of various age-related pathophysiological changes and extending both median and maximum lifespan in a wide range of organisms [7,8]. Dwarf rodents that demonstrate the suppression of growth hormone/insulin-like growth factor 1 (GH/IGF-1) signaling live longer than their wild-type (Wd) littermates [9]. Since CR suppresses GH/IGF-1 signaling, its beneficial effects are considered to be dependent on the suppression of GH/IGF-1 signaling [10]. However, CR further extends the lifespan of long-lived dwarf rodents that have GH/IGF-1 suppression [11,12]. Therefore, the beneficial effects of CR are also likely to be mediated through a GH/IGF-1-independent mechanism.

To identify the GH/IGF-1-independent mechanism involved in the effects of CR, we compared the gene expression profile of the WAT of long-living dwarf rats bearing an antisense GH transgene with that of Wd rats subjected to CR, and we found that CR upregulated the expression of genes involved in fatty acid (FA) biosynthesis in a GH/IGF-1-independent manner [13]. Sterol regulatory element binding protein-1 (SREBP-1), including its two isoforms, SREBP-1a and -1c, is a master transcriptional regulator of FA biosynthesis [14]. In WAT, SREBP-1c is predominantly expressed, rather than SREBP-1a [15]. Therefore, we applied CR to both *Srebp-1c* knockout (KO) and WT mice on a B6; 129S6 background and found that CR extended lifespan in Wd mice but not in KO mice. Moreover, CR upregulated the expression of proteins involved in FA biosynthesis and mitochondrial biogenesis in the WAT of Wd mice but not in KO mice. These findings were observed only in WAT but not in the other tissues, including liver, kidney, quadriceps femoris muscle, and heart [16]. Peroxisome proliferator-activated receptor γ coactivator-1α (PGC-1α) is a master transcriptional cofactor for mitochondrial biogenesis [17] and a key regulator of the CR-induced activation of mitochondrial biogenesis [18]. We also found that CR upregulates *Pgc-1a* mRNA in WAT of Wd mice but not in *Srebp-1c* KO mice. Moreover, a chromatin immunoprecipitation assay showed that SREBP-1 protein binds to the promoter region of the *Pgc-1a* gene, as well as the *Fasn* gene, in mouse embryonic fibroblasts (MEFs) derived from WT mice but not in those from KO mice. Therefore, we suggested that CR upregulates FA biosynthesis and mitochondrial biogenesis via SREBP-1c in WAT [16].

Fibroblast growth factor 21 (FGF21), which was initially identified as a hepatokine, is mostly secreted by the liver [19]. Circulating FGF21 binds to the FGF receptor (FGFR) and β-klotho (KLB) receptor complex in target tissues such as WAT. The binding of FGF21 to its receptors activates downstream signaling, including extracellular signal-regulated kinase (ERK) signaling, which upregulates the expression of genes involved in glucose and lipid metabolism [20–22]. FGF21 expression is negatively regulated by SREBP-1c in hepatocytes [23]. In contrast, FGF21 expression is upregulated by SREBP-1c in WAT and 3T3-L1 adipocytes [24]. FGF21 induces PGC-1α expression in the liver as an adaptation to starvation [25]. In WAT, FGF21 positively regulates PGC-1α and PPARγ expression and/or activity via feed-forward autocrine/paracrine loops [26,27]. Moreover, *Fgf21* Tg mice live longer than Wd mice and have a similar metabolic phenotype to CR mice [28]. We have also shown that the CR-associated upregulation of PGC-1α expression is partially mediated through FGF21 in WAT [29]. CR also upregulates PPARγ expression in WAT [29]. In addition, the expression of PGC-1α is increased as a result of rosiglitazone-induced PPARγ activity in WAT [30].

Leptin, which was the first substance to be identified as an adipokine, is mostly secreted by adipocytes [31]. Circulating leptin binds to the leptin receptor, which is predominantly expressed in the arcuate nucleus of the hypothalamus and reduces appetite and increases energy expenditure via the sympathetic nervous system [32]. However, the leptin receptor is also expressed in other cell types, including adipocytes [33]. It has been reported that leptin treatment downregulates the expression of SREBP-1 and its downstream targets in mouse WAT [34]. In addition, CR reduces leptin secretion by adipocytes, thereby reducing the circulating leptin concentration [35]. These findings raised the possibility that CR might suppress leptin signaling via an autocrine/paracrine loop, leading to the SREBP-1-induced upregulation of proteins involved in FA biosynthesis in WAT.

As stated above, the molecular mechanisms of CR-associated metabolic remodeling, including FA biosynthesis and mitochondrial biogenesis, are unclear. In particular, the reciprocal regulatory mechanism that involves SREBP-1, FGF21, and PGC-1α is complex. In the present study, we aimed to clarify this molecular mechanism, focusing on the expression of the master regulators of FA biosynthesis and mitochondrial biogenesis, SREBP-1 and PGC-1α, respectively, in adipocytes. To this end, we analyzed the regulation of leptin signaling, SREBP-1c, FGF21, and PGC-1α in the CR-associated metabolic remodeling of WAT and adipocytes.

2. Materials and Methods

2.1. Animals and the Collection of Mice Embryonic Fibroblasts (MEFs)

All animal experiments were approved by the Animal Experimentation Committees of Tokyo University of Science (Y17051, Y18060, Y19056) or the University of Tsukuba (19–274). We back-crossed *Srebp-1c* KO mice on a B6;129S6 background (B6; 129S6-Srebf1tm1Mbr/J; Jackson Laboratory, Bar Harbor, ME, USA) and C57Bl/6J mice (CLEA Japan, Tokyo, Japan) to obtained *Srebp-1c* KO mice on a C57Bl/6 background. All the animals were maintained under specific pathogen-free conditions. At 3 months of age, Wd and *Srebp-1c* KO mice were allocated to two groups: an *ad libitum*-fed (AL) and a CR (70% of the energy intake of AL) group. At 10 months of age, four groups of mice (WdAL, WdCR, KOAL, and KOCR) were provided with food 0.5–1 h prior to turning off the lights in the evening, then they were euthanized under isoflurane anesthesia (Mylan, Canonsburg, PA, USA) 2–3 h later, after which WAT was harvested. The time-course measurement of food intake of the Wd and KO mice that were fed AL, and the body weights of the four groups, are shown in Figure S1.

MEFs were obtained from *Srebp-1c* KO and Wd mice, and *Fgf21* KO and Wd mice on a C57Bl/6 background [36], as previously reported [16]. Briefly, 13–15-day old embryos (E13–15) were collected from pregnant mice of each KO line, minced and trypsinized. MEFs were separated by passing the tryptic digests through a cell strainer.

2.2. Cell Culture and Reagent Treatment

3T3-L1 preadipocytes were purchased from the Japanese Collection of Research Bioresources (JCRB) cell bank (Osaka, Japan) and maintained in Dulbecco's modified Eagle's medium (DMEM) containing a low glucose concentration (Wako, Osaka, Japan), 10% fetal bovine serum (FBS) (Thermo Fisher Scientific, Waltham, MA USA), and 1% penicillin/streptomycin (P/S) (Sigma-Aldrich, MO, USA). MEFs were maintained in DMEM containing a high glucose concentration (Wako), 10% FBS, 1% P/S, and 0.1 µM 2-mercaptoethanol (Sigma). The differentiation of 3T3-L1 preadipocytes or MEFs to mature adipocytes was achieved by using our previous published protocol [37]. In the present study, 3T3-L1 cells or MEFs were used as mature adipocytes 8–12 or 16 days after the induction of differentiation, respectively. PD1730741 (Funakoshi, Tokyo, Japan) was dissolved in DMSO to make a 5 mM solution, and then it was diluted in PBS. Differentiated 3T3-L1 cells (day 7) were treated with 50 nM PD1730741 for 24 h and then collected.

2.3. Retrovirus Plasmid Construction

The construction of the retrovirus plasmids for *Srebp-1c* and *Fgf21* overexpression have been described in our previous reports [16,29]. *Srebp-1a* cDNA was obtained by PCR using KOD FX Neo (Toyobo, Osaka, Japan) and the following primers: 5′-TTT GGA TCC GCC ACC ATG GAC GAG CTG GCC TT-3′ and 5′-TTT GAA TTC TTA CAG GGC CAG GCG GGA-3′. Amplified *Srebp-1a* fragments were digested with BamHI and EcoRI and subcloned into BamHI- and EcoRI-digested pBluescript II SK (+). Then, this plasmid was digested with BamHI and EcoRI, and the gene sequence was inserted into pMXs-AMNN-Puro (pMXs-AMNN-Srebp-1a-Puro) after it was also digested using the same enzymes. The target sequences of the shRNAs against *Lepr* were designed using the Public TRC Portal website (http://www.broadinstitute.org/rnai/public/seq/search), and the sequences were as follows:

5′-GCT AGG TGT AAA CTG GGA CAT CTC GAG ATG TCC CAG TTT ACA CCT AGC TTT TT-3′ and 5′-CGA AAA AGC TAG GTG TAA ACT GGG ACA TCT CGA GAT GTC CCA GTT TAC ACC TAG C-3′. The underlined letters are the sense and antisense target sequences. These oligonucleotides were inserted into BstBI- and PmeI-digested pMXs-puro-mU6 (pMXs-puro-shLeptinR).

2.4. Retrovirus Vector Preparation

Retrovirus vectors were generated as reported previously [16]. Briefly, each pMXs-AMNN-Puro plasmid or pMXs-puro-shLeptinR plasmid was transfected into Plat-E cells (kindly provided by T. Kitamura, University of Tokyo, Japan) using the calcium phosphate method. To obtain each overexpressing or shLeptinR-expressing 3T3-L1 cell line, the supernatant from each virus-containing culture was collected after 3 days. The 3T3-L1 cells were infected by incubation in the collected virus-containing supernatant for 2 days, followed by treatment with 2 μg/mL puromycin for a further 5 days. The 3T3-L1 cells overexpressing empty vectors or expressing shRNA targeting GFP (shGFP) were used as control cells.

2.5. RT-PCR and Semi-Quantitative RT-PCR

RNA was extracted from WAT and other cell types using ISOGENII (Nippon Gene, Tokyo, Japan). The purified RNA was reverse transcribed using ReverTra Ace® qPCR RT Master Mix (Toyobo) and the cDNAs were then amplified using a CFX Connect™ Real-time System, Thunderbird SYBR qPCR mix, and the primers for each gene. These procedures were performed according to the manufacturer's protocol. Since the intrinsic expression of *Srebp-1c* mRNA was very low in 3T3-L1 adipocytes, RT-PCR was not possible. Therefore, we performed conventional PCR and agarose gel electrophoresis of the PCR products, followed by ethidium bromide staining. Fluorescence of the ethidium bromide was visualized using an LAS3000 (Fujifilm, Tokyo, Japan) and data were analyzed using Multigauge software (Fujifilm). Target gene expression data were normalized to *Rps18* expression ($n = 4$). The primer pair sequences are shown in Table 1.

Table 1. List of primers for RT-PCR.

	Forward	Reverse
Adipoq	5′-TGC CGA AGA TGA CGT TAC AAC-3′	5′-CTT CAG CTC CTG TCA TTC CAA C-3′
Fasn	5′-AGC AGG CAC ACA CAA TGG AC-3′	5′-GAA GAA AGA GAG CCG GTT G-3′
Fgf21	5′-GAA GCC CAC CTG GAG ATC AG-3′	5′-CAA AGT GAG GCG ATC CAT AGA G-3′
LeptinR	5′- CAG TCT TCGG GGA TGT GAA TG-3′	5′- CAT TGT TTG GCT GTC CCA AG-3′
PeriA	5′-TGG GAA GCA TCG AGA AGG TG-3′	5′-ATG GTG TGT CGA GAA AGA GTG TTG-3′
Pgc-1α	5′-AGA CGG ATT GCC CTC ATT TG-3′	5′-CAG GGT TTG TTC TGA TCC TGT G-3′
Rps18	5′-TGC GAG TAC TCA ACA CCA ACA T-3′	5′-CTT TCC TCA ACA CCA CAT GAG C-3′
Srebp-1a	5′-GGC CGA GAT GTG CGA ACT-3′	5′-TTG TTG ATG AGC TGG AGC ATG T-3′
Srebp-1c	5′-GGA GCC ATG GAT TGC ACA TT-3′	5′-GGC CCG GGA AGT CAC TGT-3′
Tbp	5′-CCC TCA CAC TCA GAT CAT CTT CTC-3′	5′-GCC TTG TCC CTT GAA GAG AAC C-3′

2.6. Western Blotting

Cell lysis and immunoblotting were performed as previously described [37]. Briefly, the collected cells were lysed in lysis buffer (50 mM Tris-HCl (pH 6.8), 2% SDS, 3 M urea, 6% glycerol), boiled for 5 min, and sonicated. Lysates containing 15 μg protein were subjected to SDS/PAGE and the proteins were then transferred to nitrocellulose membranes. The membranes were blocked with 2.5% skim milk and 0.25% bovine serum albumin in Tris-buffered saline (50 mM Tris-HCl (pH 7.4) and 150 mM NaCl) containing 0.1% Tween 20 (TTBS) for 60 min at room temperature, then incubated with appropriate primary antibodies overnight at 4 °C. Primary antibodies against FGF21 (Abcam, Cambridge, UK), PGC-1α (Sigma, MO, USA AB3242), mitochondrial transcription factor A (TFAM) (Proteintech, Chicago, IL, USA, 19998-1-AP), SIRT3 (cell signaling technology (CST), Beverly, MA, USA, #5490), ACC (CST, #3662), p-STAT3 (Thermo Fisher Scientific, 44-3804), and LaminB1 (Medical & Biological Laboratories,

Nagoya, Japan, PM064) were used. The membranes were then incubated with an appropriate secondary antibody (a horseradish peroxidase-conjugated F(ab')2 fragment of goat anti-mouse IgG or anti-rabbit IgG; Jackson Immuno Research, West Grove, PA, USA) for 60 min at room temperature. Thereafter, they were incubated with ImmunoStar LD (Wako), specific protein bands were visualized using an LAS3000 (Fujifilm, Tokyo, Japan), and the data were analyzed using Multigauge software (Fujifilm).

2.7. Statistical Analysis

The values presented are means ± standard deviations (SDs). The data were statistically evaluated using Student's *t*-test, two-way ANOVA and/or Tukey's test, with R software (Version 3.4.1, R Foundation for Statistical Computing, Vienna, Austria). $p < 0.05$ was considered to represent statistical significance.

3. Results

3.1. Role of SREBP-1c in the Effects of CR on Gene Expression in WAT

We have reported previously that CR increases the expression of *Srebp-1c*, *Srebp-1a*, and *Pgc-1a* mRNAs in the WAT of Wd B6;129S6 mice but not in KO mice [16]. In the WAT of mice on a C57Bl/6 background, CR also increased the expression of *Srebp-1c*, *Pgc-1a*, and *Fgf21* mRNAs in Wd mice but not in KO mice (Figure 1A or Figure 1C,D). Similar findings about the CR-associated upregulation of *Fgf21* mRNA were observed in B6;129S6 mice (Figure S2). However, in contrast to mice on a B6;129S6 background, CR did not upregulate the expression of *Srebp-1a* mRNA in either Wd or KO mice on a C57Bl/6 background (Figure 1B). Overall, it is likely that the CR-associated upregulation of these factors is less exaggerated in C57bl/6 mice compared with B6;129S6 mice.

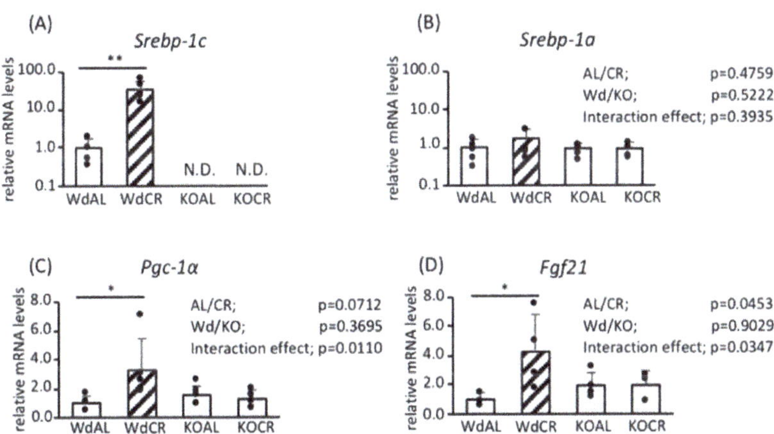

Figure 1. The effects of *Srebp-1c* KO on the expression of key regulators of CR-associated metabolic remodeling in the WAT of mice on a C57Bl/6 background. The mRNA expression levels of *Srebp-1c* (**A**), *Srebp-1a* (**B**), *Fgf21* (**C**), and *Pgc-1a* (**D**) in WAT were measured using RT-PCR and were normalized to *Tbp* expression ($n = 4$). Values are means ± SDs. * $p < 0.05$, ** $p < 0.01$ vs. AL, according to Student's *t*-test, or two-way ANOVA and Tukey's test.

3.2. Effects of SREBP-1c on the Expression of Genes and Proteins Involved in FA Biosynthesis and Mitochondrial Biogenesis

To confirm that SREBP-1c is the significant regulator of the expression of genes and proteins involved in FA biosynthesis and mitochondrial biogenesis in vitro, we generated 3T3-L1 preadipocytes that overexpressed SREBP-1c (SREBP-1c OE) using retroviral vectors. The SREBP-1c OE 3T3-L1 preadipocytes were differentiated to adipocytes (Figure 2A), and the expression levels of mRNAs and proteins of interest were analyzed.

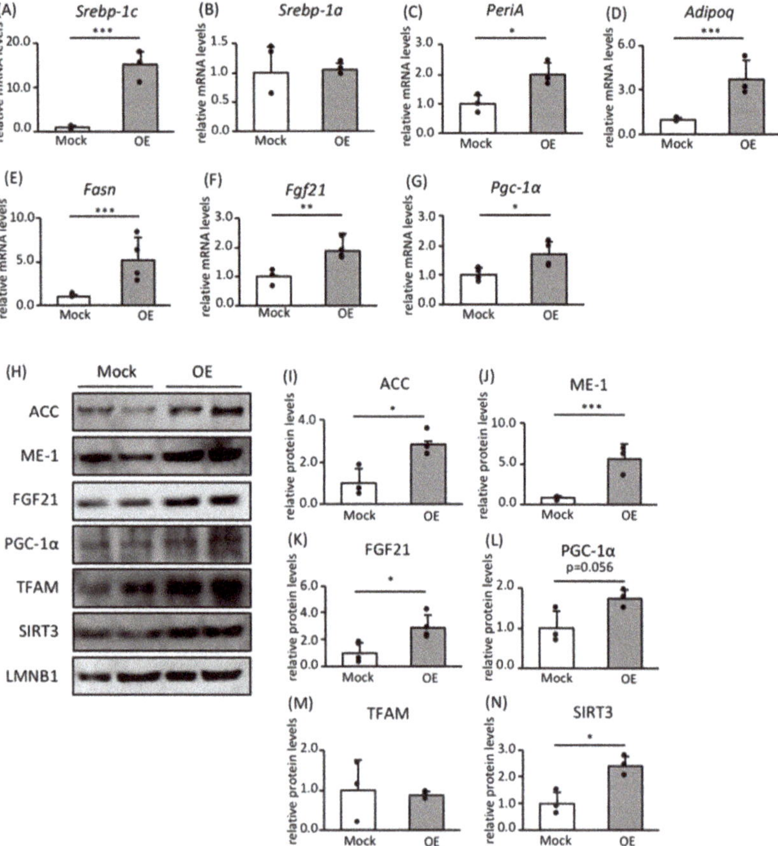

Figure 2. The effects of SREBP-1c overexpression on the expression of genes and proteins involved in fatty acid (FA) biosynthesis and mitochondrial biogenesis in mature 3T3-L1 adipocytes. Control and SREBP-1c OE preadipocytes were differentiated into mature adipocytes in four separate dishes from each phenotype. RNA was extracted and lysates were prepared from each dish. RNA was extracted and lysates were prepared from adipocytes. The mRNA expression levels of *Srebp-1c* (**A**), *Srebp-1a* (**B**), *PeriA* (**C**), *Adipoq* (**D**), *Fasn* (**E**), *Fgf21* (**F**), and *Pgc-1a* (**G**) were determined using RT-PCR and normalized to *Rps18* expression ($n = 4$). (**H**) Representative immunoblot images, showing the expression levels of proteins involved in FA biosynthesis and mitochondrial biogenesis. Quantitative analysis was performed using a chemiluminescence method. The protein expression of ACC (**I**), ME-1 (**J**), FGF21 (**K**), PGC-1α (**L**), TFAM (**M**), and SIRT3 (**N**) are shown as the relative intensities of the indicated protein divided by that of LMNB1 as an internal control ($n = 4$). Values are means ± SDs. * $p < 0.05$, ** $p < 0.01$, *** $p < 0.001$ vs. controls, according to Student's *t*-test.

In SREBP-1c OE adipocytes, the expression of *Srebp-1a* mRNA was similar to that of control cells (Figure 2B), whereas that of *perilipin A* (*PeriA*) and *adiponectin* (*Adipoq*), which are markers of adipocyte differentiation, was upregulated (Figure 2C,D). Fatty acid synthase (FASN) is a rate-limiting enzyme in FA biosynthesis. The expression of *Fasn* mRNA was high in SREBP-1c OE adipocytes (Figure 2E).

Moreover, the expression of both *Fgf21* and *Pgc-1a* mRNAs was high in OE adipocytes (Figure 2F,G). In addition, the protein expression of acetyl-CoA carboxylase (ACC) and malic enzyme 1 (ME1), which are FA biosynthetic enzymes, and FGF21 and PGC-1α were high in OE adipocytes (Figure 2H–L). With regard to mitochondrial proteins, expression of SIRT3 was also high in SREBP-1c OE adipocytes, but that of TFAM was not (Figure 2H or Figure 2M,N).

To further characterize the regulation of FA biosynthetic genes by FGF21 and PGC-1α, we generated Wd and KO adipocytes by differentiating MEFs derived from Wd and *Srebp-1c* KO mice. The expression of *Srebp-1c* was not detectable in KO adipocytes and that of *Srebp-1a* was similar to that of control adipocytes (Figure 3A,B). The expression levels of *PeriA* and *Adipoq* mRNAs in KO adipocytes did not differ from those in Wd adipocytes, suggesting that SREBP-1c deficiency did not alter differentiation (Figure 3C,D). However, the expression of *Fasn* was lower in KO adipocytes than Wd adipocytes (Figure 3E). The expression of *Fgf21* mRNA was slightly reduced, while that of *Pgc-1a* mRNA was significantly lower in KO adipocytes (Figure 3F,G). Taken together, our findings suggest that the expression of both *Pgc-1a* and *Fgf21* is positively regulated by Srebp-1c, in addition to that of *Fasn*. Moreover, SREBP-1c is the significant regulator of the expression of genes and proteins involved in FA biosynthesis and mitochondrial biogenesis in adipocytes.

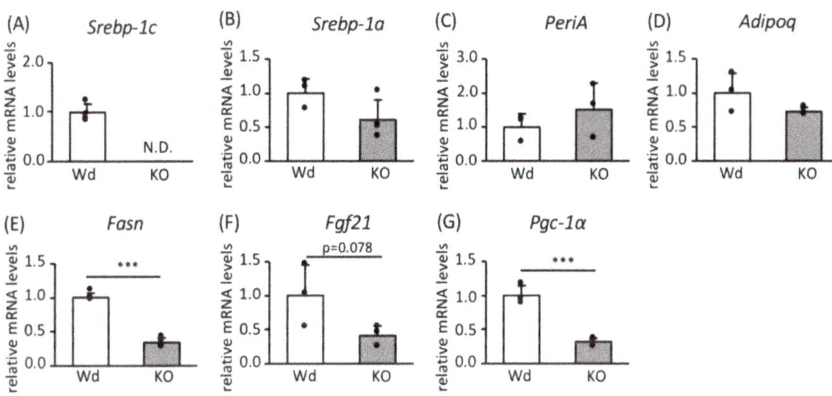

Figure 3. The effects of SREBP-1c deficiency on the expression of adipocyte differentiation markers and genes involved in FA biosynthesis and mitochondrial biogenesis in mature adipocytes. MEFs were obtained from four individual embryos of either Wd or *Srebp-1c* KO mice, differentiated to mature adipocytes, and then RNA was extracted from each dish. The mRNA expression levels of *Srebp-1c* (**A**), *Srebp-1a* (**B**), *PeriA* (**C**), *Adipoq* (**D**), *Fasn* (**E**), *Fgf21* (**F**), and *Pgc-1a* (**G**) were determined using RT-PCR and normalized to *Rps18* expression (*n* = 4). Values are means ± SDs. *** *p* < 0.001 vs. Wd, according to Student's *t*-test.

3.3. Roles of FGF21 and PGC-1α in Mitochondrial Biogenesis

CR upregulated the expression of both *Fgf21* and *Pgc-1a* mRNAs and proteins via SREBP-1c. Therefore, we next determined the roles of FGF21 and PGC-1α in mitochondrial biogenesis.

In FGF21 OE adipocytes (Figure 4A or Figure 4C,D), the expression of *Pgc-1a* mRNA and PGC-1α protein was high (Figure 4B,C or Figure 4E,F). Treatment with PD173074, an FGF receptor (FGFR) inhibitor, reduced the phosphorylation of ERK without reducing *Fgf21* mRNA and FGF21 protein expression (Figure 4A or Figure 4C–E). In addition, this treatment did not reduce the expression of *Pgc-1a* mRNA or PGC-1α protein (Figure 4B,C or Figure 4F).

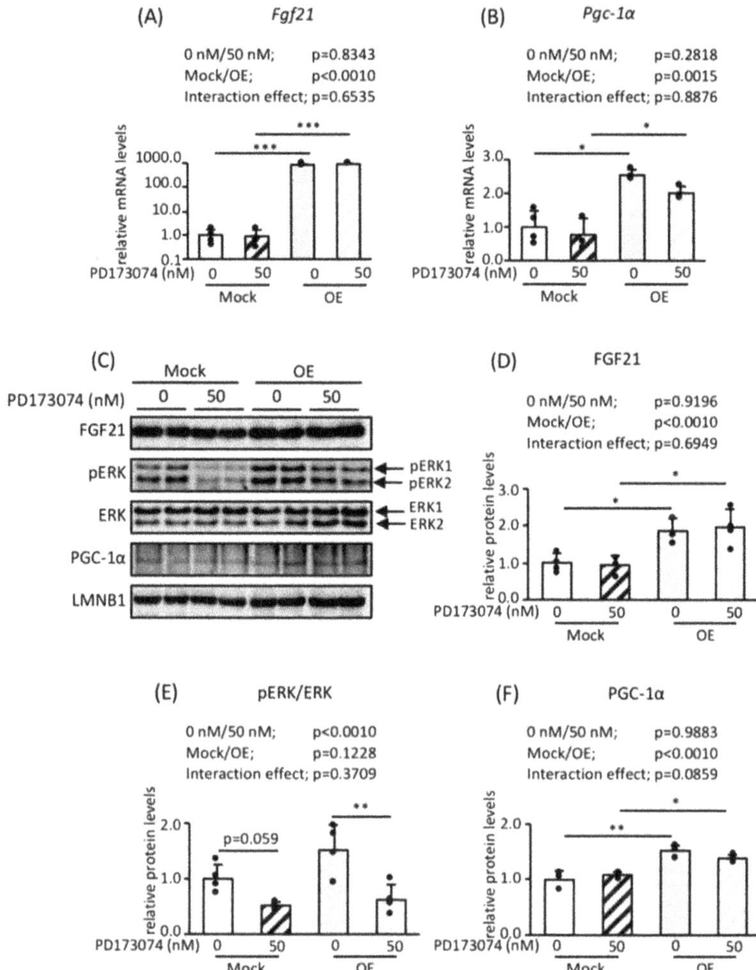

Figure 4. The effects of FGF21 overexpression and the inhibition of FGFR on the expression of genes and proteins involved in FGF21 signaling and PGC-1α in mature 3T3-L1 adipocytes. Control and FGF21 OE preadipocytes were differentiated into mature adipocytes in four separate dishes for each phenotype, and they were treated with or without 50 nM PD173074, an FGFR inhibitor, for 24 h. (**A**,**B**) RNA was extracted and lysates were prepared from each dish. The mRNA expression levels of *Fgf21* (**A**) and *Pgc-1a* (**B**) were determined using RT-PCR and normalized to *Rps18* expression (*n* = 4). (**C**) Representative immunoblot images showing the expression of proteins involved in FGF21 signaling and mitochondrial biogenesis. Quantitative analysis was performed using a chemiluminescence method. The protein expression of FGF21 (**D**) and PGC-1α (**F**) is shown as the relative intensity of the indicated protein divided by that of LMNB1 as an internal control (*n* = 4). Extracellular signal-regulated kinase (ERK) phosphorylation is expressed as the relative intensity of the phosphorylated form of ERK/total ERK (*n* = 4) (**E**). Values are means ± SDs *p < 0.05, **p < 0.01, ***p < 0.001 vs. controls administered the same treatment.

In Fgf21 KO adipocytes differentiated from MEFs, the expression levels of *PeriA*, *Adipoq*, and *Pgc-1α* mRNAs were much lower than in control cells, suggesting that the significant reduction in *Pgc-1α* mRNA expression is associated with impaired adipocyte differentiation (Figure 5A–D).

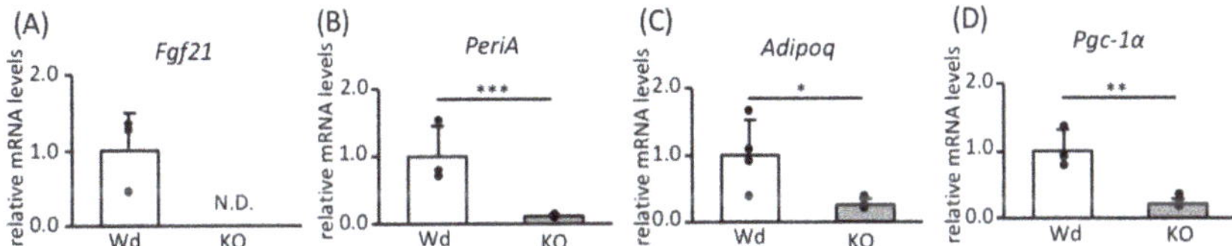

Figure 5. The effects of FGF21 deficiency on the expression of adipocyte differentiation markers and genes involved in mitochondrial biogenesis in mature adipocytes. MEFs were obtained from four individual embryos of either Wd or *Fgf21* KO mice, differentiated to mature adipocytes, and then RNA was extracted from each dish. The mRNA expression levels of *Fgf21* (**A**), *PeriA* (**B**), *Adipoq* (**C**), and *Pgc-1a* (**D**) were analyzed using RT-PCR and normalized to *Rps18* expression ($n = 4$). Values are means ± SDs. * $p < 0.05$, ** $p < 0.01$ vs. Wd, according to Student's *t*-test, *** $p < 0.001$.

Taken together, these findings indicate that FGF21 positively regulates PGC-1α expression, but ERK signaling does not have a significant effect.

3.4. Effect of Leptin Signaling on the Expression of Genes and Proteins Involved in FA Biosynthesis and Mitochondrial Biogenesis

To determine the effect of leptin signaling on FA biosynthesis and mitochondrial biogenesis in adipocytes, we knocked down leptin receptor expression in 3T3-L1 preadipocytes using a retroviral vector and analyzed the cells after differentiating them to adipocytes (Figure 6A). The activation of the leptin receptor phosphorylates STAT3, the major downstream target molecule of leptin signaling [38]. Leptin receptor knockdown (KD) reduced the phosphorylation of STAT3, confirming that leptin signaling had been inhibited (Figure 6F,G). The expression of *Srebp-1c*, *Srebp-1a*, *Fgf21*, and *Pgc-1α* mRNAs was significantly higher in leptin receptor KD adipocytes than in control cells (Figure 6B–E). Moreover, the expression of FGF21, PGC-1α, TFAM, ACC, and ME-1 proteins was higher in leptin receptor KD adipocytes (Figure 6F or Figure 6H–L). This finding suggests that a reduction in leptin signaling induces the expression of both *Srebp-1c* and *Srebp-1a* mRNAs and the expression of proteins involved in FA biosynthesis and mitochondrial biogenesis in adipocytes.

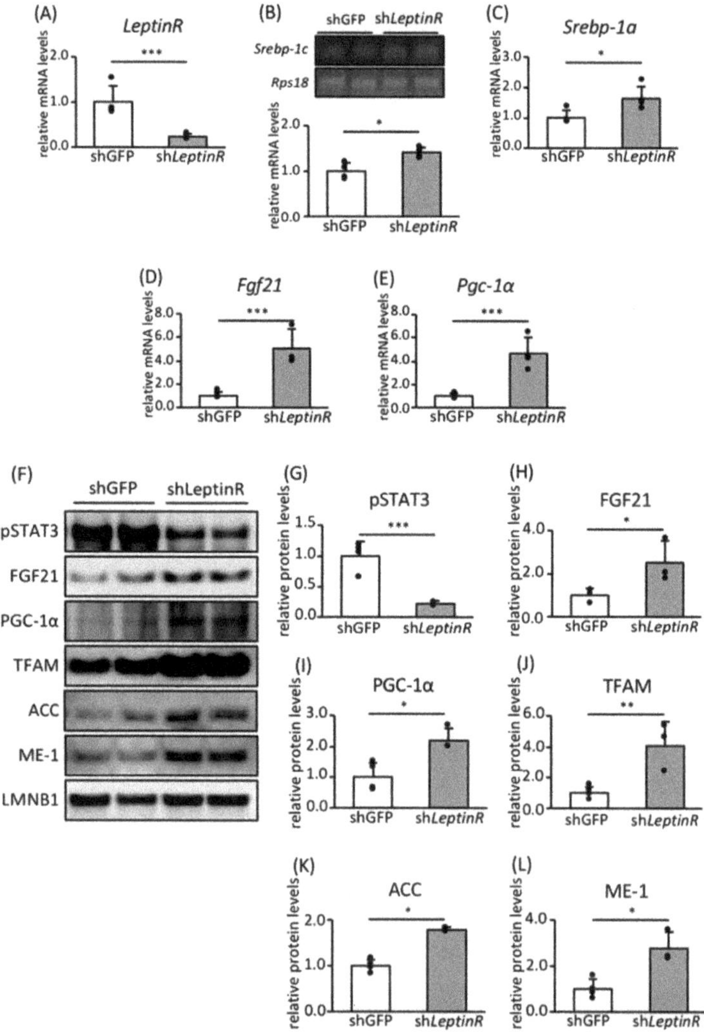

Figure 6. The effects of leptin signaling on the expression of genes and proteins involved in leptin signaling, FA biosynthesis, and mitochondrial biogenesis in mature 3T3-L1 adipocytes. LeptinR knockdown (KD) (*shLeptinR*) and control (*shGFP*) preadipocytes were differentiated into mature adipocytes in four separate dishes for each phenotype, and then RNA was extracted, and lysates were prepared from each dish. The mRNA expression levels of *LeptinR* (**A**), *Srebp-1a* (**C**), *Fgf21* (**D**), and *Pgc-1a* (**E**) were determined using RT-PCR and normalized to *Rps18* expression ($n = 4$). (**B**) Representative images of ethidium bromide-stained gels, showing fluorescence corresponding to the products of *Srebp-1c* cDNA amplification by RT-PCR. Semiquantitative analysis was performed and the data were normalized to *Rps18* expression ($n = 4$). (**F**) Representative immunoblot images showing the expression of proteins involved in leptin signaling, FA biosynthesis, and mitochondrial biogenesis. Quantitative analysis was performed using a chemiluminescence method. The protein expression of pSTAT (**G**), FGF21 (**H**), PGC-1α (**I**), TFAM (**J**), ACC (**K**), and ME-1 (**L**) is shown as the relative intensity of the indicated protein divided by that of LMNB1 as an internal control ($n = 4$). Values are means ± SDs. * p < 0.05, ** p < 0.01, *** p < 0.001 vs. shGFP, according to Student's *t*-test.

4. Discussion

SREBP-1 is a family of transcription factors that are master regulators of FA biosynthesis. It comprises two isoforms, SREBP-1c and SREBP-1a. SREBP-1c is upregulated in the livers of obese mice [39,40], and fatty liver occurs in mice with liver-specific overexpression of SREBP-1c [41]. According to previous findings, SREBP-1c functions predominantly in the liver, rather than in WAT, and is involved in hepatic steatosis [42,43]. However, we have reported previously that CR upregulates the expression of genes and/or proteins involved in FA biosynthesis and mitochondrial biogenesis, including *Pgc-1a* mRNA expression via SREBP-1c, in WAT rather than in the livers of mice on a B6;129S6 background [16]. In the present study, we have shown similar results in mice on a C57Bl/6 background and mature adipocytes differentiated from MEFs. We also demonstrated that CR upregulates FGF21 via SREBP-1c in mice on both B6;129S6 and C57Bl/6 backgrounds. *Fgf21* and *Pgc-1a* mRNA transcripts were downregulated in adipocytes differentiated from MEFs derived from SREBP1c KO mice, but these gene expressions were unchanged in WAT between WdAL and KOAL mice. We are not able to explain the discrepancy between the in vitro and in vivo findings, but these results reveal, at least, that SREBP-1c positively regulates both gene expressions. We also generated SREBP-1a OE 3T3-L1 adipocytes (Figure S3A) and the expression levels of mRNAs and proteins of interest were analyzed. Since the intrinsic expression of *Srebp-1c* mRNA was very low in 3T3-L1 adipocytes, RT-PCR was not possible. In SREBP-1a OE adipocytes, the expression of *Srebp-1c* mRNA was similar in control and SREBP-1a OE adipocytes (Figure S3B). However, the expression of *PeriA* and *Adipoq* mRNA was high in OE cells (Figure S3C,D), as was that of *Fasn* and *Fgf21* mRNAs, but this was not the case for *Pgc-1a* mRNA (Figure S3E–G). In addition, the protein expression of ACC, ME-1, FGF21, PGC-1α, and SIRT3 was unaffected or downregulated by OE (Figure S3H–N). In contrast, protein expression of TFAM was high in SREBP-1a OE adipocytes but not in SREBP-1c adipocytes. Moreover, *Fgf 21* mRNA expression was upregulated in both SREBP-1c and -1a OE 3T3-L1 adipocytes. In contrast, FGF21 protein was increased in SREBP-1c OE adipocytes but decreased in SREBP-1a OE adipocytes. We are not able to rationally explain these distorted results. However, overall, by comparing SREBP-1c and SREBP-1a OE 3T3-L1 adipocytes, we have confirmed that SREBP-1c, rather than SREBP-1a, is principally responsible for increases in the expression of genes and proteins involved in FA biosynthesis in mature adipocytes.

It has been reported that CR activates mitochondrial biogenesis in various tissues, including WAT, liver, heart, and skeletal muscle [44,45]. However, we found that CR-induced mitochondrial biogenesis is mediated by SREBP-1c only in WAT [16]. ME-1, which is upregulated in SREBP-1c OE adipocytes, is one of the enzymes involved in the pyruvate/malate cycle. Our previous proteomic analysis showed that CR upregulates the expression of proteins involved in the pyruvate/malate cycle, including ATP-citrate lyase, citrate synthase, mitochondrial pyruvate dehydrogenase E1 component subunit beta, and mitochondrial pyruvate carboxylase, as well as ME-1. Therefore, we hypothesized that CR might activate the pyruvate/malate cycle in WAT in order to switch from the use of glucose to the use of energy-dense FAs so that energy can be used more efficiently under poor food supply conditions [46]. On the basis of this hypothesis, it makes sense that CR would simultaneously upregulate the expression of proteins involved in FA biosynthesis, the pyruvate/malate cycle, and mitochondrial biogenesis via SREBP-1c. Bruss et al. has shown that CR activates de novo FA biosynthesis predominantly in WAT, rather than in the liver [47], and their findings are consistent with this hypothesis.

FGF21 positively regulates PGC-1α and PPARγ via feed-forward autocrine/paracrine loops in WAT [26,27]. It is widely accepted that PGC-1α is a master regulator of CR-associated mitochondrial biogenesis [18]. We have shown here that CR upregulates the expression of both FGF21 and PGC-1α via SREBP-1c. CR also upregulates PPARγ expression in WAT [29]. Therefore, we characterized the reciprocal regulatory mechanism involving FGF21 and PGC-1α expression in mitochondrial biogenesis. In *Fgf21* KO adipocytes, the expression of PGC-1α was low and adipocyte differentiation was impaired. However, in FGF21 OE adipocytes, the expression of *Pgc-1a* mRNA and PGF-1α protein was high [29]. FGF21 promotes the phosphorylation of ERK via the binding of FGF21 to FGFR and the β-klotho (KLB) receptor complex [21]. Treatment with an FGFR inhibitor reduced the phosphorylation of ERK and the expression of *Pgc-1a* mRNA but not that of PGC-1α protein. In SREBP-1c OE adipocytes, the expression of FGF21, PGC-1α, and SIRT3 proteins was very high. We previously demonstrated that SREBP-1c binds to the promoter of the *Pgc-1a* gene in adipocytes derived from Wd MEFs, but this did not occur in SREBP-1c KO MEFs [16]. Furthermore, in brown adipocytes, it has been reported that SREBP-1c activates the *Pgc-1a* promoter [48]. Therefore, when CR induces PGC-1α expression in adipocytes, it is likely that direct transcriptional regulation by SREBP-1c is more significant than the induction of FGF21 by SREBP-1c.

Leptin is secreted by WAT and acts as a satiety signal to the hypothalamus, activating NPY and AGRP neurons and suppressing POMC and CART neurons in the hypothalamus, subsequently activating the sympathetic nervous system and thereby lipolysis in WAT via β3-adrenergic receptors [32]. We have shown that a reduction in leptin signaling increases the expression of SREBP-1c, SREBP-1a, SREBP-1-regulated genes, FGF 21, and PGC-1α in mature adipocytes. These findings suggest that lower leptin secretion reduces leptin receptor signaling via an autocrine/paracrine loop, resulting in the greater expression of genes involved in FA biosynthesis and mitochondrial biogenesis in the WAT of CR mice. Previously, we found that the expression of proteins involved in FA biosynthesis is higher in obese fa/fa Zucker rats that have a leptin receptor mutation than in lean +/+ rats. Moreover, CR increases the expression of proteins involved in FA biosynthesis in lean +/+ rats but not in obese fa/fa Zucker rats [49]. Our present in vitro findings are consistent with these findings in Zucker rats.

Based on findings concerning the CR-associated metabolic remodeling of WAT in *Srebp-1c* KO mice, we investigated the upstream and downstream regulatory mechanisms of SREBP-1c in vitro. To confirm our results in vitro, we examined the mRNA and protein levels of most factors in both OE cells and KD or KO cells, and we were able to obtain relatively consistent data with regard to upregulated and downregulated genes. As a result, it was likely that a reduction in leptin signaling induced the expression of proteins involved in FA biosynthesis and mitochondrial biogenesis via SREBP-1c in adipocytes. PGC-1α is upregulated via both the direct transcriptional regulation of SREBP-1c and the upregulation of FGF21 indirectly regulated by SREBP-1c, but it is unlikely that the FGF21-associated upregulation of PGC-1α expression is a predominant factor in mitochondrial biogenesis induced by SREBP-1c. Therefore, we conclude that CR might downregulate an autocrine/paracrine loop involving leptin, with a reduction in leptin signaling activating de novo FA biosynthesis and mitochondrial biogenesis through the upregulation of SREBP-1c in WAT, in addition to the effects that leptin exerts via the central nervous system. SREBP-1c expression is high when the leptin concentration is low, and this is regulated in a GH/IGF-1-independent manner, but it is a key player in the CR-associated metabolic remodeling of WAT, which involves the upregulation of both FA biosynthesis and mitochondrial biogenesis. The CR-associated metabolic remodeling of WAT might be a leptin-mediated adaptive response to food shortage, causing a switch from the use of glucose to lipid as an energy substrate.

Supplementary Materials: The following is available online at http://www.mdpi.com/2072-6643/12/7/2054/s1, Figure S1: Food intake and body weight in *Srebp-1c* KO and Wd mice fed AL and subjected to CR. Figure S2: The effects of *Srebp-1c* KO on *Fgf21* expression in the WAT of mice on a C57Bl/6-129S6 background. Figure S3: The effects of SREBP-1a overexpression on the expression of genes and proteins involved in FA biosynthesis and mitochondrial biogenesis in mature 3T3-L1 adipocytes.

Author Contributions: Conceptualization, Y.H.; formal analysis, M.K.; methodology, M.K. and Y.H.; investigation, S.U., M.O., Y.D., R.T., and Y.M.; resources, M.K., Y.N., H.S., and Y.H.; writing—original draft preparation, M.K. and Y.H.; writing—review and editing, M.K. and Y.H.; supervision, M.K. and Y.H.; project administration, Y.H.; funding acquisition, Y.H. All authors have read and agreed to the published version of the manuscript.

Funding: This work was supported by Grants-in-Aid for Scientific Research (B) (No. 17H02179) from the Japan Society for the Promotion of Science.

Acknowledgments: We thank Mark Cleasby, from Edanz Group (https://en-author-services.edanzgroup.com/) for editing a draft of this manuscript.

Conflicts of Interest: All authors declare no conflict of interest.

References

1. Ouchi, N.; Parker, J.L.; Lugus, J.J.; Walsh, K. Adipokines in inflammation and metabolic disease. *Nat. Rev. Immunol.* **2011**, *11*, 85–97. [CrossRef]
2. Blüher, M.; Kahn, B.B.; Kahn, C.R. Extended longevity in mice lacking the insulin receptor in adipose tissue. *Science* **2003**, *299*, 572–574. [CrossRef] [PubMed]
3. Otabe, S.; Yuan, X.; Fukutani, T.; Wada, N.; Hashinaga, T.; Nakayama, H.; Hirota, N.; Kojima, M.; Yamada, K. Overexpression of human adiponectin in transgenic mice results in suppression of fat accumulation and prevention of premature death by high-calorie diet. *Am. J. Physiol. Endocrinol. Metab.* **2007**, *293*, 2102–2118. [CrossRef] [PubMed]
4. Farmer, S.R. Transcriptional control of adipocyte formation. *Cell Metab.* **2006**, *4*, 263–273. [CrossRef] [PubMed]
5. Chiu, C.H.; Lin, W.D.; Huang, S.Y.; Lee, Y.H. Effect of a C/EBP gene replacement on mitochondrial biogenesis in fat cells. *Genes Dev.* **2004**, *18*, 19701–19975. [CrossRef] [PubMed]
6. Argmann, C.; Dobrin, R.; Hekkinen, S.; Auburtin, A.; Pouilly, L.; Cock, T.; Koutnikova, H.; Zhu, J.; Schadt, E.E.; Auwerx, J. Pparγ2 is a key driver of longevity in the mouse. *PLoS Genet.* **2009**, *5*, e1000752. [CrossRef] [PubMed]
7. Masoro, E.J. Overview of caloric restriction and ageing. *Mech. Ageing Dev.* **2005**, *126*, 913–922. [CrossRef]
8. Sohal, R.S.; Weindruch, R. Oxidative stress, caloric restriction, and aging. *Science* **1996**, *273*, 59–63. [CrossRef]
9. Bartke, A.; Quainoo, N. Impact of growth hormone-related mutations on mammalian aging. *Front. Genet.* **2018**, *9*, 586. [CrossRef]
10. Bonkowski, M.S.; Rocha, J.S.; Masternak, M.M.; Al-Regaiey, K.A.; Bartke, A. Targeted disruption of growth hormone receptor interferes with the beneficial actions of calorie restriction. *Proc. Natl. Acad. Sci. USA* **2006**, *103*, 7901–7905. [CrossRef]
11. Sun, L.Y.; Spong, A.; Swindell, W.R.; Fang, Y.; Hill, C.; Huber, J.A.; Boehm, J.D.; Westbrook, R.; Salvatori, R.; Bartke, A. Growth hormone-releasing hormone disruption extends lifespan and regulates response to caloric restriction in mice. *Elife* **2013**, *2*, e01098. [CrossRef] [PubMed]
12. Shimokawa, I.; Higami, Y.; Tsuchiya, T.; Otani, H.; Komatsu, T.; Chiba, T.; Yamaza, H. Life span extension by reduction of the growth hormone-insulin-like growth factor-1 axis: Relation to caloric restriction. *FASEB J.* **2003**, *17*, 1108–1109. [CrossRef] [PubMed]
13. Chujo, Y.; Fujii, N.; Okita, N.; Konishi, T.; Narita, T.; Yamada, A.; Haruyama, Y.; Tashiro, K.; Chiba, T.; Shimokawa, I.; et al. Caloric restriction-associated remodeling of rat white adipose tissue: Effects on the growth hormone/insulin-like growth factor-1 axis, sterol regulatory element binding protein-1, and macrophage infiltration. *Age* **2013**, *35*, 1143–1156. [CrossRef] [PubMed]
14. Osborne, T.F.; Espenshade, P.J. Evolutionary conservation and adaptation in the mechanism that regulates SREBP action: What a long, strange tRIP it's been. *Genes Dev.* **2009**, *23*, 2578–2591. [CrossRef] [PubMed]
15. Shimomura, I.; Shimano, H.; Horton, J.D.; Goldstein, J.L.; Brown, M.S. Differential expression of exons 1a and 1c in mRNAs for sterol regulatory element binding protein-1 in human and mouse organs and cultured cells. *J. Clin. Investig.* **1997**, *99*, 838–845. [CrossRef]
16. Fujii, N.; Narita, T.; Okita, N.; Kobayashi, M.; Furuta, Y.; Chujo, Y.; Sakai, M.; Yamada, A.; Takeda, K.; Konishi, T.; et al. Sterol regulatory element-binding protein-1c orchestrates metabolic remodeling of white adipose tissue by caloric restriction. *Aging Cell* **2017**, *16*, 508–517. [CrossRef]
17. Scarpulla, R.C. Metabolic control of mitochondrial biogenesis through the PGC-1 family regulatory network. *Biochim. Biophys. Acta (BBA) Mol. Cell Res.* **2011**, *1813*, 1269–1278. [CrossRef]
18. Anderson, R.M.; Barger, J.L.; Edwards, M.G.; Braun, K.H.; O'Connor, C.E.; Prolla, T.A.; Weindruch, R. Dynamic regulation of PGC-1alpha localization and turnover implicates mitochondrial adaptation in calorie restriction and the stress response. *Aging Cell* **2008**, *7*, 101–111. [CrossRef]
19. Nishimura, T.; Nakatake, Y.; Konishi, M.; Itoh, N. Identification of a novel FGF, FGF-21, preferentially expressed in the liver. *Biochim. Biophys. Acta (BBA) Gene Struct. Expr.* **2000**, *1492*, 203–206. [CrossRef]
20. Ding, X.; Boney-Montoya, J.; Owen, B.M.; Bookout, A.L.; Coate, K.C.; Mangelsdorf, D.J.; Kliewer, S.A. βKlotho is required for fibroblast growth factor 21 effects on growth and metabolism. *Cell Metab.* **2012**, *16*, 387–393. [CrossRef]

21. Ogawa, Y.; Kurosu, H.; Yamamoto, M.; Nandi, A.; Rosenblatt, K.P.; Goetz, R.; Eliseenkova, A.V.; Mohammadi, M.; Kuro-o, M. BetaKlotho is required for metabolic activity of fibroblast growth factor 21. *Proc. Natl. Acad. Sci. USA* **2007**, *104*, 7432–7437. [CrossRef] [PubMed]

22. Markan, K.R.; Naber, M.C.; Ameka, M.K.; Anderegg, M.D.; Mangelsdorf, D.J.; Kliewer, S.A.; Mohammadi, M.; Potthoff, M.J. Circulating FGF21 is liver derived and enhances glucose uptake during refeeding and overfeeding. *Diabetes* **2014**, *63*, 4057–4063. [CrossRef]

23. Zhang, Y.; Lei, T.; Huang, J.F.; Wang, S.B.; Zhou, L.L.; Yang, Z.Q.; Chen, X.D. The link between fibroblast growth factor 21 and sterol regulatory element binding protein 1c during lipogenesis in hepatocytes. *Mol. Cell Endocrinol.* **2011**, *342*, 414–417. [CrossRef] [PubMed]

24. Véniant, M.M.; Hale, C.; Helmering, J.; Chen, M.M.; Stanislaus, S.; Busby, J.; Vonderfecht, S.; Xu, J.; Lloyd, D.J. FGF21 promotes metabolic homeostasis via white adipose and leptin in mice. *PLoS ONE* **2012**, *7*, e40164. [CrossRef]

25. Potthoff, M.J.; Inagaki, T.; Satapati, S.; Ding, X.; He, T.; Goetz, R.; Mohammadi, M.; Finck, B.N.; Mangelsdorf, D.J.; Kliewer, S.A.; et al. FGF21 induces PGC-1alpha and regulates carbohydrate and fatty acid metabolism during the adaptive starvation response. *Proc. Natl. Acad. Sci. USA* **2009**, *106*, 10853–10858. [CrossRef] [PubMed]

26. Fisher, F.M.; Kleiner, S.; Douris, N.; Fox, E.C.; Mepani, R.J.; Verdeguer, F.; Wu, J.; Kharitonenkov, A.; Flier, J.S.; Maratos-Flier, E.; et al. FGF21 regulates PGC-1α and browning of white adipose tissues in adaptive thermogenesis. *Genes Dev.* **2012**, *26*, 2712–2781. [CrossRef] [PubMed]

27. Dutchak, P.A.; Katafuchi, T.; Bookout, A.L.; Choi, J.H.; Yu, R.T.; Mangelsdorf, D.J.; Kliewer, S.A. Fibroblast growth factor-21 regulates PPARγ activity and the antidiabetic actions of thiazolidinediones. *Cell* **2012**, *148*, 556–567. [CrossRef]

28. Zhang, Y.; Xie, Y.; Berglund, E.D.; Coate, K.C.; He, T.T.; Katafuchi, T.; Xiao, G.; Potthoff, M.J.; Wei, W.; Wan, Y.; et al. The starvation hormone, fibroblast growth factor-21, extends lifespan in mice. *Elife* **2012**, *1*, e00065. [CrossRef]

29. Fujii, N.; Uta, S.; Kobayashi, M.; Sato, T.; Okita, N.; Higami, Y. Impact of aging and caloric restriction on fibroblast growth factor 21 signaling in rat white adipose tissue. *Exp. Gerontol.* **2019**, *118*, 55–64. [CrossRef]

30. Wilson-Fritch, L.; Nicoloro, S.; Chouinard, M.; Lazar, M.A.; Chui, P.C.; Leszyk, J.; Straubhaar, J.; Czech, M.P.; Corvera, S. Mitochondrial remodeling in adipose tissue associated with obesity and treatment with rosiglitazone. *J. Clin. Investig.* **2004**, *114*, 1281–1289. [CrossRef]

31. Zhang, Y.; Proenca, R.; Maffei, M.; Barone, M.; Leopold, L.; Friedman, J.M. Positional cloning of the mouse obese gene and its human homologue. *Nature* **1994**, *372*, 425–432. [CrossRef] [PubMed]

32. Ahima, R.S.; Flier, J.S. Leptin. *Annu. Rev. Physiol.* **2000**, *62*, 4134–4137. [CrossRef]

33. Campfield, L.A.; Smith, F.J.; Burn, P. The OB protein (leptin) pathway—A link between adipose tissue mass and central neural networks. *Horm. Metab. Res.* **1996**, *28*, 619–632. [CrossRef] [PubMed]

34. Soukas, A.; Cohen, P.; Socci, N.D.; Friedman, J.M. Leptin-specific patterns of gene expression in white adipose tissue. *Genes Dev.* **2000**, *14*, 9639–9680.

35. Shimokawa, I.; Higami, Y. A role for leptin in the antiaging action of dietary restriction: A hypothesis. *Aging* **1999**, *11*, 3803–3882. [CrossRef] [PubMed]

36. Hotta, Y.; Nakamura, H.; Konishi, M.; Murata, Y.; Takagi, H.; Matsumura, S.; Inoue, K.; Fushiki, T.; Itoh, N. Fibroblast growth factor 21 regulates lipolysis in white adipose tissue but is not required for ketogenesis and triglyceride clearance in liver. *Endocrinology* **2009**, *150*, 46254–46633. [CrossRef]

37. Mikami, K.; Okita, N.; Tokunaga, Y.; Ichikawa, T.; Okazaki, T.; Takemoto, K.; Nagai, W.; Matsushima, S.; Higami, Y. Autophagosomes accumulate in differentiated and hypertrophic adipocytes in a p53-independent manner. *Biochem. Biophys. Res. Commun.* **2012**, *427*, 758–763. [CrossRef]

38. Kim, Y.B.; Uotani, S.; Pierroz, D.D.; Flier, J.S.; Kahn, B.B. In vivo administration of leptin activates signal transduction directly in insulin-sensitive tissues: Overlapping but distinct pathways from insulin. *Endocrinology* **2000**, *141*, 23282–23339. [CrossRef]

39. Ferré, P.; Foufelle, F. Hepatic steatosis: A role for de novo lipogenesis and the transcription factor SREBP-1c. *Diabetes Obes. Metab.* **2010**, *12*, 83–92. [CrossRef]

40. Kammoun, H.L.; Chabanon, H.; Hainault, I.; Luquet, S.; Magnan, C.; Koike, T.; Ferré, P.; Foufelle, F. GRP78 expression inhibits insulin and ER stress-induced SREBP-1c activation and reduces hepatic steatosis in mice. *J. Clin. Investig.* **2009**, *119*, 1201–1215. [CrossRef]

41. Knebel, B.; Haas, J.; Hartwig, S.; Jacob, S.; Köllmer, C.; Nitzgen, U.; Muller-Wieland, D.; Kotzka, J. Liver-specific expression of transcriptionally active srebp-1c is associated with fatty liver and increased visceral fat mass. *PLoS ONE* **2012**, *7*, e31812. [CrossRef] [PubMed]

42. Horton, J.D.; Bashmakov, Y.; Shimomura, I.; Shimano, H. Regulation of sterol regulatory element binding proteins in livers of fasted and refed mice. *Proc. Natl. Acad. Sci. USA* **1998**, *95*, 59875–59992. [CrossRef] [PubMed]

43. Sekiya, M.; Yahagi, N.; Matsuzaka, T.; Takeuchi, Y.; Nakagawa, Y.; Takahashi, H.; Okazaki, H.; Iizuka, Y.; Ohashi, K.; Gotoda, T.; et al. SREBP-1-independent regulation of lipogenic gene expression in adipocytes. *J. Lipid Res.* **2007**, *48*, 1581–1591. [CrossRef]

44. Nisoli, E.; Tonello, C.; Cardile, A.; Cozzi, V.; Bracale, R.; Tedesco, L.; Falcone, S.; Valerio, A.; Cantoni, O.; Clementi, E.; et al. Calorie restriction promotes mitochondrial biogenesis by inducing the expression of eNOS. *Science* **2005**, *310*, 314–317. [CrossRef] [PubMed]

45. López-Lluch, G.; Hunt, N.; Jones, B.; Zhu, M.; Jamieson, H.; Hilmer, S.; Cascajo, M.V.; Allard, J.; Ingram, D.K.; Navas, P.; et al. Calorie restriction induces mitochondrial biogenesis and bioenergetic efficiency. *Proc. Natl. Acad. Sci. USA* **2006**, *103*, 1768–1773. [CrossRef] [PubMed]

46. Okita, N.; Hayashida, Y.; Kojima, Y.; Fukushima, M.; Yuguchi, K.; Mikami, K.; Yamauchi, A.; Watanabe, K.; Noguchi, M.; Nakamura, M.; et al. Differential responses of white adipose tissue and brown adipose tissue to caloric restriction in rats. *Mech. Ageing Dev.* **2012**, *133*, 255–266. [CrossRef] [PubMed]

47. Bruss, M.D.; Khambatta, C.F.; Ruby, M.A.; Aggarwal, I.; Hellerstein, M.K. Calorie restriction increases fatty acid synthesis and whole body fat oxidation rates. *Am. J. Physiol. Endocrinol. Metab.* **2010**, *298*, 108–116. [CrossRef]

48. Hao, Q.; Hansen, J.B.; Petersen, R.K.; Hallenborg, P.; Jørgensen, C.; Cinti, S.; Larsen, P.J.; Steffensen, K.R.; Wang, H.; Collins, S.; et al. ADD1/SREBP1c activates the PGC1-alpha promoter in brown adipocytes. *Biochim. Biophys. Acta* **2010**, *1801*, 4214–4229.

49. Okita, N.; Tsuchiya, T.; Fukushima, M.; Itakura, K.; Yuguchi, K.; Narita, T.; Hashizume, Y.; Sudo, Y.; China, T.; Shimokawa, I.; et al. Chronological analysis of caloric restriction-induced alteration of fatty acid biosynthesis in white adipose tissue of rats. *Exp. Gerontol.* **2015**, *63*, 59–66. [CrossRef]

© 2020 by the authors. Licensee MDPI, Basel, Switzerland. This article is an open access article distributed under the terms and conditions of the Creative Commons Attribution (CC BY) license (http://creativecommons.org/licenses/by/4.0/).

Review

The Effects of Calorie Restriction on Autophagy: Role on Aging Intervention

Ki Wung Chung [1,*] and Hae Young Chung [2,*]

[1] College of Pharmacy, Kyungsung University, Busan 48434, Korea
[2] College of Pharmacy, Pusan National University, Busan 462414, Korea
* Correspondence: kiwungc@ks.ac.kr (K.W.C.); hyjung@pusan.ac.kr (H.Y.C.);
Tel.: +82-51-663-4884 (K.W.C.); +82-51-510-2814 (H.Y.C.)

Received: 29 October 2019; Accepted: 29 November 2019; Published: 2 December 2019

Abstract: Autophagy is an important housekeeping process that maintains a proper cellular homeostasis under normal physiologic and/or pathologic conditions. It is responsible for the disposal and recycling of metabolic macromolecules and damaged organelles through broad lysosomal degradation processes. Under stress conditions, including nutrient deficiency, autophagy is substantially activated to maintain proper cell function and promote cell survival. Altered autophagy processes have been reported in various aging studies, and a dysregulated autophagy is associated with various age-associated diseases. Calorie restriction (CR) is regarded as the gold standard for many aging intervention methods. Although it is clear that CR has diverse effects in counteracting aging process, the exact mechanisms by which it modulates those processes are still controversial. Recent advances in CR research have suggested that the activation of autophagy is linked to the observed beneficial anti-aging effects. Evidence showed that CR induced a robust autophagy response in various metabolic tissues, and that the inhibition of autophagy attenuated the anti-aging effects of CR. The mechanisms by which CR modulates the complex process of autophagy have been investigated in depth. In this review, several major advances related to CR's anti-aging mechanisms and anti-aging mimetics will be discussed, focusing on the modification of the autophagy response.

Keywords: aging; autophagy; calorie restriction (CR); CR mimetic

---

1. Introduction

1.1. The Autophagy Process

Autophagy is an evolutionarily well-conserved process that occurs in all eukaryotic cells from yeast to human [1]. The highly complex autophagy-related signaling pathways have been extensively studied for the last 30 years, and they have been elucidated through the combined study of genetics and physiology in various species [2]. At least three different forms of autophagy have been identified so far: macro-autophagy, micro-autophagy, and chaperone-mediated autophagy. All three forms depend on lysosomal degradation, with macro-autophagy (hereafter referred to as autophagy) being the most prevalent form. Once activated, autophagy involves the sequestering of cytosolic components (damaged cell organelles, proteins, or other macromolecule nutrients) by phagophores that mature into autophagosomes, which are double membrane vesicles [2,3]. Autophagosomes further translocate and fuse to the acidic lysosome and form the autolysosome, where degradation and recycling occur. The diverse substrates and basal activity of these processes suggest that cells are highly dependent on it for maintaining cellular homeostasis. The importance of maintaining an adequate autophagy response has been demonstrated under both physiologic and pathologic conditions [4].

1.2. Molecular Machinery of the Autophagy Process

The molecular mechanisms and signaling pathways controlling autophagy have been extensively studied [5]. Autophagy begins with the *de novo* production of autophagosome components, followed by assembly driven by the concerted action of a group of proteins named ATG (autophagy-related genes). As the detailed molecular machinery of the autophagy process has been previously described in several review articles, only its overall features will be discussed in this review. At the start of the autophagy process, phagophore formation is initiated from the endoplasmic reticulum (ER)–mitochondrial interface, and further elongation of the phagophore depends on the Golgi and plasma membranes. The progression of autophagosome formation is largely characterized by the recruitment of ATG proteins to the phagophore [6].

The formation of the UMC-51-like kinase 1 (ULK1, homologous to yeast ATG1) complex is the earliest event in the formation of the autophagosome. ULK1 activation lies upstream of other ATG protein recruitment, and ULK1 kinase activity is required for the recruitment of the VPS34 complex (a class III PI3-kinase) to the phagophore. This is crucial for the phosphorylation of phosphatidyl inositol (PtdIns) and the subsequent production of PtdIns 3-phosphate. The further recruitment of phospholipid-binding proteins to the phagophore is important for the stabilization of protein complexes near the autophagosome formation site. Two conjugation systems are involved in the vesicle elongation process. The conjugation of ATG5 to the ATG12 complex requires the ubiquitin-like conjugation system involving ATG7 and ATG10. The conjugated ATG5–ATG12 complex is needed to further conjugate phosphoethanolamine (PE) to ATG8 (microtubule-associated protein 1 light chain 3; LC3). ATG4, ATG7, and ATG3 are required for this conjugation process. The conversion of LC3 from LC3-I (soluble form) to LC3-II (vesicle associated form) by PE conjugation is thought to be required for the closure of the expanding autophagosomal membrane. Finally, the matured autophagosome is fused with the lysosome to fulfill the main purpose of the process, culminating with the degradation and recycling of substrates in the autophagosome.

1.3. Autophagy Is Regulated by Nutrient-Sensing Signaling

A variety of physiologically important stimuli induce the autophagy process, including organelle (ER, mitochondria) damage, hypoxia, and inflammation [2]. However, nutrients and energy stress are the most powerful regulators of the autophagy process [7]. Changes in the cellular energy status such as the withdrawal of nutrients, such as glucose and amino acids, induce the activation of the autophagy process, from initiation to termination [8]. Nutrient levels can be directly recognized by the upstream signaling machinery of autophagy to regulate its initiation in response to the changing cellular energy levels (Figure 1).

Of all the nutrient-associated signaling molecules, mammalian target of rapamycin (mTOR) has been shown as one of the key upstream modulators of autophagy signaling [9,10]. mTOR is a highly conserved serine/threonine kinase that is regulated by multiple signals including energy levels, growth factors, and other cellular stressors, to coordinate cell proliferation/growth and maintain energy homeostasis. mTOR forms a complex, which is known as mTORC1 (mTOR complex 1) and mTORC2 (mTOR complex 2). mTORC1 is related to autophagy signaling changes and is activated in the presence of nutrients or growth factors. mTORC1 is usually activated under nutrient-rich conditions [11]. It can be directly activated by an increased concentration of amino acids in the cell or as downstream signaling through the action of growth factors [11,12]. Once activated, mTORC1 directly phosphorylates ULK1 [13]. Critically, the activation of mTORC1 is sufficient to inhibit autophagy in the presence of sufficient nutrients [14]. The direct repression of ULK1 kinase by mTORC1 is also well conserved across species [15]. Other components of the ATG complex directly interact with mTORC1 and repress the autophagy process [16]. Furthermore, mTORC1 can indirectly suppress autophagy by controlling lysosome biogenesis [17,18]. The transcription factor EB (TFEB) is responsible for the transcription of lysosomal and autophagy-related genes [19]. mTORC1-mediated

TFEB phosphorylation decreases its transcriptional activity, thus decreasing the overall expression of autophagy-related gene expression [20,21].

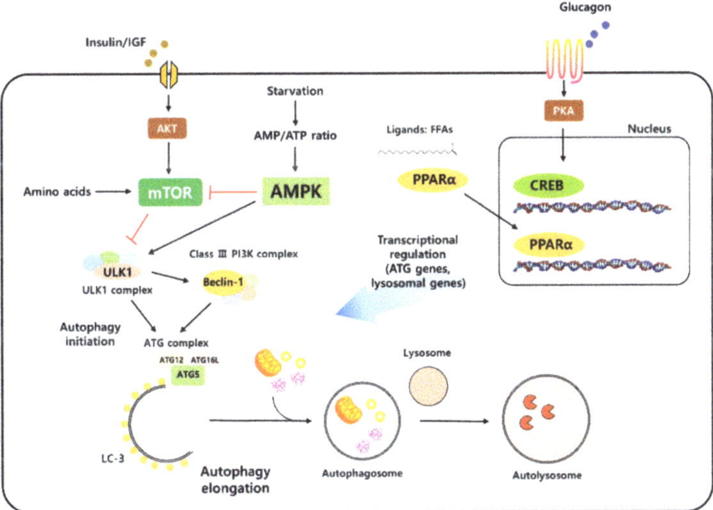

Figure 1. Autophagy is regulated by nutrient-sensing signaling. Autophagy signaling is modulated mainly by nutrient-sensing signaling pathways. Insulin and IGF (insulin-like growth factor) induce the activation of mammalian target of rapamycin (mTOR) signaling and inhibit autophagy initiation. The activation of AMP-activated protein kinase (AMPK) by an increased AMP/ATP ratio during starvation directly increases autophagy and inhibits the mTOR complex. CRE-binding protein (CREB) activation by glucagon signaling and peroxisome proliferation factor-activated receptor α (PPARα) activation by its ligands increases the gene transcription level of autophagy and lysosome-related proteins.

Under nutrient-deficient conditions, the activation of autophagy is regulated by several well-known nutrient-sensing signaling proteins. One of the most prominent players of nutrient deprivation sensing is the AMP-activated protein kinase (AMPK) [13,22]. The molecular ratio of ATP to AMP reflects the cell's energy levels, and increased levels of AMP represent an internal cell warning system that induces the cell to save energy for the maintenance of metabolic homeostasis. AMP is directly sensed by AMPK, and activated AMPK has been characterized and shown to have multiple functions in the regulation of cellular metabolism. There are several mechanisms by which AMPK induces autophagy. First, AMPK directly phosphorylates ULK, which is a process that is required for ULK1 activation and the initiation of autophagy under nutrient deprivation conditions [13]. The interaction between AMPK and ULK1 can be blocked by mTORC1-mediated ULK1 phosphorylation, indicating an intricate connection between these two pathways. Secondly, AMPK is a negative regulator of the mTOR signaling pathway [23]. Mechanistically, AMPK directly phosphorylates the tuberculosis-associated complex (TSC), which is a negative regulator of mTORC1 activation. AMPK also directly phosphorylates the Raptor subunit of the mTORC1 complex, increasing the degradation of the mTORC1 complex. These studies clearly demonstrated that AMPK, a critical regulator of nutrient availability, is able to regulate autophagy activity by coordinating mTOR-dependent and independent mechanisms.

1.4. Autophagy and Aging

Aging is associated with various changes including genomic instability, loss of proteostasis, epigenetic alterations, and deregulated nutrient-sensing pathways [24]. These changes are also associated with numerous age-related diseases including cardiovascular diseases, neurodegenerative diseases, and metabolic diseases. Among the changes that occur during aging, some are associated with autophagy-related signaling pathways [24,25]. A decline in the overall proteolytic activity and an altered nutrient-sensing signaling are directly associated with autophagy. Indeed, decreased autophagy with aging has been reported extensively in a broad range of organisms, where a progressive accumulation of damaged proteins and cellular organelles was shown to occur [26]. The decreased level of autophagy-related gene transcripts and proteins has been detected in nematodes and the fruit fly [27–29]. Aged tissues from mammals and humans also showed a lower expression of key autophagy proteins [30–32]. Consistent with changes in the levels of autophagy components, recent studies further showed a decreased overall autophagic capacity during aging in *C. elegans* [33]. Electron microscopy observations showed an age-related accumulation of autophagic vacuoles, which represents the blockage of autophagy flux. Similarly, the overall proteolysis activity is impaired during the aging process, and long-lived proteins that were not properly degraded have been detected in the liver in aged rats [34].

Further striking evidence between aging and autophagy comes from several genetic models of impaired autophagy. An unbiased screen for aging factors in yeast, nematodes, and fruit fly revealed short-lived mutants with defects in autophagy [27,35,36]. Moreover, in knockout mice, whole body deletion of autophagy-related genes led to early postnatal death, indicating an essential role of autophagy in the overall maintenance of physiological processes [37–39]. Tissue-specific conditional knockout mice models also revealed the multiple phenotypes of aging, including the aggregation and accumulation of intracellular proteins, cellular organelles, and other macromolecules [40–43]. The loss of autophagic activity in these models is likely to increasingly constrain the ability of the cells to maintain quality control, leading to the accumulation of toxic insults, and resulting in aging and age-associated pathologies [3]. On the other hand, accumulating evidence suggests that experimentally enhanced autophagy extends the lifespan and delays the aged phenotype. The overexpression of specific autophagy genes can extend the lifespan in several species. The upregulation of autophagic activity can extend longevity in *C. elegans*, as well as in the yeast, while the ubiquitous overexpression of Atg5 in mice is sufficient to stimulate autophagy and extend the lifespan [44,45]. Collectively, these observations indicate that changes in autophagic activity may be associated with longevity and that augmenting autophagic function may be an effective approach to delay aging and promote longevity in different species, including in mammals.

The mechanisms by which autophagy components or autophagic processes decrease with age remain unclear. Since the autophagy process, from initiation to completion, is complex and associated with various steps and different proteins, it is likely that the mechanisms contributing to age-associated autophagy decrease are multifactorial. The most plausible regulatory mechanism contributing to suppressed autophagy in aging is a change in the upstream signaling during autophagy initiation. Two important nutrient-sensing proteins, mTOR and AMPK, play an important role in the regulation of the initiation of autophagy [10,13]. Furthermore, these factors reflect the status of a cell, such as hormonal regulation (outside the cell) and nutrition stress (inside the cell). The nutrient sensor mTOR strongly inhibits not only the initiation of autophagy but also exerts an inhibitory effect on multiple steps in the autophagy process. It is possible that increased mTOR signaling during aging plays an important role in the age-associated suppression of autophagy. Since increased mTOR activity has been reported in various age-related diseases including metabolic and degenerative disorders, it is plausible that increased mTOR signaling is the predominant cause for the downregulation of the overall autophagy process [46]. Unlike mTOR, which is usually hyperactivated during aging, the activity or expression of AMPK is typically suppressed [47]. It is plausible that decreased AMPK might influence or suppress autophagy and act in concert with mTOR. To this end, although mechanisms disrupting autophagy signaling during aging are multifactorial, it is clear that modifications in its upstream pathways are critical for its regulation.

Another possible mechanism responsible for the decreased autophagy observed in aging is transcriptional regulation. TFEB has been previously described as a regulator of autophagy-related gene transcription; however, recent studies have revealed other important transcription factors that regulate the gene expression of autophagy-related proteins. The fasting transcriptional factor CRE-binding protein (CREB) is upregulated by glucagon under nutrient deprivation conditions, and it also upregulates autophagy gene expression including ATG7, ULK1, and TFEB. In addition to CREB, peroxisome proliferation factor-activated receptor α (PPARα), another transcription factor playing a role in starvation, also directs the transcription of autophagy genes [19,48,49]. Both transcription factors may act in concert to increase autophagy-related gene expression. The genetic deletion of both transcription factors reduced autophagy and led to an inadequate metabolic response, particularly under nutrient deprivation. Although there is no direct evidence of whether they play a role in defective autophagy during aging, there is some evidence that they are important and dysregulated during aging [50–52]. Further studies will be necessary to reveal the relationship between these transcription factors and defective autophagy during aging.

2. Calorie Restriction (CR) Modulates Autophagy Processes

2.1. Introduction to Calorie Restriction

Calorie restriction (CR) has been shown to be an established life-extension method regulating age-related diseases as well as aging itself. Although different in methodology (usually 20%–40% ad libitum intake, 40% reduction in most cases), CR showed a prolonged lifespan in a wide range of species from yeast to non-human primates, and supports healthy human aging [53]. Furthermore, CR exerts preventive effects on various age-related conditions such as cancer, neurodegenerative diseases, cardiovascular, and other metabolic diseases [54]. The diverse efficacy of CR in counteracting aging and age-related diseases has made it the golden standard of aging intervention studies. Although the anti-aging effects of CR are reproducible, the exact mechanisms of how CR exerts its anti-aging effects are debatable, because CR regulates several different aspects of physiology. These changes include modifications in the energy-sensing signaling, oxidative stress, inflammation, and other intercellular and intracellular processes. Among the many changes induced by CR, energy production and utilization is the most directly regulated signaling exerted by CR [55,56]. Since reduced energy intake and changes in nutritional status following CR may change the molecular signaling pathways associated with energy-sensing mechanisms, other mechanisms may be secondary effects to this process.

2.2. Evidence for the Beneficial Effects of CR-Mediated Autophagy

Based on the induction mechanism of autophagy and its role during starvation, it was predicted that CR might induce the autophagic process. Indeed, under many different settings of nutrient deprivation conditions, including in CR, autophagy is induced to regulate the organism's homeostasis. Although it is clear that CR represents a strong physiologically autophagic inducer, it is uncertain whether autophagy contributes to the anti-aging effects of CR. Recently, several studies have shown that autophagy induction was essential for the anti-aging effects of CR (Table 1). CR was shown to promote longevity or protect from hypoxia through a Sirtuin-1-dependent autophagy induction process [57,58]. Another study also showed that life extension through methionine restriction required autophagy activation [59]. Growing evidence supports the notion that autophagy has a substantial role in the beneficial effects of CR [60,61]. In addition to research on longevity, other studies have shown that CR robustly induces autophagy under various physiological and pathological conditions, and that it has a protective effect in the maintenance of normal functions in the organism. In the following section, the protective role of autophagy under CR conditions will be discussed.

Table 1. Studies showing protective effects of calorie restriction (CR)-induced autophagy in different organs. LC3: light chain 3.

	Liver Autophagy		
Species	**CR Methods**	**Main Results**	**Ref**
Fisher Rats	40% calorie restriction (Life-long)	CR had no substantial effect on the expression of autophagic proteins	[62]
SD Rats	Alternative day fasting (10 months)	Alternative day fasting increased autophagy at high levels	[63]
Rat	40% calorie restriction (4 months)	CR increased autophagy flux (LC3-II/LC3-I ratio) especially in the mitochondrial membrane	[64]
Mice	0%–40% calorie restriction (4 months)	A significant increase in autophagy was detected	[65]
	Muscle Autophagy		
Fisher Rats	8% calorie restriction (Life-long)	Mild CR attenuated the impairment of autophagy in rodent muscle during aging	[66]
Human	Up to 30% calorie restriction (3–15 years)	Autophagy-related genes were significantly increased in response to CR	[67]
Mice	40% calorie restriction (6–18 months)	Autophagy and mitochondrial integrity was significantly increased	[68]
	Adipose Tissue Autophagy		
Mice	40% calorie restriction (15 days)	Autophagy was significantly induced in lean mice (but not in obese mice)	[69]
Mice	40% calorie restriction (Life-long)	Autophagy activity was enhanced in CR mice compare to aged mice	[70]
	Kidney Autophagy		
Mice	40% calorie restriction (12 month)	Autophagy flux and LC3 conversion were higher in CR mice	[58]
SD Rats	40% calorie restriction (2 month)	Short-term CR increased LC3-II/LC3-I ratio and beclin-1 expression	[71]

3. Protective Effects of CR-Induced Autophagy on Different Organs

The substrates of autophagy include important macronutrients such as glycogen and lipid droplets [72,73]. Under CR conditions, it is essential for cells to use their internal nutrient stores. The breakdown products derived from autophagy provide substrates for biosynthesis and energy generation. The redistribution of nutrients, under starved or CR conditions, is essential for the cells to adapt to the changed nutritional environment. Indeed, metabolic tissues show the most dramatic changes in autophagy regulation under nutrient-starved conditions, suggesting its important role in the regulation of metabolism.

3.1. Liver

The importance and original concept of autophagy was first described in the liver, where high levels of enzymes and cellular organelles associated with lysosomal degradation are found. Liver autophagy plays an important role under physiologic and pathologic conditions by contributing to the recycling of organelles, as well as macronutrients [74]. Recent evidence showed that the role of liver autophagy under normal physiological conditions is to regulate the nutrient degradation systems, such as glycogenolysis and lipid droplet degradation [42,75]. Furthermore, lipid droplet degradation in hepatocytes (lipophagy) is particularly important under pathologic conditions such as in non-alcoholic fatty liver disease, steatohepatitis, and in hepatocellular carcinoma [75–77]. The deficient autophagic response aggravated not only lipid accumulation but also other pathologic features of liver disease. Liver autophagy is also impaired during aging. The base level of autophagy as well as autophagy induced by stress responses is impaired in the aged liver, making it vulnerable to liver damage [72].

The effects of CR on liver autophagy were assessed in several studies. Wohlgemuth et al. evaluated the effects of life-long CR in Fisher rats [62]. They found that life-long CR did not cause a substantial change in the expression of autophagic proteins in the liver. However, other studies using different CR settings found different results. Donati et al. assessed the effect of CR following alternate day fasting [63]. When studying the rate of autophagic proteolysis in the isolated livers, they found that maximum rates of autophagy were achieved in the CR groups compared to controls. A more recent study by Luevano-Martinez et al. showed the effect of CR on the induction of autophagy in liver mitochondria [64]. They isolated mitochondria from the livers of controls, and after 4 months of a CR schedule, they found an increase in the LC3-II/LC3-I ratio in CR livers, indicating enhanced liver mitochondrial autophagy. Derous et al. found similar results when they evaluated the effect of graded levels of CR on autophagy using the hepatic transcriptome [65]. Mice were subjected to a graded level of CR (from 0% to 40% CR) for 3 months, following which a significant increase in autophagy levels was observed that correlated with increased levels of CR. In the liver, the autophagy response is generally increased following CR, independently of the method used to induce CR.

In addition to CR, fasting also induced a robust hepatic autophagy. Although fasting is different from a consistent pattern of CR, they share some common features. Researchers have identified that a fasting-induced autophagy response is a fundamental process during food deprivation and is an important protective response in the regulation of metabolism [78,79].

3.2. Muscle

The skeletal muscle is the most abundant body tissue (comprising approximately 40% of the body weight) and is a dynamic tissue consistently adapting to metabolic demands. To meet the high metabolic demand, autophagy proteolytic systems engage in metabolic regulation [80]. In the muscle, autophagy regulates protein degradation and provides amino acids for energy production [43,81]. This is particularly important under nutrient-deprived or stress conditions to maintain adequate energy production. Recent studies have shown that basal autophagy is crucial for the maintenance of muscle physiology, and that a maladaptive autophagy is implicated in various muscle diseases, including muscular dystrophy, sarcopenia, and myofibril degeneration [31,66,82–84].

Several studies have shown the ability of CR to induce muscle autophagy and its beneficial effects. Wohlgemuth et al. investigated the effects of aging and mild CR on skeletal muscle autophagy and lysosome-related proteins [66]. They found LC3-I and LAMP-2 accumulation, suggesting an age-related decline in autophagic degradation. Age-related changes were inhibited by CR, concluding that mild CR attenuated the age-related impairment of autophagy in skeletal muscle in rodents. More evidence comes from a recent clinical trial study. Yang et al. showed that long-term CR enhanced the overall quality-control processes in human skeletal muscle [67]. They found that several autophagy genes, including ULK1, ATG101, beclin-1, LC3 were significantly upregulated in response to CR. Furthermore, they found decreased muscle inflammation, suggesting another beneficial role of CR on muscle biology. The study by Gutierrez-Casado et al. also showed a prominent effect of CR on autophagy in the muscle [68]. CR resulted in decreased levels of p62, suggesting a possible increase in autophagy flux. Although not experimentally demonstrated, Lee et al. suggested the importance of the role of autophagy on muscle stem cell regeneration induced by CR [85]. CR not only improved stem cell regenerative capacity but also enhanced the engraftment capacity of muscle stem cells [86]. CR-induced autophagy may prime the improvement in oxidative stress and increase mitochondrial activity in muscle stem cells, contributing to their beneficial regenerative effects in muscle.

3.3. Adipose Tissue

Adipose tissue is another important metabolic tissue that plays an important role in lipid storage during energy-sufficient conditions. A reduction in adiposity is the hallmark of CR, which is a consequence that may result from hormonal changes [87]. Although it is clear that autophagy induces lipid degradation through lipophagy in the liver, the role of autophagy in the regulation of adipose tissue lipids is more complex [88]. Singh et al. first showed that adipose tissue autophagy regulates adipose tissue mass and differentiation [89]. They found that the knockdown of Atg7, an essential autophagy gene, inhibited lipid accumulation and decreased the protein level of several adipocyte differentiation factors. Furthermore, they demonstrated that the adipocyte-specific Atg7 knockout mouse had a lean phenotype with decreased white adipose mass and enhanced insulin sensitivity. However, more recently, Cai et al. showed a protective effect of autophagy in mature adipocyte function [90]. They showed that autophagy proteins are required for adequate mitochondrial function and that the post-development ablation of autophagy caused insulin resistance.

The defective regulation of adipose tissue autophagy has been detected in mice and human obesity [69]. In mice models of obesity and in obese humans, autophagy-related genes and proteins were found to be significantly upregulated [91,92]. Although these results were interpreted as increased autophagy, at first, Soussi et al. showed that the autophagy flux was impaired in obesity [93]. This result was consistent with the conclusion derived from the work of Cai et al., showing that autophagy may play a protective role after maturation. Based on these results, it is clear that the maintenance of an appropriate activation of autophagy is needed in the adipose tissue. However, the role of CR in adipose tissue function is yet to be clarified. Nunez et al. showed that CR successfully increased autophagy in lean mice, but in obese mice, autophagy induction did not occur, suggesting that similarly to previous reports, the autophagic response is defective during obesity [69]. Ghosh et al. studied the effects of aging and CR on adipose tissue autophagy and found a diminished autophagy activity with aging, contributing to aberrant ER stress and inflammation in aged adipose tissue [70]. They also showed that autophagy activity was enhanced in the CR mice with a concomitant decrease in ER stress and inflammation. Taken together, CR has beneficial effects on adipose tissue, at least partly through the induction of the autophagy response.

3.4. Kidney

Kidneys also show beneficial effects from CR, including the induction of autophagy. In the canonical concept of metabolism, the kidney is not an active participant. However, the kidney can participate and play an important role in the metabolism of carbohydrates, proteins, and lipids [94,95]. Renal tubule cells have a high basal level of energy consumption and depend on the β-oxidation of fatty acids to generate adequate amounts of ATP [96]. Furthermore, proximal tubule cells generate glucose through gluconeogenesis, especially under nutrient-deficient conditions, and contribute to the total blood glucose level [97]. For these reasons, suitable autophagy is important for the maintenance of normal kidney physiology by regulating adequate metabolic processes and organelle quality. Defects in autophagy have been found to worsen conditions in several types of kidney diseases [98,99]. CR is known to have beneficial effects in the kidney both under physiologic and pathologic conditions [100]. In addition, CR also leads to a delayed age-associated kidney dysfunction and to structural changes [50]. Among several suggested mechanisms that explain the beneficial effects of CR in the kidneys, increased autophagy activity is an important one.

Kume et al. designed a 12-month-long CR schedule in 12-month-old mice to assess the effect of aging and CR on autophagy [58]. In comparison to the control group, CR resulted in healthy mitochondria with numerous autophagosomes in the kidney. In addition, a lower level of p62 was found in the kidney of the CR mice. The ratio of LC3 conversion and LC3 puncta were higher in the CR mice, indicating that CR-mediated autophagy increased mitochondrial integrity and protected from age-associated kidney damage. Ning et al. showed a similar result using a short-term calorie restriction model [71]. CR groups had a 40% calorie restriction for 8 weeks, and showed increased autophagy flux, autophagy-related gene expression, and reduced oxidative damage. CR also significantly decreased p62 expression and polyubiquitin aggregates.

Chung et al. also showed that short-term CR reduced age-associated renal fibrosis [50]. They found that reduced PPARα expression during aging impaired lipid metabolism and induced interstitial fibrosis in the kidney. PPARα knockout mice showed an early onset of age-associated kidney fibrosis. Although they only focused on lipid metabolism for the regulatory role of PPARα and did not check for autophagy changes in their model, PPARα plays an important role in the expression of autophagy-related genes; therefore, it is plausible that autophagy might have played a role in mediating the anti-fibrosis effects of CR in their model. Collectively, these studies strongly suggest that CR effectively induces autophagy in aging and diabetic mice and plays a protective role in these settings.

4. Benefits of Intermeal Fasting in Autophagy: Is CR the Only Solution?

Recently, an interesting study by Martinez-Lopez et al. demonstrated a pivotal role for autophagy under nutritional conditions other than CR [78]. They introduced an isocaloric twice-a-day (ITAD) feeding model with the same amount of food consumption in total as ad libitum controls. These mice were exposed to food at two short time intervals, early and late in the diurnal cycle. The concept of this model is different from calorie restriction because the total food intake is the same as the controls. ITAD still leads to intermeal fasting, which induces various physiological changes, including the autophagy process. ITAD feeding impacted autophagy flux in multiple organs including liver, adipose tissue, muscle, and neurons. ITAD feeding promoted multiple metabolic benefits in organs where autophagy was increased, and further experiments demonstrated a tissue-specific contribution of autophagy to the metabolic benefits of ITAD feeding by use of tissue-specific autophagy knockout models. Finally, in an aging and obesity model, it was concluded that consuming two meals a day without CR could prevent metabolic syndrome through the activation of autophagy. This study could easily translate to humans, as ITAD is more feasibly applied than CR. If a similar regimen was applied in humans, it could provide some beneficial effects such as autophagy induction and ultimately prevent various age-associated metabolic diseases.

More recently, Stekovic et al. showed a prominent effect of alternate day (AD) fasting on aging in non-obese humans [101]. AD fasting significantly improved physiological and molecular markers; it also improved cardiovascular markers with reduced fat mass and without any of the typical adverse effects. This study also emphasized that AD fasting can be tolerated more easily than continuous CR and lead to similar beneficial effects. Although they did not check whether the autophagic response played a role, it might be interesting to further investigate the effects of AD fasting on autophagy induction.

5. CR Mimetic as an Autophagy Inducer

CR could have beneficial effects that prolong human lifespan; however, it is challenging to implement, even in the case of short-term CR. Therefore, the development of drugs or compounds that mimic the effect of CR is an interesting topic of discussion among biologists and gerontologists [102]. Based on the pathways and proteins changed under CR conditions, many have started to investigate modulators that mimic the CR effect. Currently, several drugs and other compounds naturally occurring in the diet (nutraceuticals) have been shown to act as a CR mimetic through various mechanisms. The targets of mimetics include the glycolysis pathway, insulin/insulin-like growth factor signaling, mTOR, AMPK, sirtuins, and other pathways associated with CR. Interestingly, many well-known CR mimetics are directly or indirectly associated with autophagy regulation. The following discussion will focus on well-known CR mimetics that act through the regulation of autophagy (Figure 2).

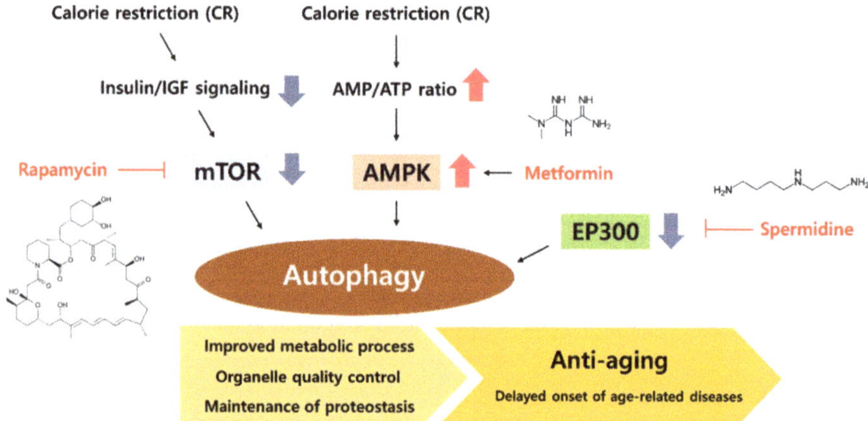

Figure 2. Calorie restriction (CR) and CR mimetics modulate the autophagy process. CR decreases mTOR signaling by reducing insulin and IGF levels. CR increases the AMP/ATP ratio and activates AMPK. Decreased mTOR and activated AMPK efficiently induce the initiation of the autophagy process. Various CR mimetics can induce the autophagy process. Rapamycin activates autophagy by inhibiting mTOR and metformin induces autophagy by activating AMPK. Spermidine enhances the overall autophagy process through the inhibition of EP300 deacetylase.

5.1. Rapamycin, an mTOR Inhibitor

Rapamycin was initially described as an immune-suppressor drug and is a commonly referred compound for CR mimetics. In later studies, it has been demonstrated that rapamycin directly binds between FKBP12 and the mTOR kinase subunits of mTORC1, causing the inhibition of mTOR and its downstream signaling pathway [103]. The mTOR inhibitory activity of rapamycin gained attention because the activity and expression of mTOR is significantly increased in aging and in age-related diseases [104]. Furthermore, CR was shown to downregulate mTOR function, leading to an increased autophagy with decreased protein synthesis [105]. Rapamycin has been documented as delaying or ameliorating age-related diseases including metabolic diseases, cardiovascular diseases, Hutchinson–Gilford progeria syndrome premature aging phenotype, and neurodegenerative diseases [104,106]. Rapamycin also showed a lifespan extension effect in various animal models including in the yeast, fruit fly, and nematode [107]. In addition, the life-extension effect of rapamycin was also verified and replicated in mice by several independent groups [108,109].

Although rapamycin activates autophagy through the inhibition of mTOR, it also shows other beneficial effects through the regulation of other signaling pathways. mTORC1 is activated not only by nutrient levels in the cell but also by cell growth hormones. mTORC1 interacts with key proteins in the anabolic process such as S6K, 4E-BP1, and SREBP1c, and activates protein, lipid, nucleotide, and organelle synthesis such as mitochondria [104]. However, evidence has also demonstrated some side effects of rapamycin such as a suppressed immune system, increased incidence of diabetes, and nephrotoxicity [110]. The safety and side effects of rapamycin in the long-term use should be carefully considered.

5.2. Metformin, an AMPK Activator

Merformin is another interesting CR mimetic. It is a guanidine-based hypoglycemic agent that is used as a drug for the treatment of type-2 diabetes, and has the ability to increase insulin sensitivity through the activation of AMPK. Although it is commonly referred as an AMPK activator, it is unlikely that metformin directly binds to either AMPK or its activator LKB1 [111]. Evidence supports that metformin may increase AMPK activation by modulating ATP production in mitochondria [112].

Since AMPK is downregulated in many types of metabolic disease, metformin showed a particular beneficial effect in various age-related metabolic diseases [113]. Further studies have shown a lifespan extension effect of metformin. During the screening of CR mimetics, Dhahbi et al. first found that metformin treatment showed a similar transcriptional profile to that of CR in mice [114]. Moreover, metformin was shown to lead to an increased lifespan in nematode and rodent models [115,116]. Interestingly, some studies showed that the beneficial effects of metformin were less pronounced under autophagy-inhibited conditions, suggesting the importance of autophagy signaling induced by metformin [117–120]. It is now clear that metformin shows its beneficial effects at least partly through the induction of autophagy. However, in some models of aging, the longevity benefit of metformin was not observed. It is clear that metformin has several beneficial effects in various metabolic diseases. However, further investigation is needed to verify whether metformin can act as a CR mimetic and consistently present anti-aging effects.

5.3. Spermidine

Unlike rapamycin and metformin, spermidine is a natural polyamine that stimulates autophagy [121]. It has been demonstrated to be involved in various cellular processes and to regulate cellular homeostasis. The external supplementation of spermidine extends the lifespan in various species including yeast, nematodes, fruit flies, and mice [121–123]. It also showed protective effects in several degenerative diseases. Importantly, many of these anti-aging and beneficial properties of spermidine were abrogated when there was a genetic impairment to autophagy [123–125]. Mechanistic studies revealed that spermidine induces autophagy through the inhibition of several acetyltransferases. EP300, one of the acetyltransferases regulated by spermidine, is a main negative regulator of autophagy [126]. Epidemiology data showed that spermidine levels decline with age, and that the increased uptake of spermidine-rich foods diminishes the overall mortality associated with cardiovascular diseases and cancer [127,128]. Interestingly, a recent report also demonstrated a similar role for aspirin, and the induction of autophagy by aspirin has been demonstrated in several species [129,130]. Collectively, these results provide new molecular mechanisms for regulating autophagy, and spermidine and aspirin could form a new type of CR mimetics with anti-aging effects.

6. Concluding Remarks

In this review, the anti-aging effects of CR-induced autophagy were discussed. Although dependent on the species and age used in the experimental models and on the duration and intensity of CR regimens, all evidence supports a role for CR in autophagy activation. CR-induced autophagy plays a pivotal role under physiological conditions by maintaining adequate homeostasis in the organism. Furthermore, in various organs and tissues under pathologic conditions including aging, CR-induced autophagy played a protective role. The underlying mechanisms of longevity extension in response to CR are not yet fully understood, but evidence supports that activated autophagy could be playing an important role. With further advances in mechanistic biology, it is interesting that autophagy-inducing CR mimetics show similar effects to CR in several organisms. While more studies are required to better understand the benefits of CR mimetics, its safety and side effects should also be carefully considered. Finally, it will be necessary to assess whether autophagy inducers are effective and can be applicable in the treatment of human diseases.

Author Contributions: K.W.C. and H.Y.C. wrote the manuscript. We thank Aging Tissue Bank (Busan, Korea) for providing research information.

Funding: This work was supported by the National Research Foundation of Korea (NRF) grant funded by the Korea government (MSIT) (2018R1A2A3075425).

Conflicts of Interest: The authors declare no conflicts of interest.

References

1. Klionsky, D.J.; Abdelmohsen, K.; Abe, A.; Abedin, M.J.; Abeliovich, H.; Acevedo Arozena, A.; Adachi, H.; Adams, C.M.; Adams, P.D.; Adeli, K.; et al. Guidelines for the use and interpretation of assays for monitoring autophagy (3rd edition). *Autophagy* **2016**, *12*, 1–222. [CrossRef]

2. Dikic, I.; Elazar, Z. Mechanism and medical implications of mammalian autophagy. *Nat. Rev. Mol. Cell Biol.* **2018**, *19*, 349–364. [CrossRef]

3. Levine, B.; Kroemer, G. Biological Functions of Autophagy Genes: A Disease Perspective. *Cell* **2019**, *176*, 11–42. [CrossRef]

4. Ravikumar, B.; Sarkar, S.; Davies, J.E.; Futter, M.; Garcia-Arencibia, M.; Green-Thompson, Z.W.; Jimenez-Sanchez, M.; Korolchuk, V.I.; Lichtenberg, M.; Luo, S.; et al. Regulation of mammalian autophagy in physiology and pathophysiology. *Physiol. Rev.* **2010**, *90*, 1383–1435. [CrossRef]

5. Yin, Z.; Pascual, C.; Klionsky, D.J. Autophagy: Machinery and regulation. *Microb. Cell* **2016**, *3*, 588–596. [CrossRef] [PubMed]

6. Parzych, K.R.; Klionsky, D.J. An overview of autophagy: Morphology, mechanism, and regulation. *Antioxid. Redox Signal.* **2014**, *20*, 460–473. [CrossRef] [PubMed]

7. He, L.; Zhang, J.; Zhao, J.; Ma, N.; Kim, S.W.; Qiao, S.; Ma, X. Autophagy: The Last Defense against Cellular Nutritional Stress. *Adv. Nutr.* **2018**, *9*, 493–504. [CrossRef] [PubMed]

8. Russell, R.C.; Yuan, H.X.; Guan, K.L. Autophagy regulation by nutrient signaling. *Cell Res.* **2014**, *24*, 42–57. [CrossRef]

9. Dunlop, E.A.; Tee, A.R. mTOR and autophagy: A dynamic relationship governed by nutrients and energy. *Semin. Cell Dev. Biol.* **2014**, *36*, 121–129. [CrossRef]

10. Kim, Y.C.; Guan, K.L. mTOR: A pharmacologic target for autophagy regulation. *J. Clin. Investig.* **2015**, *125*, 25–32. [CrossRef]

11. Bond, P. Regulation of mTORC1 by growth factors, energy status, amino acids and mechanical stimuli at a glance. *J. Int. Soc. Sports Nutr.* **2016**, *13*, 8. [CrossRef] [PubMed]

12. Bar-Peled, L.; Sabatini, D.M. Regulation of mTORC1 by amino acids. *Trends Cell Biol.* **2014**, *24*, 400–406. [CrossRef] [PubMed]

13. Kim, J.; Kundu, M.; Viollet, B.; Guan, K.L. AMPK and mTOR regulate autophagy through direct phosphorylation of Ulk1. *Nat. Cell Biol.* **2011**, *13*, 132–141. [CrossRef] [PubMed]

14. Rabanal-Ruiz, Y.; Otten, E.G.; Korolchuk, V.I. mTORC1 as the main gateway to autophagy. *Essays Biochem.* **2017**, *61*, 565–584. [CrossRef]

15. Alers, S.; Loffler, A.S.; Wesselborg, S.; Stork, B. Role of AMPK-mTOR-Ulk1/2 in the regulation of autophagy: Cross talk, shortcuts, and feedbacks. *Mol. Cell Biol.* **2012**, *32*, 2–11. [CrossRef]

16. Jung, C.H.; Ro, S.H.; Cao, J.; Otto, N.M.; Kim, D.H. mTOR regulation of autophagy. *FEBS Lett.* **2010**, *584*, 1287–1295. [CrossRef]

17. Yao, Y.; Jones, E.; Inoki, K. Lysosomal Regulation of mTORC1 by Amino Acids in Mammalian Cells. *Biomolecules* **2017**, *7*, 51. [CrossRef]

18. Puertollano, R. mTOR and lysosome regulation. *F1000Prime Rep.* **2014**, *6*, 52. [CrossRef]

19. Settembre, C.; Di Malta, C.; Polito, V.A.; Garcia Arencibia, M.; Vetrini, F.; Erdin, S.; Erdin, S.U.; Huynh, T.; Medina, D.; Colella, P.; et al. TFEB links autophagy to lysosomal biogenesis. *Science* **2011**, *332*, 1429–1433. [CrossRef]

20. Roczniak-Ferguson, A.; Petit, C.S.; Froehlich, F.; Qian, S.; Ky, J.; Angarola, B.; Walther, T.C.; Ferguson, S.M. The transcription factor TFEB links mTORC1 signaling to transcriptional control of lysosome homeostasis. *Sci. Signal.* **2012**, *5*, ra42. [CrossRef]

21. Napolitano, G.; Esposito, A.; Choi, H.; Matarese, M.; Benedetti, V.; Di Malta, C.; Monfregola, J.; Medina, D.L.; Lippincott-Schwartz, J.; Ballabio, A. mTOR-dependent phosphorylation controls TFEB nuclear export. *Nat. Commun.* **2018**, *9*, 3312. [CrossRef] [PubMed]

22. Herzig, S.; Shaw, R.J. AMPK: Guardian of metabolism and mitochondrial homeostasis. *Nat. Rev. Mol. Cell Biol.* **2018**, *19*, 121–135. [CrossRef] [PubMed]

23. Inoki, K.; Ouyang, H.; Zhu, T.; Lindvall, C.; Wang, Y.; Zhang, X.; Yang, Q.; Bennett, C.; Harada, Y.; Stankunas, K.; et al. TSC2 integrates Wnt and energy signals via a coordinated phosphorylation by AMPK and GSK3 to regulate cell growth. *Cell* **2006**, *126*, 955–968. [CrossRef] [PubMed]

24. Lopez-Otin, C.; Blasco, M.A.; Partridge, L.; Serrano, M.; Kroemer, G. The hallmarks of aging. *Cell* **2013**, *153*, 1194–1217. [CrossRef]

25. Martinez-Lopez, N.; Athonvarangkul, D.; Singh, R. Autophagy and aging. *Adv. Exp. Med. Biol.* **2015**, *847*, 73–87. [CrossRef]

26. Leidal, A.M.; Levine, B.; Debnath, J. Autophagy and the cell biology of age-related disease. *Nat. Cell Biol.* **2018**, *20*, 1338–1348. [CrossRef]

27. Simonsen, A.; Cumming, R.C.; Brech, A.; Isakson, P.; Schubert, D.R.; Finley, K.D. Promoting basal levels of autophagy in the nervous system enhances longevity and oxidant resistance in adult Drosophila. *Autophagy* **2008**, *4*, 176–184. [CrossRef]

28. Sarkis, G.J.; Ashcom, J.D.; Hawdon, J.M.; Jacobson, L.A. Decline in protease activities with age in the nematode Caenorhabditis elegans. *Mech. Ageing Dev.* **1988**, *45*, 191–201. [CrossRef]

29. Bai, H.; Kang, P.; Hernandez, A.M.; Tatar, M. Activin signaling targeted by insulin/dFOXO regulates aging and muscle proteostasis in Drosophila. *PLoS Genet.* **2013**, *9*, e1003941. [CrossRef]

30. Kaushik, S.; Arias, E.; Kwon, H.; Lopez, N.M.; Athonvarangkul, D.; Sahu, S.; Schwartz, G.J.; Pessin, J.E.; Singh, R. Loss of autophagy in hypothalamic POMC neurons impairs lipolysis. *EMBO Rep.* **2012**, *13*, 258–265. [CrossRef]

31. Carnio, S.; LoVerso, F.; Baraibar, M.A.; Longa, E.; Khan, M.M.; Maffei, M.; Reischl, M.; Canepari, M.; Loefler, S.; Kern, H.; et al. Autophagy impairment in muscle induces neuromuscular junction degeneration and precocious aging. *Cell Rep.* **2014**, *8*, 1509–1521. [CrossRef] [PubMed]

32. Cuervo, A.M.; Dice, J.F. Age-related decline in chaperone-mediated autophagy. *J. Biol. Chem.* **2000**, *275*, 31505–31513. [CrossRef] [PubMed]

33. Chang, J.T.; Kumsta, C.; Hellman, A.B.; Adams, L.M.; Hansen, M. Spatiotemporal regulation of autophagy during Caenorhabditis elegans aging. *eLife* **2017**, *6*. [CrossRef] [PubMed]

34. Xu, J.; Jiao, K.; Liu, X.; Sun, Q.; Wang, K.; Xu, H.; Zhang, S.; Wu, Y.; Wu, L.; Liu, D.; et al. Omi/HtrA2 Participates in Age-Related Autophagic Deficiency in Rat Liver. *Aging Dis.* **2018**, *9*, 1031–1042. [CrossRef]

35. Matecic, M.; Smith, D.L.; Pan, X.; Maqani, N.; Bekiranov, S.; Boeke, J.D.; Smith, J.S. A microarray-based genetic screen for yeast chronological aging factors. *PLoS Genet.* **2010**, *6*, e1000921. [CrossRef]

36. Toth, M.L.; Sigmond, T.; Borsos, E.; Barna, J.; Erdelyi, P.; Takacs-Vellai, K.; Orosz, L.; Kovacs, A.L.; Csikos, G.; Sass, M.; et al. Longevity pathways converge on autophagy genes to regulate life span in Caenorhabditis elegans. *Autophagy* **2008**, *4*, 330–338. [CrossRef]

37. Kuma, A.; Hatano, M.; Matsui, M.; Yamamoto, A.; Nakaya, H.; Yoshimori, T.; Ohsumi, Y.; Tokuhisa, T.; Mizushima, N. The role of autophagy during the early neonatal starvation period. *Nature* **2004**, *432*, 1032–1036. [CrossRef]

38. Komatsu, M.; Waguri, S.; Ueno, T.; Iwata, J.; Murata, S.; Tanida, I.; Ezaki, J.; Mizushima, N.; Ohsumi, Y.; Uchiyama, Y.; et al. Impairment of starvation-induced and constitutive autophagy in Atg7-deficient mice. *J. Cell Biol.* **2005**, *169*, 425–434. [CrossRef]

39. Saitoh, T.; Fujita, N.; Jang, M.H.; Uematsu, S.; Yang, B.G.; Satoh, T.; Omori, H.; Noda, T.; Yamamoto, N.; Komatsu, M.; et al. Loss of the autophagy protein Atg16L1 enhances endotoxin-induced IL-1beta production. *Nature* **2008**, *456*, 264–268. [CrossRef]

40. Komatsu, M.; Waguri, S.; Chiba, T.; Murata, S.; Iwata, J.; Tanida, I.; Ueno, T.; Koike, M.; Uchiyama, Y.; Kominami, E.; et al. Loss of autophagy in the central nervous system causes neurodegeneration in mice. *Nature* **2006**, *441*, 880–884. [CrossRef]

41. Stroikin, Y.; Dalen, H.; Loof, S.; Terman, A. Inhibition of autophagy with 3-methyladenine results in impaired turnover of lysosomes and accumulation of lipofuscin-like material. *Eur. J. Cell Biol.* **2004**, *83*, 583–590. [CrossRef] [PubMed]

42. Singh, R.; Kaushik, S.; Wang, Y.; Xiang, Y.; Novak, I.; Komatsu, M.; Tanaka, K.; Cuervo, A.M.; Czaja, M.J. Autophagy regulates lipid metabolism. *Nature* **2009**, *458*, 1131–1135. [CrossRef] [PubMed]

43. Masiero, E.; Agatea, L.; Mammucari, C.; Blaauw, B.; Loro, E.; Komatsu, M.; Metzger, D.; Reggiani, C.; Schiaffino, S.; Sandri, M. Autophagy is required to maintain muscle mass. *Cell Metab.* **2009**, *10*, 507–515. [CrossRef] [PubMed]

44. Lapierre, L.R.; De Magalhaes Filho, C.D.; McQuary, P.R.; Chu, C.C.; Visvikis, O.; Chang, J.T.; Gelino, S.; Ong, B.; Davis, A.E.; Irazoqui, J.E.; et al. The TFEB orthologue HLH-30 regulates autophagy and modulates longevity in Caenorhabditis elegans. *Nat. Commun.* **2013**, *4*, 2267. [CrossRef] [PubMed]

45. Pyo, J.O.; Yoo, S.M.; Ahn, H.H.; Nah, J.; Hong, S.H.; Kam, T.I.; Jung, S.; Jung, Y.K. Overexpression of Atg5 in mice activates autophagy and extends lifespan. *Nat. Commun.* **2013**, *4*, 2300. [CrossRef] [PubMed]

46. Johnson, S.C.; Rabinovitch, P.S.; Kaeberlein, M. mTOR is a key modulator of ageing and age-related disease. *Nature* **2013**, *493*, 338–345. [CrossRef] [PubMed]

47. Salminen, A.; Kaarniranta, K. AMP-activated protein kinase (AMPK) controls the aging process via an integrated signaling network. *Ageing Res. Rev.* **2012**, *11*, 230–241. [CrossRef]

48. Seok, S.; Fu, T.; Choi, S.E.; Li, Y.; Zhu, R.; Kumar, S.; Sun, X.; Yoon, G.; Kang, Y.; Zhong, W.; et al. Transcriptional regulation of autophagy by an FXR-CREB axis. *Nature* **2014**, *516*, 108–111. [CrossRef]

49. Lee, J.M.; Wagner, M.; Xiao, R.; Kim, K.H.; Feng, D.; Lazar, M.A.; Moore, D.D. Nutrient-sensing nuclear receptors coordinate autophagy. *Nature* **2014**, *516*, 112–115. [CrossRef]

50. Chung, K.W.; Lee, E.K.; Lee, M.K.; Oh, G.T.; Yu, B.P.; Chung, H.Y. Impairment of PPARα and the Fatty Acid Oxidation Pathway Aggravates Renal Fibrosis during Aging. *J. Am. Soc. Nephrol.* **2018**, *29*, 1223–1237. [CrossRef]

51. Sung, B.; Park, S.; Yu, B.P.; Chung, H.Y. Modulation of PPAR in aging, inflammation, and calorie restriction. *J. Gerontol. A Biol. Sci. Med. Sci.* **2004**, *59*, 997–1006. [CrossRef] [PubMed]

52. Morris, K.A.; Gold, P.E. Age-related impairments in memory and in CREB and pCREB expression in hippocampus and amygdala following inhibitory avoidance training. *Mech. Ageing Dev.* **2012**, *133*, 291–299. [CrossRef] [PubMed]

53. Chung, K.W.; Kim, D.H.; Park, M.H.; Choi, Y.J.; Kim, N.D.; Lee, J.; Yu, B.P.; Chung, H.Y. Recent advances in calorie restriction research on aging. *Exp. Gerontol.* **2013**, *48*, 1049–1053. [CrossRef] [PubMed]

54. Balasubramanian, P.; Howell, P.R.; Anderson, R.M. Aging and Caloric Restriction Research: A Biological Perspective with Translational Potential. *EBioMedicine* **2017**, *21*, 37–44. [CrossRef] [PubMed]

55. Libert, S.; Guarente, L. Metabolic and neuropsychiatric effects of calorie restriction and sirtuins. *Annu. Rev. Physiol.* **2013**, *75*, 669–684. [CrossRef] [PubMed]

56. Redman, L.M.; Smith, S.R.; Burton, J.H.; Martin, C.K.; Il'yasova, D.; Ravussin, E. Metabolic Slowing and Reduced Oxidative Damage with Sustained Caloric Restriction Support the Rate of Living and Oxidative Damage Theories of Aging. *Cell Metab.* **2018**, *27*, 805–815.e4. [CrossRef] [PubMed]

57. Morselli, E.; Maiuri, M.C.; Markaki, M.; Megalou, E.; Pasparaki, A.; Palikaras, K.; Criollo, A.; Galluzzi, L.; Malik, S.A.; Vitale, I.; et al. Caloric restriction and resveratrol promote longevity through the Sirtuin-1-dependent induction of autophagy. *Cell Death Dis.* **2010**, *1*, e10. [CrossRef]

58. Kume, S.; Uzu, T.; Horiike, K.; Chin-Kanasaki, M.; Isshiki, K.; Araki, S.; Sugimoto, T.; Haneda, M.; Kashiwagi, A.; Koya, D. Calorie restriction enhances cell adaptation to hypoxia through Sirt1-dependent mitochondrial autophagy in mouse aged kidney. *J. Clin. Investig.* **2010**, *120*, 1043–1055. [CrossRef]

59. Ruckenstuhl, C.; Netzberger, C.; Entfellner, I.; Carmona-Gutierrez, D.; Kickenweiz, T.; Stekovic, S.; Gleixner, C.; Schmid, C.; Klug, L.; Sorgo, A.G.; et al. Lifespan extension by methionine restriction requires autophagy-dependent vacuolar acidification. *PLoS Genet.* **2014**, *10*, e1004347. [CrossRef]

60. Rubinsztein, D.C.; Marino, G.; Kroemer, G. Autophagy and aging. *Cell* **2011**, *146*, 682–695. [CrossRef]

61. Madeo, F.; Zimmermann, A.; Maiuri, M.C.; Kroemer, G. Essential role for autophagy in life span extension. *J. Clin. Investig.* **2015**, *125*, 85–93. [CrossRef] [PubMed]

62. Wohlgemuth, S.E.; Julian, D.; Akin, D.E.; Fried, J.; Toscano, K.; Leeuwenburgh, C.; Dunn, W.A., Jr. Autophagy in the heart and liver during normal aging and calorie restriction. *Rejuvenation Res.* **2007**, *10*, 281–292. [CrossRef] [PubMed]

63. Donati, A.; Cavallini, G.; Bergamini, E. Effects of aging, antiaging calorie restriction and in vivo stimulation of autophagy on the urinary excretion of 8OHdG in male Sprague-Dawley rats. *Age* **2013**, *35*, 261–270. [CrossRef] [PubMed]

64. Luevano-Martinez, L.A.; Forni, M.F.; Peloggia, J.; Watanabe, I.S.; Kowaltowski, A.J. Calorie restriction promotes cardiolipin biosynthesis and distribution between mitochondrial membranes. *Mech. Ageing Dev.* **2017**, *162*, 9–17. [CrossRef] [PubMed]

65. Derous, D.; Mitchell, S.E.; Wang, L.; Green, C.L.; Wang, Y.; Chen, L.; Han, J.J.; Promislow, D.E.L.; Lusseau, D.; Douglas, A.; et al. The effects of graded levels of calorie restriction: XI. Evaluation of the main hypotheses underpinning the life extension effects of CR using the hepatic transcriptome. *Aging* **2017**, *9*, 1770–1824. [CrossRef] [PubMed]

66. Wohlgemuth, S.E.; Seo, A.Y.; Marzetti, E.; Lees, H.A.; Leeuwenburgh, C. Skeletal muscle autophagy and apoptosis during aging: Effects of calorie restriction and life-long exercise. *Exp. Gerontol.* **2010**, *45*, 138–148. [CrossRef] [PubMed]

67. Yang, L.; Licastro, D.; Cava, E.; Veronese, N.; Spelta, F.; Rizza, W.; Bertozzi, B.; Villareal, D.T.; Hotamisligil, G.S.; Holloszy, J.O.; et al. Long-Term Calorie Restriction Enhances Cellular Quality-Control Processes in Human Skeletal Muscle. *Cell Rep.* **2016**, *14*, 422–428. [CrossRef]

68. Gutierrez-Casado, E.; Khraiwesh, H.; Lopez-Dominguez, J.A.; Montero-Guisado, J.; Lopez-Lluch, G.; Navas, P.; de Cabo, R.; Ramsey, J.J.; Gonzalez-Reyes, J.A.; Villalba, J.M. The Impact of Aging, Calorie Restriction and Dietary Fat on Autophagy Markers and Mitochondrial Ultrastructure and Dynamics in Mouse Skeletal Muscle. *J. Gerontol. A Biol. Sci. Med. Sci.* **2019**, *74*, 760–769. [CrossRef]

69. Nunez, C.E.; Rodrigues, V.S.; Gomes, F.S.; Moura, R.F.; Victorio, S.C.; Bombassaro, B.; Chaim, E.A.; Pareja, J.C.; Geloneze, B.; Velloso, L.A.; et al. Defective regulation of adipose tissue autophagy in obesity. *Int. J. Obes.* **2013**, *37*, 1473–1480. [CrossRef]

70. Ghosh, A.K.; Mau, T.; O'Brien, M.; Garg, S.; Yung, R. Impaired autophagy activity is linked to elevated ER-stress and inflammation in aging adipose tissue. *Aging* **2016**, *8*, 2525–2537. [CrossRef]

71. Ning, Y.C.; Cai, G.Y.; Zhuo, L.; Gao, J.J.; Dong, D.; Cui, S.; Feng, Z.; Shi, S.Z.; Bai, X.Y.; Sun, X.F.; et al. Short-term calorie restriction protects against renal senescence of aged rats by increasing autophagic activity and reducing oxidative damage. *Mech. Ageing Dev.* **2013**, *134*, 570–579. [CrossRef] [PubMed]

72. Chung, K.W.; Kim, K.M.; Choi, Y.J.; An, H.J.; Lee, B.; Kim, D.H.; Lee, E.K.; Im, E.; Lee, J.; Im, D.S.; et al. The critical role played by endotoxin-induced liver autophagy in the maintenance of lipid metabolism during sepsis. *Autophagy* **2017**, *13*, 1113–1129. [CrossRef] [PubMed]

73. Zirin, J.; Nieuwenhuis, J.; Perrimon, N. Role of autophagy in glycogen breakdown and its relevance to chloroquine myopathy. *PLoS Biol.* **2013**, *11*, e1001708. [CrossRef] [PubMed]

74. Czaja, M.J.; Ding, W.X.; Donohue, T.M., Jr.; Friedman, S.L.; Kim, J.S.; Komatsu, M.; Lemasters, J.J.; Lemoine, A.; Lin, J.D.; Ou, J.H.; et al. Functions of autophagy in normal and diseased liver. *Autophagy* **2013**, *9*, 1131–1158. [CrossRef] [PubMed]

75. Madrigal-Matute, J.; Cuervo, A.M. Regulation of Liver Metabolism by Autophagy. *Gastroenterology* **2016**, *150*, 328–339. [CrossRef] [PubMed]

76. Ueno, T.; Komatsu, M. Autophagy in the liver: Functions in health and disease. *Nat. Rev. Gastroenterol. Hepatol.* **2017**, *14*, 170–184. [CrossRef]

77. Czaja, M.J. Function of Autophagy in Nonalcoholic Fatty Liver Disease. *Dig. Dis. Sci.* **2016**, *61*, 1304–1313. [CrossRef]

78. Martinez-Lopez, N.; Tarabra, E.; Toledo, M.; Garcia-Macia, M.; Sahu, S.; Coletto, L.; Batista-Gonzalez, A.; Barzilai, N.; Pessin, J.E.; Schwartz, G.J.; et al. System-wide Benefits of Intermeal Fasting by Autophagy. *Cell Metab.* **2017**, *26*, 856–871. [CrossRef]

79. Chen, L.; Wang, K.; Long, A.; Jia, L.; Zhang, Y.; Deng, H.; Li, Y.; Han, J.; Wang, Y. Fasting-induced hormonal regulation of lysosomal function. *Cell Res.* **2017**, *27*, 748–763. [CrossRef]

80. Neel, B.A.; Lin, Y.; Pessin, J.E. Skeletal muscle autophagy: A new metabolic regulator. *Trends Endocrinol. Metab.* **2013**, *24*, 635–643. [CrossRef]

81. Meijer, A.J.; Lorin, S.; Blommaart, E.F.; Codogno, P. Regulation of autophagy by amino acids and MTOR-dependent signal transduction. *Amino Acids* **2015**, *47*, 2037–2063. [CrossRef] [PubMed]

82. Jiao, J.; Demontis, F. Skeletal muscle autophagy and its role in sarcopenia and organismal aging. *Curr. Opin. Pharmacol.* **2017**, *34*, 1–6. [CrossRef]

83. Carter, H.N.; Kim, Y.; Erlich, A.T.; Zarrin-Khat, D.; Hood, D.A. Autophagy and mitophagy flux in young and aged skeletal muscle following chronic contractile activity. *J. Physiol.* **2018**, *596*, 3567–3584. [CrossRef] [PubMed]

84. Di Rienzo, M.; Antonioli, M.; Fusco, C.; Liu, Y.; Mari, M.; Orhon, I.; Refolo, G.; Germani, F.; Corazzari, M.; Romagnoli, A.; et al. Autophagy induction in atrophic muscle cells requires ULK1 activation by TRIM32 through unanchored K63-linked polyubiquitin chains. *Sci. Adv.* **2019**, *5*, eaau8857. [CrossRef] [PubMed]

85. Lee, D.E.; Bareja, A.; Bartlett, D.B.; White, J.P. Autophagy as a Therapeutic Target to Enhance Aged Muscle Regeneration. *Cells* **2019**, *8*, 183. [CrossRef]

86. Cerletti, M.; Jang, Y.C.; Finley, L.W.; Haigis, M.C.; Wagers, A.J. Short-term calorie restriction enhances skeletal muscle stem cell function. *Cell Stem Cell* **2012**, *10*, 515–519. [CrossRef]

87. Miller, K.N.; Burhans, M.S.; Clark, J.P.; Howell, P.R.; Polewski, M.A.; DeMuth, T.M.; Eliceiri, K.W.; Lindstrom, M.J.; Ntambi, J.M.; Anderson, R.M. Aging and caloric restriction impact adipose tissue, adiponectin, and circulating lipids. *Aging Cell* **2017**, *16*, 497–507. [CrossRef]

88. Ferhat, M.; Funai, K.; Boudina, S. Autophagy in Adipose Tissue Physiology and Pathophysiology. *Antioxid. Redox Signal.* **2019**, *31*, 487–501. [CrossRef]

89. Singh, R.; Xiang, Y.; Wang, Y.; Baikati, K.; Cuervo, A.M.; Luu, Y.K.; Tang, Y.; Pessin, J.E.; Schwartz, G.J.; Czaja, M.J. Autophagy regulates adipose mass and differentiation in mice. *J. Clin. Investig.* **2009**, *119*, 3329–3339. [CrossRef]

90. Cai, J.; Pires, K.M.; Ferhat, M.; Chaurasia, B.; Buffolo, M.A.; Smalling, R.; Sargsyan, A.; Atkinson, D.L.; Summers, S.A.; Graham, T.E.; et al. Autophagy Ablation in Adipocytes Induces Insulin Resistance and Reveals Roles for Lipid Peroxide and Nrf2 Signaling in Adipose-Liver Crosstalk. *Cell Rep.* **2018**, *25*, 1708–1717.e5. [CrossRef]

91. Haim, Y.; Bluher, M.; Slutsky, N.; Goldstein, N.; Kloting, N.; Harman-Boehm, I.; Kirshtein, B.; Ginsberg, D.; Gericke, M.; Guiu Jurado, E.; et al. Elevated autophagy gene expression in adipose tissue of obese humans: A potential non-cell-cycle-dependent function of E2F1. *Autophagy* **2015**, *11*, 2074–2088. [CrossRef] [PubMed]

92. Jansen, H.J.; van Essen, P.; Koenen, T.; Joosten, L.A.; Netea, M.G.; Tack, C.J.; Stienstra, R. Autophagy activity is up-regulated in adipose tissue of obese individuals and modulates proinflammatory cytokine expression. *Endocrinology* **2012**, *153*, 5866–5874. [CrossRef] [PubMed]

93. Soussi, H.; Reggio, S.; Alili, R.; Prado, C.; Mutel, S.; Pini, M.; Rouault, C.; Clement, K.; Dugail, I. DAPK2 Downregulation Associates With Attenuated Adipocyte Autophagic Clearance in Human Obesity. *Diabetes* **2015**, *64*, 3452–3463. [CrossRef] [PubMed]

94. Bobulescu, I.A. Renal lipid metabolism and lipotoxicity. *Curr. Opin. Nephrol. Hypertens.* **2010**, *19*, 393–402. [CrossRef] [PubMed]

95. Stumvoll, M.; Meyer, C.; Mitrakou, A.; Gerich, J.E. Important role of the kidney in human carbohydrate metabolism. *Med. Hypotheses* **1999**, *52*, 363–366. [CrossRef] [PubMed]

96. Kang, H.M.; Ahn, S.H.; Choi, P.; Ko, Y.A.; Han, S.H.; Chinga, F.; Park, A.S.; Tao, J.; Sharma, K.; Pullman, J.; et al. Defective fatty acid oxidation in renal tubular epithelial cells has a key role in kidney fibrosis development. *Nat. Med.* **2015**, *21*, 37–46. [CrossRef]

97. Gerich, J.E. Role of the kidney in normal glucose homeostasis and in the hyperglycaemia of diabetes mellitus: Therapeutic implications. *Diabet. Med.* **2010**, *27*, 136–142. [CrossRef]

98. Huber, T.B.; Edelstein, C.L.; Hartleben, B.; Inoki, K.; Jiang, M.; Koya, D.; Kume, S.; Lieberthal, W.; Pallet, N.; Quiroga, A.; et al. Emerging role of autophagy in kidney function, diseases and aging. *Autophagy* **2012**, *8*, 1009–1031. [CrossRef]

99. Deng, X.; Xie, Y.; Zhang, A. Advance of autophagy in chronic kidney diseases. *Ren. Fail.* **2017**, *39*, 306–313. [CrossRef]

100. Xu, X.M.; Cai, G.Y.; Bu, R.; Wang, W.J.; Bai, X.Y.; Sun, X.F.; Chen, X.M. Beneficial Effects of Caloric Restriction on Chronic Kidney Disease in Rodent Models: A Meta-Analysis and Systematic Review. *PLoS ONE* **2015**, *10*, e0144442. [CrossRef]

101. Stekovic, S.; Hofer, S.J.; Tripolt, N.; Aon, M.A.; Royer, P.; Pein, L.; Stadler, J.T.; Pendl, T.; Prietl, B.; Url, J.; et al. Alternate Day Fasting Improves Physiological and Molecular Markers of Aging in Healthy, Non-obese Humans. *Cell Metab.* **2019**, *30*, 462–476. [CrossRef] [PubMed]

102. Shintani, H.; Shintani, T.; Ashida, H.; Sato, M. Calorie Restriction Mimetics: Upstream-Type Compounds for Modulating Glucose Metabolism. *Nutrients* **2018**, *10*, 1821. [CrossRef] [PubMed]

103. Yang, H.; Rudge, D.G.; Koos, J.D.; Vaidialingam, B.; Yang, H.J.; Pavletich, N.P. mTOR kinase structure, mechanism and regulation. *Nature* **2013**, *497*, 217–223. [CrossRef] [PubMed]

104. Saxton, R.A.; Sabatini, D.M. mTOR signaling in Growth, Metabolism, and Disease. *Cell* **2017**, *169*, 361–371. [CrossRef]

105. Escobar, K.A.; Cole, N.H.; Mermier, C.M.; VanDusseldorp, T.A. Autophagy and aging: Maintaining the proteome through exercise and caloric restriction. *Aging Cell* **2019**, *18*, e12876. [CrossRef]

106. Li, J.; Kim, S.G.; Blenis, J. Rapamycin: One drug, many effects. *Cell Metab.* **2014**, *19*, 373–379. [CrossRef]

107. Ehninger, D.; Neff, F.; Xie, K. Longevity, aging and rapamycin. *Cell. Mol. Life Sci.* **2014**, *71*, 4325–4346. [CrossRef]

108. Bitto, A.; Ito, T.K.; Pineda, V.V.; LeTexier, N.J.; Huang, H.Z.; Sutlief, E.; Tung, H.; Vizzini, N.; Chen, B.; Smith, K.; et al. Transient rapamycin treatment can increase lifespan and healthspan in middle-aged mice. *eLife* **2016**, *5*, e16351. [CrossRef]

109. Harrison, D.E.; Strong, R.; Sharp, Z.D.; Nelson, J.F.; Astle, C.M.; Flurkey, K.; Nadon, N.L.; Wilkinson, J.E.; Frenkel, K.; Carter, C.S.; et al. Rapamycin fed late in life extends lifespan in genetically heterogeneous mice. *Nature* **2009**, *460*, 392–395. [CrossRef]

110. Pallet, N.; Legendre, C. Adverse events associated with mTOR inhibitors. *Expert Opin. Drug Saf.* **2013**, *12*, 177–186. [CrossRef]

111. Rena, G.; Hardie, D.G.; Pearson, E.R. The mechanisms of action of metformin. *Diabetologia* **2017**, *60*, 1577–1585. [CrossRef] [PubMed]

112. Vial, G.; Detaille, D.; Guigas, B. Role of Mitochondria in the Mechanism(s) of Action of Metformin. *Front. Endocrinol.* **2019**, *10*, 294. [CrossRef] [PubMed]

113. Valencia, W.M.; Palacio, A.; Tamariz, L.; Florez, H. Metformin and ageing: Improving ageing outcomes beyond glycaemic control. *Diabetologia* **2017**, *60*, 1630–1638. [CrossRef] [PubMed]

114. Dhahbi, J.M.; Mote, P.L.; Fahy, G.M.; Spindler, S.R. Identification of potential caloric restriction mimetics by microarray profiling. *Physiol. Genom.* **2005**, *23*, 343–350. [CrossRef]

115. Cabreiro, F.; Au, C.; Leung, K.Y.; Vergara-Irigaray, N.; Cocheme, H.M.; Noori, T.; Weinkove, D.; Schuster, E.; Greene, N.D.; Gems, D. Metformin retards aging in *C. elegans* by altering microbial folate and methionine metabolism. *Cell* **2013**, *153*, 228–239. [CrossRef]

116. Martin-Montalvo, A.; Mercken, E.M.; Mitchell, S.J.; Palacios, H.H.; Mote, P.L.; Scheibye-Knudsen, M.; Gomes, A.P.; Ward, T.M.; Minor, R.K.; Blouin, M.J.; et al. Metformin improves healthspan and lifespan in mice. *Nat. Commun.* **2013**, *4*, 2192. [CrossRef]

117. Song, Y.M.; Lee, Y.H.; Kim, J.W.; Ham, D.S.; Kang, E.S.; Cha, B.S.; Lee, H.C.; Lee, B.W. Metformin alleviates hepatosteatosis by restoring SIRT1-mediated autophagy induction via an AMP-activated protein kinase-independent pathway. *Autophagy* **2015**, *11*, 46–59. [CrossRef]

118. Shi, W.Y.; Xiao, D.; Wang, L.; Dong, L.H.; Yan, Z.X.; Shen, Z.X.; Chen, S.J.; Chen, Y.; Zhao, W.L. Therapeutic metformin/AMPK activation blocked lymphoma cell growth via inhibition of mTOR pathway and induction of autophagy. *Cell Death Dis.* **2012**, *3*, e275. [CrossRef]

119. Wang, Y.; Xu, W.; Yan, Z.; Zhao, W.; Mi, J.; Li, J.; Yan, H. Metformin induces autophagy and G0/G1 phase cell cycle arrest in myeloma by targeting the AMPK/mTORC1 and mTORC2 pathways. *J. Exp. Clin. Cancer Res.* **2018**, *37*, 63. [CrossRef]

120. Wang, G.Y.; Bi, Y.G.; Liu, X.D.; Zhao, Y.; Han, J.F.; Wei, M.; Zhang, Q.Y. Autophagy was involved in the protective effect of metformin on hyperglycemia-induced cardiomyocyte apoptosis and Connexin43 downregulation in H9c2 cells. *Int. J. Med. Sci.* **2017**, *14*, 698–704. [CrossRef]

121. Madeo, F.; Bauer, M.A.; Carmona-Gutierrez, D.; Kroemer, G. Spermidine: A physiological autophagy inducer acting as an anti-aging vitamin in humans? *Autophagy* **2019**, *15*, 165–168. [CrossRef] [PubMed]

122. Yue, F.; Li, W.; Zou, J.; Jiang, X.; Xu, G.; Huang, H.; Liu, L. Spermidine Prolongs Lifespan and Prevents Liver Fibrosis and Hepatocellular Carcinoma by Activating MAP1S-Mediated Autophagy. *Cancer Res.* **2017**, *77*, 2938–2951. [CrossRef] [PubMed]

123. Eisenberg, T.; Abdellatif, M.; Schroeder, S.; Primessnig, U.; Stekovic, S.; Pendl, T.; Harger, A.; Schipke, J.; Zimmermann, A.; Schmidt, A.; et al. Cardioprotection and lifespan extension by the natural polyamine spermidine. *Nat. Med.* **2016**, *22*, 1428–1438. [CrossRef] [PubMed]

124. Minois, N.; Carmona-Gutierrez, D.; Bauer, M.A.; Rockenfeller, P.; Eisenberg, T.; Brandhorst, S.; Sigrist, S.J.; Kroemer, G.; Madeo, F. Spermidine promotes stress resistance in Drosophila melanogaster through autophagy-dependent and -independent pathways. *Cell Death Dis.* **2012**, *3*, e401. [CrossRef]

125. Morselli, E.; Galluzzi, L.; Kepp, O.; Criollo, A.; Maiuri, M.C.; Tavernarakis, N.; Madeo, F.; Kroemer, G. Autophagy mediates pharmacological lifespan extension by spermidine and resveratrol. *Aging* **2009**, *1*, 961–970. [CrossRef]

126. Pietrocola, F.; Lachkar, S.; Enot, D.P.; Niso-Santano, M.; Bravo-San Pedro, J.M.; Sica, V.; Izzo, V.; Maiuri, M.C.; Madeo, F.; Marino, G.; et al. Spermidine induces autophagy by inhibiting the acetyltransferase EP300. *Cell Death Differ.* **2015**, *22*, 509–516. [CrossRef]

127. Minois, N.; Carmona-Gutierrez, D.; Madeo, F. Polyamines in aging and disease. *Aging* **2011**, *3*, 716–732. [CrossRef]

128. Kiechl, S.; Pechlaner, R.; Willeit, P.; Notdurfter, M.; Paulweber, B.; Willeit, K.; Werner, P.; Ruckenstuhl, C.; Iglseder, B.; Weger, S.; et al. Higher spermidine intake is linked to lower mortality: A prospective population-based study. *Am. J. Clin. Nutr.* **2018**, *108*, 371–380. [CrossRef]
129. Pietrocola, F.; Castoldi, F.; Markaki, M.; Lachkar, S.; Chen, G.; Enot, D.P.; Durand, S.; Bossut, N.; Tong, M.; Malik, S.A.; et al. Aspirin Recapitulates Features of Caloric Restriction. *Cell Rep.* **2018**, *22*, 2395–2407. [CrossRef]
130. Pietrocola, F.; Castoldi, F.; Maiuri, M.C.; Kroemer, G. Aspirin-another caloric-restriction mimetic. *Autophagy* **2018**, *14*, 1162–1163. [CrossRef]

© 2019 by the authors. Licensee MDPI, Basel, Switzerland. This article is an open access article distributed under the terms and conditions of the Creative Commons Attribution (CC BY) license (http://creativecommons.org/licenses/by/4.0/).

 *nutrients*

Review

Shedding Light on the Effects of Calorie Restriction and Its Mimetics on Skin Biology

Yeon Ja Choi

Department of Biopharmaceutical Engineering, Division of Chemistry and Biotechnology, Dongguk University, Gyeongju 38066, Korea; yjchoi@dongguk.ac.kr; Tel.: +82-54-770-2223

Received: 6 May 2020; Accepted: 22 May 2020; Published: 24 May 2020

Abstract: During the aging process of an organism, the skin gradually loses its structural and functional characteristics. The skin becomes more fragile and vulnerable to damage, which may contribute to age-related diseases and even death. Skin aging is aggravated by the fact that the skin is in direct contact with extrinsic factors, such as ultraviolet irradiation. While calorie restriction (CR) is the most effective intervention to extend the lifespan of organisms and prevent age-related disorders, its effects on cutaneous aging and disorders are poorly understood. This review discusses the effects of CR and its alternative dietary intake on skin biology, with a focus on skin aging. CR structurally and functionally affects most of the skin and has been reported to rescue both age-related and photo-induced changes. The anti-inflammatory, anti-oxidative, stem cell maintenance, and metabolic activities of CR contribute to its beneficial effects on the skin. To the best of the author's knowledge, the effects of fasting or a specific nutrient-restricted diet on skin aging have not been evaluated; these strategies offer benefits in wound healing and inflammatory skin diseases. In addition, well-known CR mimetics, including resveratrol, metformin, rapamycin, and peroxisome proliferator-activated receptor agonists, show CR-like prevention against skin aging. An overview of the role of CR in skin biology will provide valuable insights that would eventually lead to improvements in skin health.

Keywords: skin aging; calorie restriction; intermittent fasting; CR mimetic; photoaging; skin appendages

1. Introduction

The skin is the largest organ of the body and provides important protection from life-threatening environmental factors. The skin undergoes physiological and functional deterioration as organisms age, which manifests as visible changes that are clearly apparent. The esthetic implication of skin aging has been a motivating factor behind numerous studies investigating this phenomenon. In addition, skin aging is largely influenced by extrinsic factors owing to its location; this process is called extrinsic skin aging, which is a separate process to intrinsic and chronological skin aging.

Calorie restriction (CR) is the most effective intervention to extend the lifespan of various organisms and has been used as a benchmark for longevity in anti-aging research [1,2]. There are numerous reports demonstrating the preventive effects of CR on aging and associated diseases, such as chronic nephropathies, cardiomyopathies, diabetes, autoimmune conditions, respiratory diseases, and neurological degeneration [3,4]. However, the effects of CR on the skin are poorly understood. The lifespan-prolonging changes induced by CR in the skin are believed to be less pronounced than in other major organs, including the liver, heart, and brain. This review serves to summarize the characteristics of skin aging and discuss current studies focusing on the effects of CR and alternative approaches to CR on cutaneous physiology and aging.

2. Skin Aging

2.1. Extrinsic Factors of Skin Aging

The aging process involves physiological and functional deterioration that progresses throughout the lifetime of an organism, induced by various genetic and non-genetic environmental factors. Eventually, homeostasis is disrupted, and susceptibility to disease or death is increased [5]. Skin aging is affected by both intrinsic and extrinsic factors, with extrinsic environmental factors contributing to the effects of chronological aging.

The concept of "exposome" was developed by American cancer epidemiologist Christopher Wild in 2005 [6] and refers to the totality of exposures to which an individual is subjected to from conception to death. The skin aging exposome, proposed by Krutmann et al., suitably describes the external and internal factors, and their interactions, that affect humans from conception to death, as well as the response of the human body to these factors that lead to skin aging [7]. The environmental factors of a skin aging exposome can be categorized into solar radiation (ultraviolet (UV) radiation, visible light, and infrared radiation), air pollution, tobacco smoke, nutrition, cosmetic products, and miscellaneous factors [7].

Among the extrinsic factors, UV radiation is an established accelerator of skin aging, through a process termed photoaging [8]. While UVC (200–280 nm) is filtered out by the ozone layer, UVB (280–320 nm) and UVA (320–400 nm) are the principal wavebands responsible for photoaging. An organism is exposed to sunlight UV comprising approximately 5% UVB and 95% UVA, but the degree of UVB exposure is expected to increase with ozone layer depletion [9]. UVB generally induces DNA damage, which results in skin tumorigenesis and causes excess melanin production and sunburn at high doses. UVA causes little DNA damage but generates significant oxidative stress, which mediates oxidative damage to DNA and non-DNA targets [10]. UVB and UVA both contribute to the characteristic features of photoaging [11].

Unlike experimental animals, humans who have lived strictly indoors for a lifetime are rare. In humans, photoaging is superimposed onto intrinsic aging. However, the difference between intrinsic aging and photoaging can be evaluated by comparing the features of UV-exposed skin sites, such as the face and dorsal side of the forearm, with non-exposed sites, e.g., the buttock skin. For example, while the loss of extracellular matrix (ECM) is a distinct feature of intrinsically aged skin, photoaged skin contains abundant elastin and collagen fibers, which are fragmented and disorganized [11,12]. Ultimately, the morphological changes and functional loss of skin during the aging process result from a combination of intrinsic and extrinsic aging.

2.2. Structural and Functional Alterations in Aged Skin

The skin is a complex organ with multiple cell types. It consists of two primary layers, the epidermis and dermis, as well as skin appendages, including hair follicles, sebaceous glands, sweat glands, and nails [13]. The physiological and functional alterations of each part of the skin during aging are reviewed in the following sections.

2.2.1. Epidermal Changes in Aged Skin

The outermost layer of the skin, a highly specialized and multilayered epithelium, is called the epidermis. The mature epidermis is a stratified squamous epithelium that is composed of numerous keratinocyte layers, including the "stratum basale" (or the basal layer), "stratum spinosum" (or the spinous layer), "stratum granulosum" (or the granular layer), and the "stratum corneum" (or the corneal layer). Keratinocytes proliferate symmetrically and asymmetrically and differentiate, slowly moving towards the surface replacing old cells. During the terminal differentiation process, the cells become more flattened and water impermeable.

Changes in the thickness of the epidermis with age are varied, but mostly appear to be sustained or decreasing thickness. While the epidermis on the upper inner arm was reported to be thinner in the elderly [14], highlighting the alteration of the epidermis during intrinsic skin aging, other reports showed that epidermal thickness was not correlated with age [15–17]. In a murine study, older animals had a thinner epidermis at four different skin areas (dorsal, ventral, pinna, and footpad) of laboratory-raised CBA mice [18], but no differences were observed in the thickness of the dorsal and ventral epidermis of C57BL/6 mice [19]. In another study, photoaged skin appeared hypertrophic, atrophic, or unaltered, and histologically, the stratum corneum showed hyperkeratosis [8]. Cigarette smoking, another extrinsic factor of skin aging, was found to be negatively correlated with the thickness of the stratum corneum [16].

A vital function of the epidermis is to act as a protective interface between the body and the external environment, by preventing infection and the loss of body fluids, resisting mechanical stress, and participating in immune responses. The skin barrier function is also partially influenced by age. In menopausal women, measurements of transepidermal water loss showed a minor change in the hydration of the cornified layer, decreased sebum production, and a significantly higher skin surface pH [20]. Chronic itching is also common in aging skin. This may be due to the age-related decline in Merkel cell numbers, which act as mechanoreceptors [21], which causes the sense of touch to turn to an itching sensation [22].

A basement membrane lies beneath the epidermis at the dermo-epidermal junction, which adheres the epidermis to the dermis through a connection of the basal keratinocytes to the basement membrane by hemidesmosomes, and the fibroblasts in the dermis attach to the basement membrane by anchoring fibrils [23,24]. The junction is undulating. The downward folds of the epidermis are called epidermal ridges or rete ridges, and the upward projections of the dermis are called the dermal papillae. Significant flattening of the rete ridges was consistently observed in independent studies of different areas of skin [17,25]. This change is believed to reduce the interface between the epidermis and dermis, decreasing epidermal resistance to shearing stress, which makes the epidermis more fragile [26].

2.2.2. Dermal and Hypodermal Changes in Aged Skin

The dermis is comprised of a connective tissue layer of mesenchymal origin and is subdivided into the following three layers, in order of proximity to the epidermis: papillary, reticular, and hypodermis. Fibroblasts, the most representative cells resident in the dermis, produce and secrete ECM proteins. Collagen is an essential ECM protein found in the dermis; type I and type III collagen are particularly abundant there. Other dermal ECM components include elastin fibers, proteoglycans, and hyaluronic acid, which provide strength, support, and flexibility.

Structurally, the dermis in humans and mice becomes thinner and loses elasticity with age [15,19,27]. Solar elastosis is a hallmark of human photoaging, characterized by an accumulation of partially degraded elastin fibers in the upper dermis. Although this alteration is not typically observed in intrinsic skin aging, an abnormal elastin network is also detected with age in sun-protected skin [8,26]. Massive deposition of other components of the ECM, such as glycosaminoglycans and interstitial collagen, has been observed in photoaged skin. The number of dermal fibroblasts and their capability to produce ECM are lower in intrinsically aged skin. Fibril collagen is degraded markedly with age, as well as by UV irradiation [28].

Matrix metalloproteinases (MMPs), a class of proteolytic enzymes, are considered the leading physiological cause of the breakdown of dermal ECM proteins. The expression and activities of MMPs are increased in aged skin and senescent fibroblasts, whereas the expression of tissue inhibitor of metalloproteinases (TIMPs) is decreased [29]. UVB irradiation also contributes to the activation MMPs [30]. Cathepsin K is a lysosomal protease that plays a vital role in clearing elastin that has been partially degraded by MMPs in the ECM [31]. Codriansky et al. reported that Cathepsin K was induced in young dermal fibroblasts as a response to UVA irradiation but not in fibroblasts from old donors [32].

Two interesting studies recently reported the cellular and molecular mechanisms underlying age-related dermal functions. Aged upper dermal fibroblasts gradually acquire the characteristics of the lower dermis, with reduced expression of ECM proteins and increased adipogenic traits [33].

Marsh et al. revealed that fibroblast positions are stably maintained over time, but clusters of fibroblasts are lost, and the membrane extends to fill the space of lost neighboring fibroblasts in aged skin. These findings provide a mechanism for a loss of cellularity in aged fibroblasts [34].

2.2.3. Changes in Hair Follicles during Skin Aging

Hair follicles are comprised of an outer root sheath (ORS) and inner root sheath (IRS), which enclose the hair shaft. The hair bulb is at the base of the follicle, which contains proliferating matrix cells that grow to form the hair shaft and surround the dermal papilla at the bottom of the hair follicle. The dermal papilla consists of specialized mesenchymal cells [35]. The hair bulge, part of the ORS and located at the insertion site of the erector pili muscle of the hair follicle, is where epidermal stem cells reside. In adult mammals, hair grows in a regenerative cycle of phases, namely anagen (growth phase), catagen (regression phase), telogen (resting phase), and exogen (hair shaft shedding phase) [36]; this cycle is tightly regulated by the integrated action of multiple signaling pathways.

Elderly hairs become thinner, weaker, dry, dull, and sparse, due to hair follicle miniaturization and hair shaft weathering. Senescent alopecia is prevalent in the aged population, which is a diffuse and non-patterned type of hair loss that differs from androgenetic alopecia [37]. A lack of correlation between age and the total follicle number has been reported [38,39]. Old C57/Bl6 mice exhibited swelling hair follicles and a variable loss of normal hair follicle triplet patterning compared to young animals [40]. Furthermore, increased levels of inhibitors of Wnt, an activator of hair growth, such as dickkopf Wnt signaling pathway inhibitor 1 (DKK1) and secreted frizzled-related protein 4 (Sfrp4), were found in aged mice [41]. Aged hair follicle stem cells are believed to be a major contributor to a slow hair cycle and loss of hair during skin aging, which will be discussed in depth in Section 3.

Gray hair is one of the most noticeable signs of aging [37]. The number of melanocytes in the hair matrix decreases in aged hair follicles [42]. Melanocyte stem cells (MSCs) are maintained in the hair bulge area, and the mature melanocytes reside in the hair bulb. The frequency of melanocyte-inducing transcription factor (MITF)-positive melanocytes per basal keratinocytes in the hair bulge decreases significantly with age [43]. The abnormal maintenance of MSCs, together with a loss of differentiated progeny, contributes to physiological hair graying [43].

2.2.4. Changes in Sweat Glands during Skin Aging

The sweat glands are small tubular structures in the skin, producing and excreting sweat. Functionally, sweat glands remove excess micronutrients, metabolic waste, and toxins from the body, and are involved in thermoregulation. Sweat glands can be divided into three types: eccrine, apocrine, and apoeccrine [44]. The eccrine sweat glands are the most numerous, distributed across almost the entire body surface area, and are smaller than the other two glands. Apocrine and apoeccrine are limited to specific regions of the body, such as axilla.

A decrease in the number of eccrine sweat glands and a shrunken morphology was detected in the scalp skin of old males (83.8 ± 2.8 years old) compared to younger males (33 ± 6.3 years old) [15]. The responsiveness of the eccrine sweat gland to pharmacological stimuli was estimated in different age groups of men. The results revealed a comparable density of activated glands but a lower sweat gland output per active gland in the old group (age > 58 years old) [45], which implies a functional decline in the sweat glands during aging. Recent data suggested that epithelial autophagy contributes to the homeostasis of sweat glands, showing a significant decrease in the number of functional sweat glands in conditionally lacking Atg7 in K14-positive precursor cells [46]. In another study, the age-related reduction of sweat gland function was found to be regionally different [47]. In addition, sweating between old and young adults during exercise in the heat was comparable, indicating that the ability to regulate body core temperature during heat stress was retained in older adults [44].

2.2.5. Changes in Sebaceous Glands during Skin Aging

Sebaceous glands are unique microscopic gland structures that accompany hair follicles. In humans, sebaceous glands are distributed throughout all skin sites but show a high abundance on the face and scalp. These glands secrete a complex oily and waxy mixture called sebum, which lubricates and waterproofs the skin and hair [48] and also participates in the immunity of mammals through the production of antimicrobial peptides, cytokines, and chemokines, such as interleukin (IL)-1β, IL-6, IL-8/CXCL-8, and tumor necrosis factors (TNFs) [49]. Sebum is comprised of triglycerides, wax esters, cholesterol esters, squalene, and free fatty acids [48]. Sebocytes are the major cells within the sebaceous glands. Fully mature sebocytes act in a holocrine manner; this is a unique secretion process that destroys the cell and results in the secretion of the product into the lumen [50].

During skin aging, the size and secretory activity of sebocytes decrease, which results in a decreased level of the surface lipid and dry skin [51]. Aged sebocytes were found to express more growth-regulated protein alpha (GRO-α), a CXC chemokine, which was attributed to the constitutive activation of NF-κB [52]. Cigarette smoke was found to decrease the level of scavenger receptor B1 (SRB1) [53], which is an oxidative stress-sensitive, transmembrane receptor that is well known for cholesterol uptake from high-density lipoprotein (HDL) [54]. Reduced SRB1 levels due to cigarette smoking compromised the cholesterol uptake of sebocytes, leading to an alteration of the sebocyte lipid content [53]. Moreover, age-related hormonal changes contribute to decreased lipid synthesis and changes in the gene expression profiles of sebocytes [55].

3. Effects of CR on Skin Aging

3.1. Effects of CR on Wound Healing

CR is universally believed to be a remarkable dietary manipulation of aging and age-related diseases, but its effects on the skin are poorly understood. Although CR has beneficial effects in other organs, it had an insignificant influence on age-related cutaneous phenotypes and an association with adverse outcomes in wound healing. Previous studies showed that CR retarded wound healing and collagen production [56–59]. During wound healing, activated fibroblasts transform to myofibroblasts and migrate to the area of the lesion, where they assist in closing the wound by promoting the synthesis and secretion of collagen. In different studies, the capacity of wound repair in animals with food ad libitum (AL), CR, and CR followed by refeeding for one month prior to wounding was compared [60,61]. Slower wound healing was observed in AL and CR aged animals compared to young subjects, which is consistent with previous data, but it was reported that CR animals with refeeding before the wound healed showed similar healing to that of the young animals, with enhanced synthesis of type I collagen. CR reduced collagen glycation, which are abnormal protein adducts detected in diabetic or aged skin, in older Rhesus monkeys by 30% [59]. These studies suggested that CR assisted in preserving the proliferative capacity required for wound repair.

3.2. Effects of CR on Morphological and Structural Changes in the Skin

A recent study on female 8-week-old Swiss mice fed a 60% reduced diet for six months revealed a thicker epidermis and reduced dermal white adipose tissue, demonstrating that tissue ultrastructure is modified by prolonged CR [62]. In addition, CR induced dermal vasculature development, accompanied by higher levels of vascular endothelial growth factor (VEGF), compared to AL mice [62,63]. Abdominal skin from 4-, 12-, and 24-month-old Fisher male rats that were fed a CR diet showed that age-related increase in the thickness of dermis and hypodermis was rescued. In contrast to previous data which showed thinner or age-independent changes in aged skin, the epidermis layer increased according to age and was comparable with AL and CR animals [63]. Interestingly, CR induced morphological changes to the fur coats of laboratory rats [62]. CR animals displayed significantly more and longer guard hairs in their skin fur coats than AL animals, whereas other types of hairs (Awl, Auchene, and Zigzag) remained unchanged. These changes provided a fur coat with better thermoregulatory properties. This same study also showed higher hair follicle growth in CR animals, which is associated with an increase in interfollicular and hair follicle stem cells.

A study that investigated the effect of CR on photoaging showed that CR reduced wrinkle formation when compared to AL animals, both with and without UVB irradiation [64]. Epidermal thickness increased after UVB radiation, which was accelerated by CR, as observed by corresponding epidermal proliferating cell nuclear antigen (PCNA) levels. CR also influenced the histological alteration upon UVB radiation, but further molecular and mechanistic evaluations are required to determine the precise effect of CR on photoaging.

3.3. Effects of CR on Skin Stem Cells

Changes in stem cells have been implicated primarily in aging, as well as skin aging, because adult stem cells in tissues are essential for organ homeostasis and repair. In the epidermis, a range of stem cell populations are located in different regions, and each stem cell compartment produces a subset of differentiated epidermal cells. Interfollicular epidermal (IFE) stem cells are localized in the basal layer of the epidermis, and hair follicle stem cells (HFSCs) and MSCs are in the bulge and the hair germ [65].

Age-related decline in the renewal capacity of the hair cycle fully involves HFSC aging. DNA damage accumulates in the HFSCs during repetitive hair cycling, which leads to proteolysis of COL17A1, an important component of the follicle stem cell niche [66]. Deficiency in COL17A1 results in HFSC loss of stemness and differentiation into an epidermal lineage. A comparison between young and aged murine epidermal stem cells (ESCs) showed that they have similar in vitro growth and differentiation potentials, but local environmental factors influence skin aging [67]. UV has been demonstrated to induce stem cell apoptosis in the basal layer and hair bulge, which contributed partially to epidermal atrophy, slow wound healing, and depigmentation. Intrinsically aged murine skin had a comparable abundance of CD34$^+$ epidermal stem cells [40].

The anti-aging capability of CR is related to its ability to reprogram stemness and boost the regenerative capacity of stem cells. Previously, CR improved the functioning of various stem cell populations, including hematopoietic and intestinal stem cells in mice and germline stem cells in flies [68]. However, there is limited research investigating the effect of CR on skin-residing stem cells. One study showed that CR expanded pools of IFE stem cells and HFSCs in CR animals, which promoted the growth and maintenance of their fur coats [62]. Furthermore, stem cells are under the control of a rhythmic circadian machinery; CR reversed the reprogrammed daily rhythms to adapt to tissue-specific stress in aged epidermal stem cells [69].

3.4. Effects of CR on Carcinogenesis

CR, by the restriction of fats or carbohydrates, delayed the rate and reduced the incidence of papilloma development [70]. Additionally, CR prevented UV-mediated skin tumor formation [71]. CR decreased the expression of oncogenic H-Ras and significantly activated Ras-GTP in skin stimulated with 12-0-tetradecanoylphorbol-13-acetate (TPA) [72]. In addition, the TPA-induced activation of PI3K/Akt and p42/p44-MAPK signaling was reduced in CR skin. Chemically induced ulcerative skin was observed to be more infrequent in CR than AL skin. Furthermore, the decrease in p53 gene expression in p53$^{+/-}$ mice may have reduced the beneficial effects of CR in these circumstances [73]. Together, these results suggest that CR prevents skin carcinogenesis.

3.5. Metabolic Effect of CR on Skin Aging

Molecular alterations to metabolically adapt to limited calorie intake mediates the beneficial effects of CR. CR stimulates respiratory rates by enhancing mitochondrial biogenesis and stimulating uncoupling between oxygen consumption and oxidative phosphorylation. A metabolic shift to a more oxidative phenotype has been reported in the dermal compartment [62]. Metabolomic analysis revealed that UV exposure induced catabolism of biomolecules and increased oxidative stress [74].

The metabolome data showed altered activity in upper glycolysis and glycerolipid biosynthesis and decreased protein and polyamine biosynthesis in aged skin [75]. CR-mediated metabolic alterations might be employed through changes in cellular signaling, epidermal barrier function, and skin structure during skin aging; further investigation is necessary to better determine the underlying molecular activities.

4. Effects of Alternative Ways of Dietary Restriction on Skin Aging

The food intake of experimental animals under CR in aging research is severely restricted; overall, calorie intake or food intake in CR models is reduced by approximately 10%–50%, without malnutrition, compared to AL controls [76], which could be challenging for humans to practice and sustain. Therefore, several practical approaches, such as intermittent (e.g., alternate day fasting) and periodic (fasting that lasts three days or longer, every two or more weeks) fasting, or alternative methods of dietary restriction, have been suggested for humans [77]. The restriction of specific nutrients rather than the decrease in total food intake has shown beneficial effects on lifespan extension and prevention of age-related diseases. In addition, various pharmacological interventions, from natural products to synthetic compounds, have been developed and studied to mimic the benefits of CR as an anti-aging strategy. However, the effects of these alternative ways of CR on skin biology, including the skin aging process and skin disorders, have been paid less attention. The following sections of this review describe current research into the effects of alternative dietary restriction approaches and CR mimetics on skin biology and aging.

4.1. Effects of Fasting on Skin Biology

A recent review summarized current literature on the impact of fasting on skin biology [78]. Most of the study focused on the efficacy of fasting on wound healing. While fasting for three days delayed wound healing [78], short-term, repeated fasting (four consecutive days, every two weeks) for two months before the wound, improved wound healing with increases in epithelialization, contraction, healing, collagen levels, and hydroxyproline [58]. This is consistent with the increased capacity of wound repair in the animals of the caloric restricted-refed group. In another study, four days of a diet that mimics fasting (FMD) reduced severe ulcerating dermatitis in C57BL/6 mice, which indicates that FMD protects against inflammation and inflammation-associated skin lesions [79]. Bragazzi et al., the author of the recent review [78], emphasized the need for evidence-based and standardized protocols of fasting and qualitative improvement in research on fasting and skin.

4.2. Effects of Specific Macronutrient Restriction on Skin Biology

Previous studies have shown that a decrease in either dietary protein or sugar can reduce mortality and extend the life span of *Drosophila* [80] and mice [81], independently of the calorie intake. Furthermore, the reduced intake of specific essential amino acids, such as methionine, tryptophan, or branched-chain amino acids, had beneficial effects on delaying aging or improving health [82,83]. There has been little research on the impact of macronutrient restriction on skin biology and aging.

Protein restriction (PR, 0% kcal protein of total calorie) or methionine restriction (MR, 14% kcal protein containing 0.05% methionine) regimens were tested in the context of wound healing in normal and diabetic animals [84]. The mice preconditioned with PR for one week or MR for two weeks before surgery showed comparable wound healing to mice fed a complete diet. Under diabetic conditions, PR or MR improved perioperative glucose tolerance and perioperative hyperglycemia, without any impairment in wound healing. These results lessen the concerns of poor wound healing or susceptibility to infection associated with the typical CR method during surgery, suggesting the potential clinical application of these regimens.

On the other hand, a carbohydrate-restricted diet promotes skin senescence in senescence-accelerated prone mice. Histologically, the epidermis and dermis were thinner in the carbohydrate-restricted group, and cutaneous expression of the senescence markers p16 and p21 and lipid peroxidation was increased by long-term carbohydrate restriction [85,86]. Considering that this group also showed a significant progression of visible aging and decreased survival rate, the duration of nutrient restriction should be carefully considered.

5. Protective Effect of CR Mimetics on Skin Aging and Skin Disorders

CR mimetics have attracted considerable attention for many years because of their health-promoting effects [87]. Notwithstanding the limitations of some mimetics partially mimicking the effect of CR and the unclear mechanisms of action of CR mimetics, CR mimetics still have numerous advantages including convenience of application such as its potential use as a health food supplement [88–90]. In skin aging research, many pharmacological compounds have been studied as potential skin aging interventions. In this section, the current knowledge about several well-known CR mimicking compounds in terms of skin aging and disorders will be discussed (Figure 1).

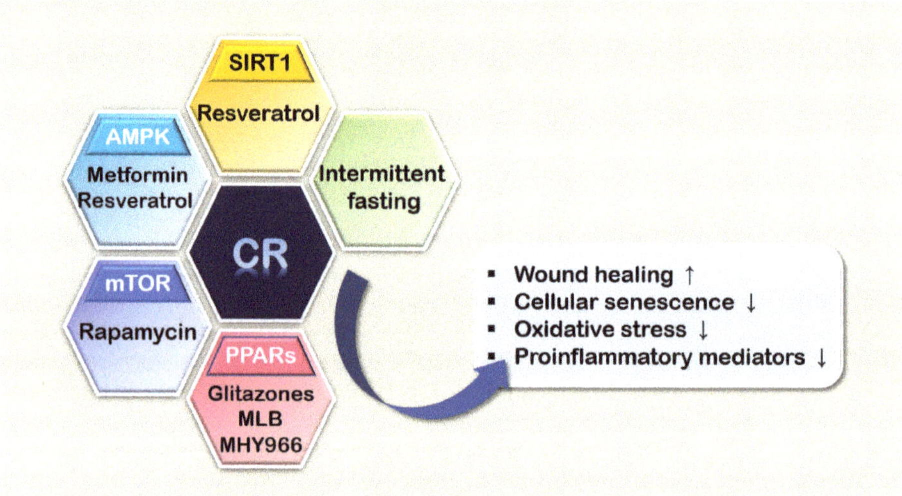

Figure 1. Summary of CR mimetics and their effects on skin aging. AMPK, AMP-activated protein kinase; CR, calorie restriction; MLB, magnesium lithospermate B; mTOR, the mammalian target of rapamycin; SIRT1, the mammalian homolog of SIR2; PPAR, peroxisome proliferator-activated receptors.

5.1. Sirtuin and Resveratrol

The activation of sirtuins, which are nicotinamide dinucleotide (NAD^+)-dependent deacetylases, has been reported to extend the lifespan of various organisms, including yeast, worms, fruit flies, and mice [91–93] and has been identified as a mediator of the beneficial effects of CR. More than 14,000 compounds that activate sirtuin have been identified [94]. Resveratrol (3,5,4′-trihydroxystilbene) was identified as the first potent activator of Sirtuin and has been studied extensively as a CR mimetic [95–100].

SIRT1, the mammalian homolog of yeast SIR2, is expressed ubiquitously throughout the skin. Immunohistochemical staining of elderly skin showed that the level of SIRT1 decreased and there was a steady reduction in the proliferation of dermal fibroblasts [101]. UVA and UVB irradiation also induced a decrease in gene expression or activity of SIRT1 in dermal fibroblasts [102,103], keratinocytes [104], and melanocytes. The overexpression of SIRT1 prevented human skin fibroblast senescence through deacetylation of forkhead box O3α (FOXO3α) and p53 [105]. Epidermis-specific SIRT1 deletion inhibited the regeneration of both the epidermis and dermal stroma, which shows that epidermal SIRT1 is essential for wound repair [106]. The strong correlation between skin aging and sirtuin expression and extensive mechanistic studies have supported SIRT1 as a pharmacological target and resveratrol as a powerful prevention therapy of skin aging.

The topical application of 2% resveratrol increased the repair of tissue wounds more so than for vehicle-treated rats and was associated with the induction of angiogenesis, fibroplasia, and collagen organization [107]. The administration of resveratrol to wounds improved epithelization, hair follicle regeneration, and collagen deposition in both young and old rodents [108]. Resveratrol stimulates the production of collagen types I and II, reduces the expression of AP-1 and NF-kB factors, and slows down the process of skin photoaging in human keratinocytes and mouse skin [109]. Oxidative stress-induced senescence was ameliorated by resveratrol in primary human keratinocytes [110]. Resveratrol was initially characterized as a SIRT activator, but other types of signaling, such as AMP-activated protein kinase (AMPK) and FOXO3, also contribute to its actions.

Resveratrol has been used increasingly in cosmetology and dermatology because of its antioxidant, anti-inflammatory, anti-proliferative, and anti-pigmentation properties [111]. Epidermal permeation of resveratrol has been assessed in vitro and in vivo. Most of the resveratrol was detected in the stratum corneum, and resveratrol penetrated the porcine skin at 20–49 μm, corresponding to the viable epidermis, at a constant concentration [112]. Dietary supplements and various types of cosmetics, such as sunscreen and ampules containing resveratrol for skin rejuvenation, are currently available, studies are underway to improve the delivery of the topical application of resveratrol to enhance its efficacy and stability [113–116].

5.2. AMPK and Metformin

Metformin (N',N'-dimethylbiguanide) is one of the first-line drugs for treating type 2 diabetes [117]; its anti-aging property was illustrated by lifespan extension in *C. elegans* [118–120], *Drosophila* [121], and mice [122,123]. Although the precise molecular mechanisms of the effects of metformin remain unclear, it is known that metformin activates AMPK, which serves as an energy sensor and regulator of glucose homeostasis [124]. AMPK signaling intersects with the mammalian target of rapamycin (mTOR) [125], extracellular signal-regulated kinase (ERK) [126], and SIRT3 [127] and is also involved in mitochondrial biogenesis and activating autophagy [128].

AMPK activation in human skin reportedly decreases during the aging process [110]. Cutaneous AMPK activity is also downregulated by UVB irradiation in humans and mice [129]. AMPKα deletion in keratin 14-expressing ESCs resulted in hyperactive mTOR signaling leading to extensive hyperproliferation after acute wounding, UVB exposure, and phorbol ester application. These findings suggest that the essential role of ESC-specific AMPK is in the control of ESC proliferation and physiological skin repair [130]. Additionally, the activation of AMPK has a beneficial effect on oxidative stress-mediated UV-induced cellular senescence.

Both systemic and topical application of metformin successfully attenuated UVB-induced epidermal hyperplasia and skin tumorigenesis [129]. Metformin reversed the diminished collagen I production induced by UVA and suppressed MMP-1 expression, which substantiates the potential use of metformin in prevention against dermal aging. Metformin was reported to have a prominent effect on wound healing. Metformin-regulated AMPK/mTOR signaling resulted in M2 macrophage polarization by inhibiting NLRP3 inflammasome activation [131].

Zhao et al. estimated the efficacies of popular anti-aging agents, including resveratrol, metformin, and rapamycin; topical application of resveratrol and metformin, but not rapamycin, improved wound healing in young mice, and metformin exerted strong regenerative efficacy in aged skin [108]. Some of the beneficial effects of metformin have been achieved through cutaneous application. For example, a transdermal formulation for metformin, such as a cream and transdermal patch, was developed for patients who could not tolerate the oral dose or could not swallow large tablets. A recent paper reported a significant improvement in skin delivery by incorporating metformin into solid lipid nanoparticles and subsequently formulating an effective topical gel [132]. The future development of more advanced metformin delivery systems will allow for expanded application.

5.3. mTOR and Rapamycin

Rapamycin was first introduced as an inhibitor of mTOR, which is a serine-threonine kinase that regulates cell survival, growth, proliferation, motility, protein synthesis, transcription [133], and autophagy [134–136]. Rapamycin has also significantly increased the life span of various experimental models [137–140] and has potential as a CR mimetic. However, there has been limited research on the role of mTOR in skin aging. An age-related and UVB-induced increase in the activity of mTOR and RICTOR protein, which is a major component of mTOR complex 2 (mTORC2) [141], has been reported. In turn, the activation of mTORC2 signaling was found to mediate NF-kB activation during skin aging. The epidermal deficiency of mTORC2 signaling caused moderate tissue hypoplasia, reduced keratinocyte proliferation, and attenuated the hyperplastic response to TPA [142]. In an in vitro study, mTORC2 activity-deficient keratinocytes displayed a longer lifespan, less senescence, and an enhanced tolerance to cellular stressors. While this study was not performed in aged animals, it highlights the potential implications of mTOR signaling in skin aging and the therapeutic resistance of epithelial tumors. Moreover, mTOR has been implicated in the pathogenesis of various skin disorders, such as psoriasis [143,144], which strongly suggests the dermatological application of rapamycin.

Rapamycin effectively suppressed UVB-induced oxidative stress and collagen degradation in skin fibroblasts [145]. A recent clinical trial (ClinicalTrials.gov Identifier: NCT03103893) showed that topical rapamycin reduced the senescence and age-related features in human skin [146]. This study showed that the p16^{INK4A} level, a marker of cellular senescence, and solar elastosis was decreased in rapamycin-treated skin. Moreover, collagen VII, a critical component of the basement membrane, was increased and disorganized collagen was restored. These histological and molecular observations in rapamycin-treated skin highlight rapamycin as a potential anti-aging therapy with efficacy in humans.

5.4. PPAR Agonists

Peroxisome proliferator-activated receptors (PPARs) are nuclear receptors with diverse biological effects in the promotion of cellular proliferation and differentiation, lipid and carbohydrate metabolism, inflammatory responses, and tissue remodeling [147,148]. There are three PPAR isoforms, PPARα, PPARβ/δ, and PPARγ, which are distributed in different tissues and have selectivity and responsiveness to specific ligands [149]. Among them, PPARα and γ have been well investigated in aging research. The expression of PPARα and PPARγ genes was found to be decreased during aging, which was rescued by CR [147,150]. UV irradiation reduced PPARγ levels, which resulted in dysregulation in epidermal lipids in human skin, contributing to the development of skin photoaging [151].

PPARγ is also expressed throughout the skin and in most types of skin cells [149]. In the epidermis, PPARγ has an essential role in skin barrier regulation [152] and negatively regulates the gene expression of proinflammatory genes through the antagonization of inflammatory transcription factors NF-κB and AP-1. Therefore, synthetic ligand "glitazones," which are a class of oral antidiabetic drugs, were applied; they had protective effects on inflammatory skin disorders, such as atopic dermatitis and psoriasis [153,154].

In addition to well-known agonists of PPARs, the beneficial effects of potential novel ligands in skin aging have been reported. Abietic acid, as a PPARα/γ dual ligand, decreased UVB-induced MMP-1 expression significantly by downregulating UVB-induced MAPK and NF-κB signaling in dermal fibroblasts [155]. Magnesium lithospermate B activated PPARβ/δ, leading to the upregulation of collagen expression in aged murine skin [156]. Treatment with the synthetic compound MHY966 (2-bromo-4-(5-chloro-benzo[d]thiazol-2-yl) phenol), a novel PPARα/γ dual agonist, protected UVB-exposed hairless mice from lipid peroxidation and elevated cutaneous proinflammatory mediators, including NF-κB, iNOS, and COX-2 [157].

6. Conclusions

Dietary restrictions affect the structure and function of skin. CR has beneficial effects on skin aging in terms of wound repair, stem maintenance, and carcinogenesis. While the physiological and pathological features of aged skin are relatively well characterized, the full effects of CR on skin physiology remain to be elucidated. Investigations pertaining to the effects of CR and CR alternatives on skin aging are limited, and extensive research is necessary to resolve these gaps in knowledge. Skin disease is increasing in prevalence among the elderly, and while many age-related skin disorders are not lethal, they are integral to general health status and overall quality of life. Currently, hormone therapy, antioxidant intervention, and the therapeutic application of stem cells are used to treat skin aging. The use of CR and CR mimetics has great potential to rejuvenate and maintain healthy skin, as well as improve age-related skin disorders. This review substantiates the need for further investigation into CR and related mimetics as potential therapeutic agents for skin aging and age-related disorders.

Funding: This study was supported by the Korean National Research Foundation (NRF) funded by the Korean government (NRF-2019R1F1A1057138) and the Dongguk University Research Fund of 2019.

Acknowledgments: I would like to thank Editage (www.editage.co.kr) for English language editing.

Conflicts of Interest: The author declares no conflicts of interest.

References

1. Yu, B.P. Aging and oxidative stress: Modulation by dietary restriction. *Free Radic. Biol. Med.* **1996**, *21*, 651–668. [CrossRef]
2. Yu, B.P. Why dietary restriction may extend life: A hypothesis. *Geriatrics* **1989**, *44*, 87–90. [PubMed]
3. Yu, B.P.; Chung, H.Y. Stress resistance by caloric restriction for longevity. *Ann. N. Y. Acad. Sci.* **2001**, *928*, 39–47. [CrossRef] [PubMed]
4. Demetrius, L. Of mice and men. When it comes to studying ageing and the means to slow it down, mice are not just small humans. *EMBO Rep.* **2005**, *6*, S39–S44. [CrossRef] [PubMed]
5. Kim, D.H.; Bang, E.; Jung, H.J.; Noh, S.G.; Yu, B.P.; Choi, Y.J.; Chung, H.Y. Anti-aging Effects of Calorie Restriction (CR) and CR Mimetics based on the Senoinflammation Concept. *Nutrients* **2020**, *12*, 422. [CrossRef] [PubMed]
6. Wild, C.P. Complementing the genome with an "exposome": The outstanding challenge of environmental exposure measurement in molecular epidemiology. *Cancer Epidemiol. Biomark. Prev.* **2005**, *14*, 1847–1850. [CrossRef] [PubMed]
7. Krutmann, J.; Bouloc, A.; Sore, G.; Bernard, B.A.; Passeron, T. The skin aging exposome. *J. Dermatol. Sci.* **2017**, *85*, 152–161. [CrossRef]
8. Berneburg, M.; Plettenberg, H.; Krutmann, J. Photoaging of human skin. *Photodermatol. Photoimmunol. Photomed.* **2000**, *16*, 239–244. [CrossRef]
9. Rousseaux, M.C.; Ballare, C.L.; Giordano, C.V.; Scopel, A.L.; Zima, A.M.; Szwarcberg-Bracchitta, M.; Searles, P.S.; Caldwell, M.M.; Diaz, S.B. Ozone depletion and UVB radiation: Impact on plant DNA damage in southern South America. *Proc. Natl. Acad. Sci. USA* **1999**, *96*, 15310–15315. [CrossRef]
10. Karran, P.; Brem, R. Protein oxidation, UVA and human DNA repair. *DNA Repair (Amst.)* **2016**, *44*, 178–185. [CrossRef]
11. Gilchrest, B.A. Photoaging. *J. Invest. Dermatol.* **2013**, *133*, E2–E6. [CrossRef] [PubMed]
12. Eckhart, L.; Tschachler, E.; Gruber, F. Autophagic Control of Skin Aging. *Front. Cell Dev. Biol.* **2019**, *7*, 143. [CrossRef] [PubMed]
13. Watt, F.M. Mammalian skin cell biology: At the interface between laboratory and clinic. *Science* **2014**, *346*, 937–940. [CrossRef] [PubMed]
14. Branchet, M.C.; Boisnic, S.; Frances, C.; Robert, A.M. Skin thickness changes in normal aging skin. *Gerontology* **1990**, *36*, 28–35. [CrossRef] [PubMed]
15. Jo, S.J.; Kim, J.Y.; Jang, S.; Choi, S.J.; Kim, K.H.; Kwon, O. Decrease of versican levels in the follicular dermal papilla is a remarkable aging-associated change of human hair follicles. *J. Dermatol. Sci.* **2016**, *84*, 354–357. [CrossRef] [PubMed]

16. Sandby-Moller, J.; Poulsen, T.; Wulf, H.C. Epidermal thickness at different body sites: Relationship to age, gender, pigmentation, blood content, skin type and smoking habits. *Acta Derm. Venereol.* **2003**, *83*, 410–413. [CrossRef] [PubMed]

17. Giangreco, A.; Goldie, S.J.; Failla, V.; Saintigny, G.; Watt, F.M. Human skin aging is associated with reduced expression of the stem cell markers beta1 integrin and MCSP. *J. Investig. Dermatol.* **2010**, *130*, 604–608. [CrossRef]

18. Bhattacharyya, T.K.; Thomas, J.R. Histomorphologic changes in aging skin: Observations in the CBA mouse model. *Arch. Facial Plast. Surg.* **2004**, *6*, 21–25. [CrossRef]

19. Meador, W.D.; Sugerman, G.P.; Story, H.M.; Seifert, A.W.; Bersi, M.R.; Tepole, A.B.; Rausch, M.K. The regional-dependent biaxial behavior of young and aged mouse skin: A detailed histomechanical characterization, residual strain analysis, and constitutive model. *Acta Biomater.* **2020**, *101*, 403–413. [CrossRef]

20. Luebberding, S.; Krueger, N.; Kerscher, M. Age-related changes in skin barrier function—Quantitative evaluation of 150 female subjects. *Int. J. Cosmet. Sci.* **2013**, *35*, 183–190. [CrossRef]

21. Xiao, Y.; Williams, J.S.; Brownell, I. Merkel cells and touch domes: More than mechanosensory functions? *Exp. Dermatol.* **2014**, *23*, 692–695. [CrossRef] [PubMed]

22. Feng, J.; Luo, J.; Yang, P.; Du, J.; Kim, B.S.; Hu, H. Piezo2 channel-Merkel cell signaling modulates the conversion of touch to itch. *Science* **2018**, *360*, 530–533. [CrossRef] [PubMed]

23. Simpson, C.L.; Patel, D.M.; Green, K.J. Deconstructing the skin: Cytoarchitectural determinants of epidermal morphogenesis. *Nat. Rev. Mol. Cell Biol.* **2011**, *12*, 565–580. [CrossRef] [PubMed]

24. Tsuruta, D.; Hashimoto, T.; Hamill, K.J.; Jones, J.C. Hemidesmosomes and focal contact proteins: Functions and cross-talk in keratinocytes, bullous diseases and wound healing. *J. Dermatol. Sci.* **2011**, *62*, 1–7. [CrossRef] [PubMed]

25. Smith, L. Histopathologic characteristics and ultrastructure of aging skin. *Cutis* **1989**, *43*, 414–424.

26. Rittie, L.; Fisher, G.J. Natural and sun-induced aging of human skin. *Cold Spring Harb. Perspect. Med.* **2015**, *5*, a015370. [CrossRef]

27. Haydont, V.; Bernard, B.A.; Fortunel, N.O. Age-related evolutions of the dermis: Clinical signs, fibroblast and extracellular matrix dynamics. *Mech. Ageing Dev.* **2019**, *177*, 150–156. [CrossRef]

28. Varani, J.; Schuger, L.; Dame, M.K.; Leonard, C.; Fligiel, S.E.; Kang, S.; Fisher, G.J.; Voorhees, J.J. Reduced fibroblast interaction with intact collagen as a mechanism for depressed collagen synthesis in photodamaged skin. *J. Investig. Dermatol.* **2004**, *122*, 1471–1479. [CrossRef]

29. Gu, Y.; Han, J.; Jiang, C.; Zhang, Y. Biomarkers, oxidative stress and autophagy in skin aging. *Ageing Res. Rev.* **2020**, *59*, 101036. [CrossRef]

30. Fisher, G.J.; Kang, S.; Varani, J.; Bata-Csorgo, Z.; Wan, Y.; Datta, S.; Voorhees, J.J. Mechanisms of photoaging and chronological skin aging. *Arch. Dermatol.* **2002**, *138*, 1462–1470. [CrossRef]

31. Quintanilla-Dieck, M.J.; Codriansky, K.; Keady, M.; Bhawan, J.; Runger, T.M. Expression and regulation of cathepsin K in skin fibroblasts. *Exp. Dermatol.* **2009**, *18*, 596–602. [CrossRef] [PubMed]

32. Codriansky, K.A.; Quintanilla-Dieck, M.J.; Gan, S.; Keady, M.; Bhawan, J.; Runger, T.M. Intracellular degradation of elastin by cathepsin K in skin fibroblasts–a possible role in photoaging. *Photochem. Photobiol.* **2009**, *85*, 1356–1363. [CrossRef] [PubMed]

33. Salzer, M.C.; Lafzi, A.; Berenguer-Llergo, A.; Youssif, C.; Castellanos, A.; Solanas, G.; Peixoto, F.O.; Stephan-Otto Attolini, C.; Prats, N.; Aguilera, M.; et al. Identity Noise and Adipogenic Traits Characterize Dermal Fibroblast Aging. *Cell* **2018**, *175*, 1575–1590.e22. [CrossRef] [PubMed]

34. Marsh, E.; Gonzalez, D.G.; Lathrop, E.A.; Boucher, J.; Greco, V. Positional Stability and Membrane Occupancy Define Skin Fibroblast Homeostasis In Vivo. *Cell* **2018**, *175*, 1620–1633.e13. [CrossRef] [PubMed]

35. Fuchs, E.; Raghavan, S. Getting under the skin of epidermal morphogenesis. *Nat. Rev. Genet.* **2002**, *3*, 199–209. [CrossRef]

36. Chuong, C.M.; Randall, V.A.; Widelitz, R.B.; Wu, P.; Jiang, T.X. Physiological regeneration of skin appendages and implications for regenerative medicine. *Physiology (Bethesda)* **2012**, *27*, 61–72. [CrossRef]

37. Ji, J.; Ho, B.S.; Qian, G.; Xie, X.M.; Bigliardi, P.L.; Bigliardi-Qi, M. Aging in hair follicle stem cells and niche microenvironment. *J. Dermatol.* **2017**, *44*, 1097–1104. [CrossRef]

38. Sinclair, R.; Chapman, A.; Magee, J. The lack of significant changes in scalp hair follicle density with advancing age. *Br. J. Dermatol.* **2005**, *152*, 646–649. [CrossRef]

39. Whiting, D.A. How real is senescent alopecia? A histopathologic approach. *Clin. Dermatol.* **2011**, *29*, 49–53. [CrossRef]

40. Giangreco, A.; Qin, M.; Pintar, J.E.; Watt, F.M. Epidermal stem cells are retained in vivo throughout skin aging. *Aging Cell* **2008**, *7*, 250–259. [CrossRef]

41. Chen, C.C.; Murray, P.J.; Jiang, T.X.; Plikus, M.V.; Chang, Y.T.; Lee, O.K.; Widelitz, R.B.; Chuong, C.M. Regenerative hair waves in aging mice and extra-follicular modulators follistatin, dkk1, and sfrp4. *J. Investig. Dermatol.* **2014**, *134*, 2086–2096. [CrossRef] [PubMed]

42. Tobin, D.J.; Paus, R. Graying: Gerontobiology of the hair follicle pigmentary unit. *Exp. Gerontol.* **2001**, *36*, 29–54. [CrossRef]

43. Nishimura, E.K.; Granter, S.R.; Fisher, D.E. Mechanisms of hair graying: Incomplete melanocyte stem cell maintenance in the niche. *Science* **2005**, *307*, 720–724. [CrossRef] [PubMed]

44. Baker, L.B. Physiology of sweat gland function: The roles of sweating and sweat composition in human health. *Temperature (Austin)* **2019**, *6*, 211–259. [CrossRef] [PubMed]

45. Kenney, W.L.; Fowler, S.R. Methylcholine-activated eccrine sweat gland density and output as a function of age. *J. Appl. Physiol. (1985)* **1988**, *65*, 1082–1086. [CrossRef] [PubMed]

46. Sukseree, S.; Bergmann, S.; Pajdzik, K.; Sipos, W.; Gruber, F.; Tschachler, E.; Eckhart, L. Suppression of Epithelial Autophagy Compromises the Homeostasis of Sweat Glands during Aging. *J. Investig. Dermatol.* **2018**, *138*, 2061–2063. [CrossRef]

47. Inoue, Y.; Nakao, M.; Araki, T.; Murakami, H. Regional differences in the sweating responses of older and younger men. *J. Appl. Physiol. (1985)* **1991**, *71*, 2453–2459. [CrossRef]

48. Zouboulis, C.C.; Picardo, M.; Ju, Q.; Kurokawa, I.; Torocsik, D.; Biro, T.; Schneider, M.R. Beyond acne: Current aspects of sebaceous gland biology and function. *Rev. Endocr. Metab. Disord.* **2016**, *17*, 319–334. [CrossRef]

49. Kabashima, K.; Honda, T.; Ginhoux, F.; Egawa, G. The immunological anatomy of the skin. *Nat. Rev. Immunol.* **2019**, *19*, 19–30. [CrossRef]

50. Schneider, M.R.; Paus, R. Sebocytes, multifaceted epithelial cells: Lipid production and holocrine secretion. *Int. J. Biochem. Cell Biol.* **2010**, *42*, 181–185. [CrossRef]

51. Makrantonaki, E.; Zouboulis, C.C.; German National Genome Research. The skin as a mirror of the aging process in the human organism—State of the art and results of the aging research in the German National Genome Research Network 2 (NGFN-2). *Exp. Gerontol.* **2007**, *42*, 879–886. [CrossRef] [PubMed]

52. Fimmel, S.; Devermann, L.; Herrmann, A.; Zouboulis, C. GRO-alpha: A potential marker for cancer and aging silenced by RNA interference. *Ann. N. Y. Acad. Sci.* **2007**, *1119*, 176–189. [CrossRef] [PubMed]

53. Crivellari, I.; Sticozzi, C.; Belmonte, G.; Muresan, X.M.; Cervellati, F.; Pecorelli, A.; Cavicchio, C.; Maioli, E.; Zouboulis, C.C.; Benedusi, M.; et al. SRB1 as a new redox target of cigarette smoke in human sebocytes. *Free Radic. Biol. Med.* **2017**, *102*, 47–56. [CrossRef] [PubMed]

54. Valacchi, G.; Davis, P.A.; Khan, E.M.; Lanir, R.; Maioli, E.; Pecorelli, A.; Cross, C.E.; Goldkorn, T. Cigarette smoke exposure causes changes in Scavenger Receptor B1 level and distribution in lung cells. *Int. J. Biochem. Cell Biol.* **2011**, *43*, 1065–1070. [CrossRef] [PubMed]

55. Makrantonaki, E.; Adjaye, J.; Herwig, R.; Brink, T.C.; Groth, D.; Hultschig, C.; Lehrach, H.; Zouboulis, C.C. Age-specific hormonal decline is accompanied by transcriptional changes in human sebocytes in vitro. *Aging Cell* **2006**, *5*, 331–344. [CrossRef] [PubMed]

56. Reiser, K.; McGee, C.; Rucker, R.; McDonald, R. Effects of aging and caloric restriction on extracellular matrix biosynthesis in a model of injury repair in rats. *J. Gerontol. A Biol. Sci. Med. Sci.* **1995**, *50A*, B40–B47. [CrossRef] [PubMed]

57. Roth, G.S.; Kowatch, M.A.; Hengemihle, J.; Ingram, D.K.; Spangler, E.L.; Johnson, L.K.; Lane, M.A. Effect of age and caloric restriction on cutaneous wound closure in rats and monkeys. *J. Gerontol. A Biol. Sci. Med. Sci.* **1997**, *52*, B98–B102. [CrossRef]

58. Hayati, F.; Maleki, M.; Pourmohammad, M.; Sardari, K.; Mohri, M.; Afkhami, A. Influence of Short-term, Repeated Fasting on the Skin Wound Healing of Female Mice. *Wounds* **2011**, *23*, 38–43.

59. Sell, D.R.; Lane, M.A.; Obrenovich, M.E.; Mattison, J.A.; Handy, A.; Ingram, D.K.; Cutler, R.G.; Roth, G.S.; Monnier, V.M. The effect of caloric restriction on glycation and glycoxidation in skin collagen of nonhuman primates. *J. Gerontol. A Biol. Sci. Med. Sci.* **2003**, *58*, 508–516. [CrossRef]

60. Reed, M.J.; Penn, P.E.; Li, Y.; Birnbaum, R.; Vernon, R.B.; Johnson, T.S.; Pendergrass, W.R.; Sage, E.H.; Abrass, I.B.; Wolf, N.S. Enhanced cell proliferation and biosynthesis mediate improved wound repair in refed, caloric-restricted mice. *Mech. Ageing Dev.* **1996**, *89*, 21–43. [CrossRef]

61. Hunt, N.D.; Li, G.D.; Zhu, M.; Miller, M.; Levette, A.; Chachich, M.E.; Spangler, E.L.; Allard, J.S.; Hyun, D.H.; Ingram, D.K.; et al. Effect of calorie restriction and refeeding on skin wound healing in the rat. *Age (Dordr.)* **2012**, *34*, 1453–1458. [CrossRef] [PubMed]

62. Forni, M.F.; Peloggia, J.; Braga, T.T.; Chinchilla, J.E.O.; Shinohara, J.; Navas, C.A.; Camara, N.O.S.; Kowaltowski, A.J. Caloric Restriction Promotes Structural and Metabolic Changes in the Skin. *Cell Rep.* **2017**, *20*, 2678–2692. [CrossRef] [PubMed]

63. Bhattacharyya, T.K.; Merz, M.; Thomas, J.R. Modulation of cutaneous aging with calorie restriction in Fischer 344 rats: A histological study. *Arch. Facial Plast. Surg.* **2005**, *7*, 12–16. [CrossRef] [PubMed]

64. Bhattacharyya, T.K.; Hsia, Y.; Weeks, D.M.; Dixon, T.K.; Lepe, J.; Thomas, J.R. Association of Diet With Skin Histological Features in UV-B-Exposed Mice. *JAMA Facial Plast. Surg.* **2017**, *19*, 399–405. [CrossRef] [PubMed]

65. Hsu, Y.C.; Li, L.; Fuchs, E. Emerging interactions between skin stem cells and their niches. *Nat. Med.* **2014**, *20*, 847–856. [CrossRef]

66. Matsumura, H.; Mohri, Y.; Binh, N.T.; Morinaga, H.; Fukuda, M.; Ito, M.; Kurata, S.; Hoeijmakers, J.; Nishimura, E.K. Hair follicle aging is driven by transepidermal elimination of stem cells via COL17A1 proteolysis. *Science* **2016**, *351*, aad4395. [CrossRef]

67. Stern, M.M.; Bickenbach, J.R. Epidermal stem cells are resistant to cellular aging. *Aging Cell* **2007**, *6*, 439–452. [CrossRef]

68. Schultz, M.B.; Sinclair, D.A. When stem cells grow old: Phenotypes and mechanisms of stem cell aging. *Development* **2016**, *143*, 3–14. [CrossRef]

69. Solanas, G.; Peixoto, F.O.; Perdiguero, E.; Jardi, M.; Ruiz-Bonilla, V.; Datta, D.; Symeonidi, A.; Castellanos, A.; Welz, P.S.; Caballero, J.M.; et al. Aged Stem Cells Reprogram Their Daily Rhythmic Functions to Adapt to Stress. *Cell* **2017**, *170*, 678–692.e20. [CrossRef]

70. Birt, D.F.; Pinch, H.J.; Barnett, T.; Phan, A.; Dimitroff, K. Inhibition of skin tumor promotion by restriction of fat and carbohydrate calories in SENCAR mice. *Cancer Res.* **1993**, *53*, 27–31.

71. Hopper, B.D.; Przybyszewski, J.; Chen, H.W.; Hammer, K.D.; Birt, D.F. Effect of ultraviolet B radiation on activator protein 1 constituent proteins and modulation by dietary energy restriction in SKH-1 mouse skin. *Mol. Carcinog.* **2009**, *48*, 843–852. [CrossRef] [PubMed]

72. Xie, L.; Jiang, Y.; Ouyang, P.; Chen, J.; Doan, H.; Herndon, B.; Sylvester, J.E.; Zhang, K.; Molteni, A.; Reichle, M.; et al. Effects of dietary calorie restriction or exercise on the PI3K and Ras signaling pathways in the skin of mice. *J. Biol. Chem.* **2007**, *282*, 28025–28035. [CrossRef] [PubMed]

73. Perkins, S.N.; Hursting, S.D.; Phang, J.M.; Haines, D.C. Calorie restriction reduces ulcerative dermatitis and infection-related mortality in p53-deficient and wild-type mice. *J. Investig. Dermatol.* **1998**, *111*, 292–296. [CrossRef] [PubMed]

74. Randhawa, M.; Sangar, V.; Tucker-Samaras, S.; Southall, M. Metabolic signature of sun exposed skin suggests catabolic pathway overweighs anabolic pathway. *PLoS ONE* **2014**, *9*, e90367. [CrossRef]

75. Kuehne, A.; Hildebrand, J.; Soehle, J.; Wenck, H.; Terstegen, L.; Gallinat, S.; Knott, A.; Winnefeld, M.; Zamboni, N. An integrative metabolomics and transcriptomics study to identify metabolic alterations in aged skin of humans in vivo. *BMC Genom.* **2017**, *18*, 169. [CrossRef]

76. Simpson, S.J.; Le Couteur, D.G.; Raubenheimer, D.; Solon-Biet, S.M.; Cooney, G.J.; Cogger, V.C.; Fontana, L. Dietary protein, aging and nutritional geometry. *Ageing Res. Rev.* **2017**, *39*, 78–86. [CrossRef]

77. Longo, V.D.; Mattson, M.P. Fasting: Molecular mechanisms and clinical applications. *Cell Metab.* **2014**, *19*, 181–192. [CrossRef]

78. Bragazzi, N.L.; Sellami, M.; Salem, I.; Conic, R.; Kimak, M.; Pigatto, P.D.M.; Damiani, G. Fasting and Its Impact on Skin Anatomy, Physiology, and Physiopathology: A Comprehensive Review of the Literature. *Nutrients* **2019**, *11*, 249. [CrossRef]

79. Brandhorst, S.; Choi, I.Y.; Wei, M.; Cheng, C.W.; Sedrakyan, S.; Navarrete, G.; Dubeau, L.; Yap, L.P.; Park, R.; Vinciguerra, M.; et al. A Periodic Diet that Mimics Fasting Promotes Multi-System Regeneration, Enhanced Cognitive Performance, and Healthspan. *Cell Metab.* **2015**, *22*, 86–99. [CrossRef]

80. Mair, W.; Piper, M.D.; Partridge, L. Calories do not explain extension of life span by dietary restriction in Drosophila. *PLoS Biol.* **2005**, *3*, e223. [CrossRef]

81. Solon-Biet, S.M.; McMahon, A.C.; Ballard, J.W.; Ruohonen, K.; Wu, L.E.; Cogger, V.C.; Warren, A.; Huang, X.; Pichaud, N.; Melvin, R.G.; et al. The ratio of macronutrients, not caloric intake, dictates cardiometabolic health, aging, and longevity in ad libitum-fed mice. *Cell Metab.* **2014**, *19*, 418–430. [CrossRef] [PubMed]

82. Grandison, R.C.; Piper, M.D.; Partridge, L. Amino-acid imbalance explains extension of lifespan by dietary restriction in Drosophila. *Nature* **2009**, *462*, 1061–1064. [CrossRef] [PubMed]

83. Fontana, L.; Cummings, N.E.; Arriola Apelo, S.I.; Neuman, J.C.; Kasza, I.; Schmidt, B.A.; Cava, E.; Spelta, F.; Tosti, V.; Syed, F.A.; et al. Decreased Consumption of Branched-Chain Amino Acids Improves Metabolic Health. *Cell Rep.* **2016**, *16*, 520–530. [CrossRef] [PubMed]

84. Trocha, K.; Kip, P.; MacArthur, M.R.; Mitchell, S.J.; Longchamp, A.; Trevino-Villarreal, J.H.; Tao, M.; Bredella, M.A.; De Amorim Bernstein, K.; Mitchell, J.R.; et al. Preoperative Protein or Methionine Restriction Preserves Wound Healing and Reduces Hyperglycemia. *J. Surg. Res.* **2019**, *235*, 216–222. [CrossRef] [PubMed]

85. Wu, Q.; Shuang, E.; Yamamoto, K.; Tsuduki, T. Carbohydrate-restricted diet promotes skin senescence in senescence-accelerated prone mice. *Biogerontology* **2019**, *20*, 71–82. [CrossRef] [PubMed]

86. Okouchi, R.; Sakanoi, Y.; Tsuduki, T. The Effect of Carbohydrate-Restricted Diets on the Skin Aging of Mice. *J. Nutr. Sci. Vitaminol. (Tokyo)* **2019**, *65*, S67–S71. [CrossRef]

87. Singh, S.; Singh, A.K.; Garg, G.; Rizvi, S.I. Fisetin as a caloric restriction mimetic protects rat brain against aging induced oxidative stress, apoptosis and neurodegeneration. *Life Sci.* **2018**, *193*, 171–179. [CrossRef]

88. Choi, K.M.; Lee, H.L.; Kwon, Y.Y.; Kang, M.S.; Lee, S.K.; Lee, C.K. Enhancement of mitochondrial function correlates with the extension of lifespan by caloric restriction and caloric restriction mimetics in yeast. *Biochem. Biophys. Res. Commun.* **2013**, *441*, 236–242. [CrossRef]

89. Shintani, H.; Shintani, T.; Ashida, H.; Sato, M. Calorie Restriction Mimetics: Upstream-Type Compounds for Modulating Glucose Metabolism. *Nutrients* **2018**, *10*, 1821. [CrossRef]

90. Madeo, F.; Carmona-Gutierrez, D.; Hofer, S.J.; Kroemer, G. Caloric Restriction Mimetics against Age-Associated Disease: Targets, Mechanisms, and Therapeutic Potential. *Cell Metab.* **2019**, *29*, 592–610. [CrossRef]

91. Kaeberlein, M.; McVey, M.; Guarente, L. The SIR2/3/4 complex and SIR2 alone promote longevity in Saccharomyces cerevisiae by two different mechanisms. *Genes Dev.* **1999**, *13*, 2570–2580. [CrossRef] [PubMed]

92. Rogina, B.; Helfand, S.L. Sir2 mediates longevity in the fly through a pathway related to calorie restriction. *Proc. Natl. Acad. Sci. USA* **2004**, *101*, 15998–16003. [CrossRef] [PubMed]

93. Kanfi, Y.; Naiman, S.; Amir, G.; Peshti, V.; Zinman, G.; Nahum, L.; Bar-Joseph, Z.; Cohen, H.Y. The sirtuin SIRT6 regulates lifespan in male mice. *Nature* **2012**, *483*, 218–221. [CrossRef] [PubMed]

94. Lee, S.H.; Lee, J.H.; Lee, H.Y.; Min, K.J. Sirtuin signaling in cellular senescence and aging. *BMB Rep.* **2019**, *52*, 24–34. [CrossRef] [PubMed]

95. Howitz, K.T.; Bitterman, K.J.; Cohen, H.Y.; Lamming, D.W.; Lavu, S.; Wood, J.G.; Zipkin, R.E.; Chung, P.; Kisielewski, A.; Zhang, L.L.; et al. Small molecule activators of sirtuins extend Saccharomyces cerevisiae lifespan. *Nature* **2003**, *425*, 191–196. [CrossRef]

96. Baur, J.A.; Pearson, K.J.; Price, N.L.; Jamieson, H.A.; Lerin, C.; Kalra, A.; Prabhu, V.V.; Allard, J.S.; Lopez-Lluch, G.; Lewis, K.; et al. Resveratrol improves health and survival of mice on a high-calorie diet. *Nature* **2006**, *444*, 337–342. [CrossRef]

97. Collins, J.J.; Evason, K.; Kornfeld, K. Pharmacology of delayed aging and extended lifespan of Caenorhabditis elegans. *Exp. Gerontol.* **2006**, *41*, 1032–1039. [CrossRef]

98. Bass, T.M.; Weinkove, D.; Houthoofd, K.; Gems, D.; Partridge, L. Effects of resveratrol on lifespan in Drosophila melanogaster and Caenorhabditis elegans. *Mech. Ageing Dev.* **2007**, *128*, 546–552. [CrossRef]

99. Rascon, B.; Hubbard, B.P.; Sinclair, D.A.; Amdam, G.V. The lifespan extension effects of resveratrol are conserved in the honey bee and may be driven by a mechanism related to caloric restriction. *Aging* **2012**, *4*, 499–508. [CrossRef]

100. Wang, C.; Wheeler, C.T.; Alberico, T.; Sun, X.; Seeberger, J.; Laslo, M.; Spangler, E.; Kern, B.; de Cabo, R.; Zou, S. The effect of resveratrol on lifespan depends on both gender and dietary nutrient composition in Drosophila melanogaster. *Age (Dordr. Neth.)* **2013**, *35*, 69–81. [CrossRef]

101. Golubtsova, N.N.; Filippov, F.N.; Gunin, A.G. Age-Related Changes in the Content of Sirtuin 1 in Human Dermal Fibroblasts. *Adv. Gerontol.* **2017**, *7*, 302–306. [CrossRef]

102. Lago, J.C.; Puzzi, M.B. The effect of aging in primary human dermal fibroblasts. *PLoS ONE* **2019**, *14*, e0219165. [CrossRef] [PubMed]

103. Lei, D.; Huang, Y.; Xie, H.; Yi, Y.; Long, J.; Lin, S.; Huang, C.; Jian, D.; Li, J. Fluorofenidone inhibits UV-A induced senescence in human dermal fibroblasts via the mammalian target of rapamycin-dependent SIRT1 pathway. *J. Dermatol.* **2018**, *45*, 791–798. [CrossRef] [PubMed]

104. Kim, H.K. Protective Effect of Garlic on Cellular Senescence in UVB-Exposed HaCaT Human Keratinocytes. *Nutrients* **2016**, *8*, 464. [CrossRef] [PubMed]

105. Chung, K.W.; Choi, Y.J.; Park, M.H.; Jang, E.J.; Kim, D.H.; Park, B.H.; Yu, B.P.; Chung, H.Y. Molecular Insights into SIRT1 Protection Against UVB-Induced Skin Fibroblast Senescence by Suppression of Oxidative Stress and p53 Acetylation. *J. Gerontol. A Biol. Sci. Med. Sci.* **2015**, *70*, 959–968. [CrossRef] [PubMed]

106. Qiang, L.; Sample, A.; Liu, H.; Wu, X.; He, Y.Y. Epidermal SIRT1 regulates inflammation, cell migration, and wound healing. *Sci. Rep.* **2017**, *7*, 14110. [CrossRef] [PubMed]

107. Christovam, A.C.; Theodoro, V.; Mendonca, F.A.S.; Esquisatto, M.A.M.; Dos Santos, G.M.T.; do Amaral, M.E.C. Activators of SIRT1 in wound repair: An animal model study. *Arch. Dermatol. Res.* **2019**, *311*, 193–201. [CrossRef]

108. Zhao, P.; Sui, B.D.; Liu, N.; Lv, Y.J.; Zheng, C.X.; Lu, Y.B.; Huang, W.T.; Zhou, C.H.; Chen, J.; Pang, D.L.; et al. Anti-aging pharmacology in cutaneous wound healing: Effects of metformin, resveratrol, and rapamycin by local application. *Aging Cell* **2017**, *16*, 1083–1093. [CrossRef]

109. Ratz-Lyko, A.; Arct, J. Resveratrol as an active ingredient for cosmetic and dermatological applications: A review. *J. Cosmet. Laser Ther.* **2019**, *21*, 84–90. [CrossRef]

110. Ido, Y.; Duranton, A.; Lan, F.; Weikel, K.A.; Breton, L.; Ruderman, N.B. Resveratrol prevents oxidative stress-induced senescence and proliferative dysfunction by activating the AMPK-FOXO3 cascade in cultured primary human keratinocytes. *PLoS ONE* **2015**, *10*, e0115341. [CrossRef]

111. Boo, Y.C. Human Skin Lightening Efficacy of Resveratrol and Its Analogs: From in Vitro Studies to Cosmetic Applications. *Antioxidants (Basel)* **2019**, *8*, 332. [CrossRef] [PubMed]

112. Alonso, C.; Marti, M.; Barba, C.; Carrer, V.; Rubio, L.; Coderch, L. Skin permeation and antioxidant efficacy of topically applied resveratrol. *Arch. Dermatol. Res.* **2017**, *309*, 423–431. [CrossRef] [PubMed]

113. Nastiti, C.; Ponto, T.; Mohammed, Y.; Roberts, M.S.; Benson, H.A.E. Novel Nanocarriers for Targeted Topical Skin Delivery of the Antioxidant Resveratrol. *Pharmaceutics* **2020**, *12*, 108. [CrossRef] [PubMed]

114. Ravikumar, P.; Katariya, M.; Patil, S.; Tatke, P.; Pillai, R. Skin delivery of resveratrol encapsulated lipidic formulation for melanoma chemoprevention. *J. Microencapsul.* **2019**, *36*, 535–551. [CrossRef] [PubMed]

115. Negi, P.; Aggarwal, M.; Sharma, G.; Rathore, C.; Sharma, G.; Singh, B.; Katare, O.P. Niosome-based hydrogel of resveratrol for topical applications: An effective therapy for pain related disorder(s). *Biomed. Pharmacother.* **2017**, *88*, 480–487. [CrossRef]

116. Pentek, T.; Newenhouse, E.; O'Brien, B.; Chauhan, A.S. Development of a Topical Resveratrol Formulation for Commercial Applications Using Dendrimer Nanotechnology. *Molecules* **2017**, *22*, 137. [CrossRef]

117. Kirpichnikov, D.; McFarlane, S.I.; Sowers, J.R. Metformin: An update. *Ann. Intern. Med.* **2002**, *137*, 25–33. [CrossRef]

118. Cabreiro, F.; Au, C.; Leung, K.Y.; Vergara-Irigaray, N.; Cocheme, H.M.; Noori, T.; Weinkove, D.; Schuster, E.; Greene, N.D.; Gems, D. Metformin retards aging in C. elegans by altering microbial folate and methionine metabolism. *Cell* **2013**, *153*, 228–239. [CrossRef]

119. De Haes, W.; Frooninckx, L.; Van Assche, R.; Smolders, A.; Depuydt, G.; Billen, J.; Braeckman, B.P.; Schoofs, L.; Temmerman, L. Metformin promotes lifespan through mitohormesis via the peroxiredoxin PRDX-2. *Proc. Natl. Acad. Sci. USA* **2014**, *111*, E2501–E2509. [CrossRef]

120. Wu, L.; Zhou, B.; Oshiro-Rapley, N.; Li, M.; Paulo, J.A.; Webster, C.M.; Mou, F.; Kacergis, M.C.; Talkowski, M.E.; Carr, C.E.; et al. An Ancient, Unified Mechanism for Metformin Growth Inhibition in C. elegans and Cancer. *Cell* **2016**, *167*, 1705–1718.e1713. [CrossRef]

121. Slack, C.; Foley, A.; Partridge, L. Activation of AMPK by the putative dietary restriction mimetic metformin is insufficient to extend lifespan in Drosophila. *PLoS ONE* **2012**, *7*, e47699. [CrossRef] [PubMed]

122. Onken, B.; Driscoll, M. Metformin induces a dietary restriction-like state and the oxidative stress response to extend C. elegans Healthspan via AMPK, LKB1, and SKN-1. *PLoS ONE* **2010**, *5*, e8758. [CrossRef] [PubMed]

123. Martin-Montalvo, A.; Mercken, E.M.; Mitchell, S.J.; Palacios, H.H.; Mote, P.L.; Scheibye-Knudsen, M.; Gomes, A.P.; Ward, T.M.; Minor, R.K.; Blouin, M.J.; et al. Metformin improves healthspan and lifespan in mice. *Nat. Commun.* **2013**, *4*, 2192. [CrossRef] [PubMed]

124. Faure, M.; Bertoldo, M.J.; Khoueiry, R.; Bongrani, A.; Brion, F.; Giulivi, C.; Dupont, J.; Froment, P. Metformin in Reproductive Biology. *Front. Endocrinol. (Lausanne)* **2018**, *9*, 675. [CrossRef]

125. Wu, K.; Tian, R.; Huang, J.; Yang, Y.; Dai, J.; Jiang, R.; Zhang, L. Metformin alleviated endotoxemia-induced acute lung injury via restoring AMPK-dependent suppression of mTOR. *Chem. Biol. Interact.* **2018**, *291*, 1–6. [CrossRef]

126. Mihaylova, M.M.; Shaw, R.J. The AMPK signalling pathway coordinates cell growth, autophagy and metabolism. *Nat. Cell Biol.* **2011**, *13*, 1016–1023. [CrossRef]

127. Karnewar, S.; Neeli, P.K.; Panuganti, D.; Kotagiri, S.; Mallappa, S.; Jain, N.; Jerald, M.K.; Kotamraju, S. Metformin regulates mitochondrial biogenesis and senescence through AMPK mediated H3K79 methylation: Relevance in age-associated vascular dysfunction. *Biochim. Biophys. Acta Mol. Basis Dis.* **2018**, *1864*, 1115–1128. [CrossRef]

128. Kuang, Y.; Hu, B.; Feng, G.; Xiang, M.; Deng, Y.; Tan, M.; Li, J.; Song, J. Metformin prevents against oxidative stress-induced senescence in human periodontal ligament cells. *Biogerontology* **2020**, *21*, 13–27. [CrossRef]

129. Wu, C.L.; Qiang, L.; Han, W.; Ming, M.; Viollet, B.; He, Y.Y. Role of AMPK in UVB-induced DNA damage repair and growth control. *Oncogene* **2013**, *32*, 2682–2689. [CrossRef]

130. Crane, E.D.W.; Zhang, H.W.; O'Neil, G.L.; Crane, J.D. AMPK Restrains Epidermal Stem Cell Aging by Inhibiting mTORC1-Induced Hyper-Proliferation. *CELL-STEM-CELL-D-19-00666* **2019**. [CrossRef]

131. Qing, L.; Fu, J.; Wu, P.; Zhou, Z.; Yu, F.; Tang, J. Metformin induces the M2 macrophage polarization to accelerate the wound healing via regulating AMPK/mTOR/NLRP3 inflammasome singling pathway. *Am. J. Transl. Res.* **2019**, *11*, 655–668. [PubMed]

132. Rostamkalaei, S.S.; Akbari, J.; Saeedi, M.; Morteza-Semnani, K.; Nokhodchi, A. Topical gel of Metformin solid lipid nanoparticles: A hopeful promise as a dermal delivery system. *Colloids Surf. B Biointerfaces* **2019**, *175*, 150–157. [CrossRef] [PubMed]

133. Tokunaga, C.; Yoshino, K.; Yonezawa, K. mTOR integrates amino acid- and energy-sensing pathways. *Biochem. Biophys. Res. Commun.* **2004**, *313*, 443–446. [CrossRef] [PubMed]

134. Salminen, A.; Kaarniranta, K. Regulation of the aging process by autophagy. *Trends Mol. Med.* **2009**, *15*, 217–224. [CrossRef]

135. Kim, Y.C.; Guan, K.L. mTOR: A pharmacologic target for autophagy regulation. *J. Clin. Investig.* **2015**, *125*, 25–32. [CrossRef]

136. Munson, M.J.; Ganley, I.G. MTOR, PIK3C3, and autophagy: Signaling the beginning from the end. *Autophagy* **2015**, *11*, 2375–2376. [CrossRef]

137. Dodds, S.G.; Livi, C.B.; Parihar, M.; Hsu, H.K.; Benavides, A.D.; Morris, J.; Javors, M.; Strong, R.; Christy, B.; Hasty, P.; et al. Adaptations to chronic rapamycin in mice. *Pathobiol. Aging Age Relat. Dis.* **2016**, *6*, 31688. [CrossRef]

138. Swindell, W.R. Rapamycin in mice. *Aging* **2017**, *9*, 1941–1942. [CrossRef]

139. Blagosklonny, M.V. Fasting and rapamycin: Diabetes versus benevolent glucose intolerance. *Cell Death Dis.* **2019**, *10*, 607. [CrossRef] [PubMed]

140. Papadopoli, D.; Boulay, K.; Kazak, L.; Pollak, M.; Mallette, F.; Topisirovic, I.; Hulea, L. mTOR as a central regulator of lifespan and aging. *F1000Research* **2019**, *8*. [CrossRef] [PubMed]

141. Choi, Y.J.; Moon, K.M.; Chung, K.W.; Jeong, J.W.; Park, D.; Kim, D.H.; Yu, B.P.; Chung, H.Y. The underlying mechanism of proinflammatory NF-kappaB activation by the mTORC2/Akt/IKKalpha pathway during skin aging. *Oncotarget* **2016**, *7*, 52685–52694. [CrossRef] [PubMed]

142. Tassone, B.; Saoncella, S.; Neri, F.; Ala, U.; Brusa, D.; Magnuson, M.A.; Provero, P.; Oliviero, S.; Riganti, C.; Calautti, E. Rictor/mTORC2 deficiency enhances keratinocyte stress tolerance via mitohormesis. *Cell Death Differ.* **2017**, *24*, 731–746. [CrossRef] [PubMed]

143. Buerger, C. Epidermal mTORC1 Signaling Contributes to the Pathogenesis of Psoriasis and Could Serve as a Therapeutic Target. *Front. Immunol.* **2018**, *9*, 2786. [CrossRef] [PubMed]

144. Burger, C.; Shirsath, N.; Lang, V.; Diehl, S.; Kaufmann, R.; Weigert, A.; Han, Y.Y.; Ringel, C.; Wolf, P. Blocking mTOR Signalling with Rapamycin Ameliorates Imiquimod-induced Psoriasis in Mice. *Acta Derm. Venereol.* **2017**, *97*, 1087–1094. [CrossRef]

145. Qin, D.; Ren, R.; Jia, C.; Lu, Y.; Yang, Q.; Chen, L.; Wu, X.; Zhu, J.; Guo, Y.; Yang, P.; et al. Rapamycin Protects Skin Fibroblasts from Ultraviolet B-Induced Photoaging by Suppressing the Production of Reactive Oxygen Species. *Cell Physiol. Biochem.* **2018**, *46*, 1849–1860. [CrossRef]

146. Chung, C.L.; Lawrence, I.; Hoffman, M.; Elgindi, D.; Nadhan, K.; Potnis, M.; Jin, A.; Sershon, C.; Binnebose, R.; Lorenzini, A.; et al. Topical rapamycin reduces markers of senescence and aging in human skin: An exploratory, prospective, randomized trial. *Geroscience* **2019**, *41*, 861–869. [CrossRef]

147. Sung, B.; Park, S.; Yu, B.P.; Chung, H.Y. Modulation of PPAR in aging, inflammation, and calorie restriction. *J. Gerontol. A Biol. Sci. Med. Sci.* **2004**, *59*, 997–1006. [CrossRef]

148. Bishop-Bailey, D. Peroxisome proliferator-activated receptors in the cardiovascular system. *Br. J. Pharmacol.* **2000**, *129*, 823–834. [CrossRef]

149. Ramot, Y.; Mastrofrancesco, A.; Camera, E.; Desreumaux, P.; Paus, R.; Picardo, M. The role of PPARgamma-mediated signalling in skin biology and pathology: New targets and opportunities for clinical dermatology. *Exp. Dermatol.* **2015**, *24*, 245–251. [CrossRef]

150. Masternak, M.M.; Bartke, A. PPARs in Calorie Restricted and Genetically Long-Lived Mice. *PPAR Res.* **2007**, *2007*, 28436. [CrossRef]

151. Kim, E.J.; Jin, X.J.; Kim, Y.K.; Oh, I.K.; Kim, J.E.; Park, C.H.; Chung, J.H. UV decreases the synthesis of free fatty acids and triglycerides in the epidermis of human skin in vivo, contributing to development of skin photoaging. *J. Dermatol. Sci.* **2010**, *57*, 19–26. [CrossRef] [PubMed]

152. Adachi, Y.; Hatano, Y.; Sakai, T.; Fujiwara, S. Expressions of peroxisome proliferator-activated receptors (PPARs) are directly influenced by permeability barrier abrogation and inflammatory cytokines and depressed PPARalpha modulates expressions of chemokines and epidermal differentiation-related molecules in keratinocytes. *Exp. Dermatol.* **2013**, *22*, 606–608. [CrossRef] [PubMed]

153. Rizos, C.V.; Kei, A.; Elisaf, M.S. The current role of thiazolidinediones in diabetes management. *Arch. Toxicol.* **2016**, *90*, 1861–1881. [CrossRef] [PubMed]

154. Yki-Jarvinen, H. Thiazolidinediones. *N. Engl. J. Med.* **2004**, *351*, 1106–1118. [CrossRef] [PubMed]

155. Jeon, Y.; Jung, Y.; Youm, J.K.; Kang, K.S.; Kim, Y.K.; Kim, S.N. Abietic acid inhibits UVB-induced MMP-1 expression in human dermal fibroblast cells through PPARalpha/gamma dual activation. *Exp. Dermatol.* **2015**, *24*, 140–145. [CrossRef]

156. Jung, Y.R.; Lee, E.K.; Kim, D.H.; Park, C.H.; Park, M.H.; Jeong, H.O.; Yokozawa, T.; Tanaka, T.; Im, D.S.; Kim, N.D.; et al. Upregulation of Collagen Expression via PPARbeta/delta Activation in Aged Skin by Magnesium Lithospermate B from Salvia miltiorrhiza. *J. Nat. Prod.* **2015**, *78*, 2110–2115. [CrossRef]

157. Park, M.H.; Park, J.Y.; Lee, H.J.; Kim, D.H.; Chung, K.W.; Park, D.; Jeong, H.O.; Kim, H.R.; Park, C.H.; Kim, S.R.; et al. The novel PPAR alpha/gamma dual agonist MHY 966 modulates UVB-induced skin inflammation by inhibiting NF-kappaB activity. *PLoS ONE* **2013**, *8*, e76820. [CrossRef]

© 2020 by the author. Licensee MDPI, Basel, Switzerland. This article is an open access article distributed under the terms and conditions of the Creative Commons Attribution (CC BY) license (http://creativecommons.org/licenses/by/4.0/).

Article

Myocardial Dysfunction after Severe Food Restriction Is Linked to Changes in the Calcium-Handling Properties in Rats

Adriana Fernandes de Deus [1], Vítor Loureiro da Silva [1], Sérgio Luiz Borges de Souza [1], Gustavo Augusto Ferreira Mota [1], Paula Grippa Sant'Ana [1], Danielle Fernandes Vileigas [1], Ana Paula Lima-Leopoldo [2], André Soares Leopoldo [2], Dijon Henrique Salomé de Campos [1], Loreta Casquel de Tomasi [1], Carlos Roberto Padovani [3], Stephen C. Kolwicz Jr. [4] and Antonio Carlos Cicogna [1,*]

[1] Department of Internal Medicine, Botucatu Medical School, São Paulo State University, Botucatu 18618687, Brazil

[2] Department of Sports, Center of Physical Education and Sports, Federal University of Espírito Santo, Vitória 29075-910, Brazil

[3] Department of Biostatistics, Institute of Biosciences, São Paulo State University, Botucatu 18618970, Brazil

[4] Department of Health and Exercise Physiology, Ursinus College, Collegeville, PA 19426, USA

* Correspondence: ac.cicogna@unesp.br; Tel.: +55-(14)-3880-1618

Received: 18 July 2019; Accepted: 14 August 2019; Published: 22 August 2019

Abstract: Severe food restriction (FR) impairs cardiac performance, although the causative mechanisms remain elusive. Since proteins associated with calcium handling may contribute to cardiac dysfunction, this study aimed to evaluate whether severe FR results in alterations in the expression and activity of Ca^{2+}-handling proteins that contribute to impaired myocardial performance. Male 60-day-old Wistar–Kyoto rats were fed a control or restricted diet (50% reduction in the food consumed by the control group) for 90 days. Body weight, body fat pads, adiposity index, as well as the weights of the soleus muscle and lung, were obtained. Cardiac remodeling was assessed by morphological measures. The myocardial contractile performance was analyzed in isolated papillary muscles during the administration of extracellular Ca^{2+} and in the absence or presence of a sarcoplasmic reticulum Ca^{2+}-ATPase (SERCA2a) specific blocker. The expression of Ca^{2+}-handling regulatory proteins was analyzed via Western Blot. Severe FR resulted in a 50% decrease in body weight and adiposity measures. Cardiac morphometry was substantially altered, as heart weights were nearly twofold lower in FR rats. Papillary muscles isolated from FR hearts displayed mechanical dysfunction, including decreased developed tension and reduced contractility and relaxation. The administration of a SERCA2a blocker led to further decrements in contractile function in FR hearts, suggesting impaired SERCA2a activity. Moreover, the FR rats presented a lower expression of L-type Ca^{2+} channels. Therefore, myocardial dysfunction induced by severe food restriction is associated with changes in the calcium-handling properties in rats.

Keywords: malnutrition; heart impairment; papillary muscle assay; calcium transient proteins; SERCA2a; L-type calcium channel

---

1. Introduction

Previous literature demonstrated that food restriction (FR) from 10% to 40% increases longevity and prevents aging-related diseases such as diabetes mellitus, hypertension, and cancer [1–5]. It is suggested that an excessive availability of macronutrients, rather than a specific nutrient, results in increased oxidative stress [6]. Therefore, a reduction of total caloric intake via FR may result in beneficial adaptations that promote overall health. However, the FR associated to malnutrition in humans has detrimental health effects, particularly on cardiac function [7–10]. In experimental models, our previous studies [11–20] and studies from other laboratories [21–24] suggest that severe dietary restriction, around 50%, leads to malnutrition and depresses cardiac performance. In young and mature rats subjected to long-term severe FR, myocardial dysfunction was associated with mitochondrial failure [25], changes in ultrastructure, including myofibril density reduction and myofilament/Z line disorganization [17,20,26,27], as well as collagen accumulation [27,28] and β-adrenergic system changes [28]. Despite these findings, the intrinsic factors responsible for the impairment of cardiac function have not been elucidated.

Calcium handling is an essential process that facilitates myocardial contraction and relaxation. Previous work examined whether alterations in Ca^{2+} handling participates in the decline of myocardial function in response to malnutrition, but the results are divergent [12–17,29,30]. While one study suggested that cardiac ultrastructural changes were responsible for deleterious cardiac outcomes [17], several studies demonstrated that abnormal activity of Ca^{2+} handling specific regulatory proteins has a fundamental role in the negative repercussions of severe FR on myocardial function [13,15,16,19]. Although malnutrition was shown to alter the gene transcription process that forms proteins involved in Ca^{2+} handling [14,15,19,31], limited studies evaluated changes at the protein level. Work by De Tomasi et al. [15] focused on the L-type calcium channel, and found only a significant reduction in its protein expression in the hearts of FR rats. However, as it is known that the levels of transcripts and proteins are not always in direct correlation [32,33] and no study evaluated all the proteins involved in Ca^{2+} handling, additional work is required to understand the role that this mechanism plays in the development of cardiac dysfunction during long-term FR.

Thus, the aim of this study is to test the hypothesis that malnutrition induced by restricting the food intake causes the deterioration of myocardial function due to alterations in the expression of calcium-handling proteins and sarcoplasmic reticulum Ca^{2+}-ATPase (SERCA2a) activity. The results of our study demonstrate that 90 days of FR severely diminishes cardiac function in rats, which is in part due to decreased intracellular calcium influx through the L-type channel and depressed SERCA2a activity.

2. Materials and Methods

2.1. Animal Model and Experimental Protocol

Sixty-day-old male Wistar–Kyoto rats were obtained from UNICAMP—the State University of Campinas, Brazil. The animals were randomized into control (C) and food restriction (FR) groups. C animals (n = 14) received commercial chow (3.76% fat, 20.96% protein, 52.28% carbohydrate, 9.60% fiber, and 13.40% moisture) and water ad libitum. The FR group (n = 13) was subjected to a severe food restriction equivalent to 50% of the average amount consumed by group C. Food intake for the control group was measured daily and used to calculate food quantity for the FR group. Rats were maintained on this dietary regimen for 90 days and were weighed once a week. The rats were kept in individual cages at a 23 °C room temperature, with light cycles of 12 h and relative humidity of 60%. At the end of experimental protocol, the animals were anesthetized with an intramuscular injection of ketamine (50 mg/kg; Dopalen®, Sespo Indústria e Comércio Ltd.a—Divisão Vetbrands, Jacareí, São Paulo, Brasil) and xylazine (10 mg/kg; Dopalen®, Sespo Indústria e Comércio Ltd.a—Divisão Vetbrands, Jacareí, São Paulo, Brasil), decapitated, and thoracotomized.

The project was approved by the "Committee for Experimental Research Ethics of the Faculty of Medicine in Botucatu—UNESP" (CEUA 755-2009), in accordance with the "Guide for the Care and Use of Laboratory Animals".

2.2. General Characteristics

Initial (IBW) and final body weights (FBW), total body fat (BF), adiposity index (AI), naso-anal and tibia length, and soleus muscle and lung weight were measured to assess the effects of food restriction on body parameters of rats. The adipose tissue fat pads (epididymal, retroperitoneal, and visceral) were dissected and weighed. Adiposity index (AI) was calculated using the formula: AI = [total body fat (BF)/FBW] × 100. BF was measured from the sum of the individual fat pad weights as follows: BF = epididymal fat + retroperitoneal fat + visceral fat.

2.3. Cardiac Morphological Post Mortem Study

Cardiac remodeling was determined by macroscopic analysis of the following parameters: total weight of the heart (HW) and left ventricle (LVW), right ventricle (RVW), and atria (ATW), and normalized to body weight.

2.4. Myocardial Function

The cardiac contractile performance was evaluated by studying isolated papillary muscles from the LV, as previously described [34,35]. The following mechanical parameters were measured during isometric contraction: developed tension (DT; g/mm^2), positive tension derivative (+dT/dt; $g/mm^2/s$) and negative tension derivative (−dT/dt; $g/mm^2/s$). The mechanical behavior of papillary muscles was assessed by increasing extracellular Ca^{2+} concentrations (0.5 mM, 1.0 mM, 1.5 mM, 2.0 mM, and 2.5 mM) in the presence and absence of cyclopiazonic acid (CPA, 30 mM), which is a highly specific blocker of SERCA2a. All the variables were normalized per cross-sectional area of papillary muscle (CSA). To avoid the central core hypoxia and impaired functional performance, papillary muscles with CSA >1.5 mm^2 were excluded from analysis.

2.5. Expression of Calcium-Handling Protein

Protein expression was analyzed by Western blot. Fragments of the LV were frozen in liquid nitrogen and stored in a freezer at −80 °C. Frozen samples were homogenized with a Polytron apparatus (IKA T25 Basic Ultra Turrax TM, Wilmington, NC, USA) in hypotonic lysis buffer (50 mM of potassium phosphate pH 7.0, 0.3 M of sucrose, 0.5 mM of dithiothreitol [DTT], 1 mM of ethylenediamine tetraacetic acid [EDTA] buffer pH 8.0, 0.3 mM phenylmethylsulfonyl fluoride [PMSF], 10 mM of sodium fluoride [NaF], and phosphatase inhibitor). The homogenization product was centrifuged (5804R Eppendorf, Hamburg, Germany) at 12.000 rpm for 20 min at 4 °C, and the supernatant was transferred to tubes and stored in a freezer at −80 °C tubes. Protein concentration was analyzed by the method of Bradford [36] using the curves of Bovine Serum Albumin (BSA) Protein Standard (Bio-Rad, Hercules, CA, USA). Samples were diluted in Laemmli buffer (240 mM of Tris-HCL, 0.8% sodium dodecyl sulfate [SDS], 40% glycerol, 0.02% bromophenol blue, and 200 mM of β-mercaptoethanol) and separated via electrophoresis using Mini-Protean 3 Electrophoresis Cell (Bio-Rad, Hercules, CA, USA). Electrophoresis was performed with a biphasic gel; stacking (240 mM Tris-HCl pH 6.8, 30% polyacrylamide, 10% ammonium persulfate [APS], and tetramethylethylenediamine [TEMED]) and resolution (240 mM of Tris-HCl pH 8.8, 30% polyacrylamide, 10% APS, and TEMED) at a concentration of 6% to 10%, depending on the molecular weight of the analyzed protein. The Kaleidoscope Prestained Protein Standard (Bio-Rad, Hercules, CA, USA) was used to identify the size of the bands. Electrophoresis was performed at 120 V (Power Pac HC 3.0A, Bio-Rad, Hercules, CA, USA), for 3 h with running buffer (0.25 M of Tris, 192 mM of glycine, and 1% SDS). Proteins were transferred to a nitrocellulose membrane using a Mini Trans-Blot (Bio-Rad, Hercules, CA, USA) system with transfer buffer (25 mM of Tris, 192 mM of glycine, 20% methanol, and 0.1% SDS). Membranes were washed twice with TBS-T buffer (20 mM of Tris-HCl pH 7.4, 137 mM of NaCl, and 0.1% of Tween 20). Membranes were blocked with 5% nonfat dry milk in TBS-T for 120 min at room temperature under constant agitation. The membrane was washed three times with TBS-T and incubated with primary antibody diluted in blocking solution under constant agitation for 12 h. After incubation with the primary antibody, membranes were washed three times with TBS-T and incubated with the secondary antibody in blocking solution for 2 h under constant stirring. Then, the membrane was washed three times with TBS-T to remove secondary antibody excess. Immunodetection was performed using chemiluminescence according to the manufacturer's instructions (Amersham Biosciences, Piscataway, NJ, USA) and detected by autoradiography. Quantification analysis of blots was performed using Scion Image software (Scion Corporation, Frederick, Maryland, EUA). Antibodies used were: Ryanodine (1:1000; ABR, Affinity BioReagents, Golden, CO, USA); Calsequestrin (1:2000; ABR, Affinity BioReagents); Exchanger Na$^+$/Ca^{2+} (1:2000; Upstate, Lake Placid, NY, USA); SERCA2a (1:2500; ABR, Affinity BioReagents); Phospholamban (1:5000; ABR, Affinity BioReagents); Phospho-Phospholamban (Ser16) (1:5000; Badrilla, Leeds, West Yorkshire, UK); Phospho-Phospholamban (Thr17) (1:5000; Badrilla); L-type calcium channel (anti-α1C) (1:1000; Santa Cruz Biotechonology Inc., Santa Cruz, CA, USA); and β-Actin (1:1000; Santa Cruz Biotechnology Inc.). Secondary antibodies that were linked to peroxidase were used (IgG anti-mouse or IgG anti-rabbit, depending on the protein; 1:5000-1:10,000; Santa Cruz Biotechnology Inc.).

2.6. Statistical Analysis

All the data were tested for normality before statistical analysis using the Shapiro–Wilk test. Data from physical characteristics, cardiac post-mortem morphology, and Western blot of calcium-handling proteins were reported by descriptive measures of position and variability and subjected to the "t" test for independent samples. For isolated papillary muscle experiments, the adjustment of the DT, +dT/dt, and −dT/dt response linear model, in the presence and absence of CPA as a function of the extracellular Ca^{2+} concentration elevation, expressed by *Response Variable = a + b[Ca^{2+}]$_{extracellular}$*, was performed by the minimum squares technique complemented with the comparative test of average profiles of responses regarding the parallelism and coincidence of the adjusted models to the group. The statistical analyses were performed using SigmaPlot 12.0 (Systat Software, Inc., San Jose, CA, USA). The level of significance for all variables was $p < 0.05$.

3. Results

3.1. Physical Characteristics

Severe food restriction (FR) resulted in substantial changes to the physical phenotype of the rats. As shown in Table 1, 90 days of FR significantly reduced body weight and fat mass measures, resulting in an approximate 12-fold decrease in the adiposity index. Furthermore, masses of the soleus muscle, lung, and tibia were significantly lower in FR rats. These results show that 90 days of FR severely alters physical phenotype, including the impairment and/or regression of organ growth.

Table 1. Physical characteristics of animals.

	Groups	
	C (*n* = 14)	FR (*n* = 13)
IBW (g)	313 ± 41.2	301 ± 35.2
FBW (g)	445 ± 39.1	228 ± 19.1 *
Food intake (g/day)	21.1 ± 2.2	10.6 ± 1.1
Epididymal fat (g)	9.60 ± 3.42	0.90 ± 0.56 *
Retroperitoneal fat (g)	7.00 ± 2.80	0.19 ± 0.12 *
Visceral fat (g)	5.24 ± 1.68	0.67 ± 0.33 *
Total body fat (g)	21.8 ± 6.90	1.75 ± 0.70 *
Adiposity index	4.86 ± 1.45	0.76 ± 0.27 *
Naso-anal length (cm)	27.5 ± 0.70	24.6 ± 0.80 *
Soleus muscle (g)	0.19 ± 0.03	0.10 ± 0.01 *
Lung (g)	2.00 ± 0.41	1.19 ± 0.11 *
Tibia length (cm)	4.38 ± 0.07	4.15 ± 0.03 *
IBW/FBW ratio (g/g)	0.71 ± 0.10	1.33 ± 0.19 *

Data expressed as mean ± standard deviation. C: control group; FR: restriction food group; IBW: initial body weight; FBW: Final body weight. Student "t" test for independent samples. * $p < 0.05$ vs. C.

3.2. Macroscopic Cardiac Morphology

As shown in Table 2, the weight of the left ventricle (LV), right ventricle (RV), and atria (AT) were approximately 50% lower in FR rats compared to their controls. However, the data normalized by FBW were similar between the groups.

Table 2. Cardiac morphological post-mortem data.

	Groups	
	C ($n = 14$)	FR ($n = 13$)
LV (g)	0.84 ± 0.10	0.44 ± 0.05 *
RV (g)	0.25 ± 0.04	0.12 ± 0.01 *
AT (g)	0.10 ± 0.02	0.05 ± 0.01 *
Total heart (g)	1.20 ± 0.15	0.62 ± 0.07 *
LV/FBW (mg/g)	1.90 ± 0.22	1.93 ± 0.13
RV/FBW (mg/g)	0.56 ± 0.08	0.53 ± 0.03
AT/FBW (mg/g)	0.24 ± 0.04	0.24 ± 0.02
Heart/FBW (mg/g)	2.70 ± 0.32	2.70 ± 0.16

Data expressed as mean ± standard deviation. C: control; FR: food restriction; LV: left ventricular weight, RV: right ventricle weight, ATW: atrium weight; Heart: heart weight; FBW: final body weight. Student "t" test for independent samples. * $p < 0.05$ vs. C.

3.3. Assessment of Papillary Muscle Function

To determine whether abnormal heart size was associated with myocardial dysfunction, the contractile performance of LV papillary muscles isolated from control and FR hearts was measured. In response to increasing extracellular Ca^{2+} concentration, the developed tension (DT) was significantly reduced in papillary muscles from FR hearts (Figure 1A). Papillary muscles from FR hearts also had significantly lower values for positive tension derivative (+dT/dt) and negative tension derivative (−dT/dt) (Figure 1B,C), which is consistent with impaired contractility and relaxation. Additionally, the administration of cyclopiazonic acid (CPA), which is a specific inhibitor of SERCA2a, further decreased +dT/dt and −dT/dt (Figure 1B,C) in FR hearts. Moreover, Figure 2 shows that the blocking percentage by CPA was lower in the FR than the control group. This result suggests a prior impairment of SERCA2a activity under caloric restriction, which would reduce the magnitude of the effect of cyclopiazonic acid inhibition. Overall, these findings show that 90 days of FR severely impairs myocardial contraction and relaxation, which is likely due to depressed SERCA2a activity.

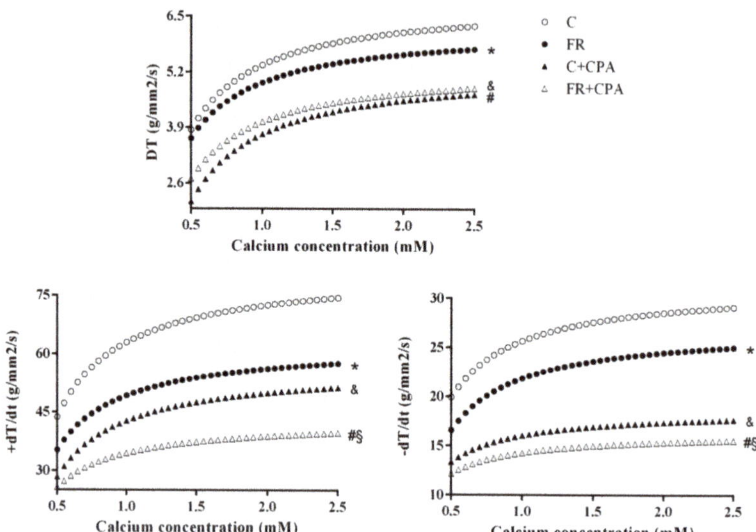

Figure 1. Food restriction (FR) animals presented altered functional responses in the papillary muscle assay in the following parameters: (**A**) developed tension (DT), (**B**) positive tension derivative (+dT/dt), and (**C**) negative tension derivative (−dT/dt) C: control group. CPA: Cyclopiazonic acid. C + CPA: control submitted to CPA; FR + CPA: food restriction submitted to CPA. Data presented as mean ± standard deviation. Minimum squares technique complemented with the comparative test of average profiles of responses. $p < 0.05$; * C vs. FR; § C + CPA vs. FR + CPA; # FR vs. FR + CPA; and & C vs. C + CPA.

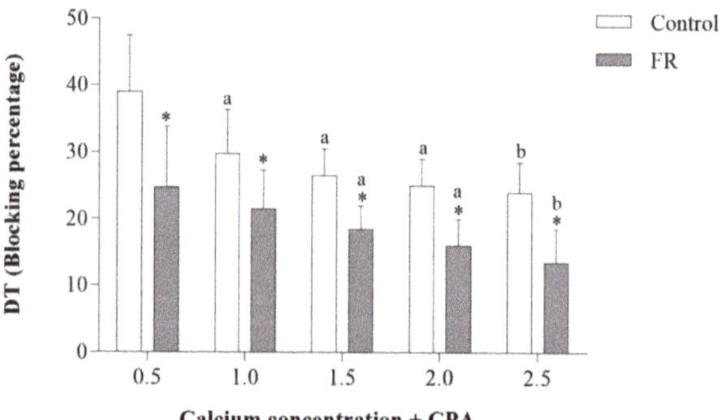

Figure 2. Blocking percentage under cyclopiazonic acid in papillary muscle preparation. C: control; FR: food restriction; CPA: cyclopiazonic acid; DT: developed tension. Data are expressed as mean ± standard deviation. Repeated measures two-way ANOVA complemented with Bonferroni post-hoc test. $p < 0.05$. [a] vs. 0.5 mM Ca^{2+} concentration; [b] vs. 0.5 and 1.0 mM Ca^{2+} concentration; * vs. control. ($n = 13$–14 each group).

3.4. Expression of Calcium-Handling Protein

Calcium-handling proteins are integral in facilitating myocardial contraction and relaxation. Therefore, a Western blot analysis was conducted focusing on the primary proteins involved in calcium handling, including SERCA2a, phospholamban (PLB), ryanodine receptor (RYR), calsequestrin (CSQ), sodium-calcium exchanger (NCX), and the L-type calcium channel (L channel). As shown in Figure 3B,F–H, no statistical differences were noted in SERCA2a, RYR, CSQ, or NCX. Additionally, no differences in PLB or the phosphorylation sites at PLB Ser[16] or PLB Thr[17] were observed (Figure 3C–E).

However, 90 days of FR did result in a significant lowering of L-channel protein levels (Figure 3I). Furthermore, the ratios of SERCA2a/PLB, phosphorylated PLB (pPLB) Ser16/PLB, pPLB Thr17/PLB were similar in both groups (Figure 3J–L).

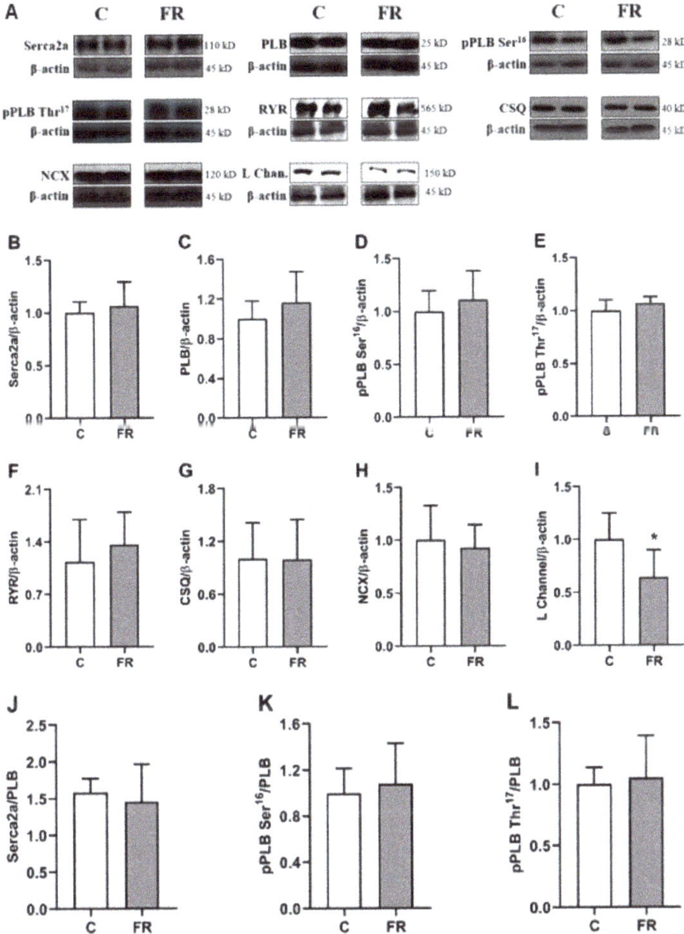

Figure 3. Protein expression of calcium-handling regulators evaluated by Western blot in the myocardium from control (C) and food restriction groups (FR) (n = 6 in each group). (**A**) Representative bands of the proteins. Quantification of myocardial (**B**) sarcoplasmic reticulum calcium-ATPase (SERCA2a), (**C**) total phospholamban (PLB), (**D**) phosphorylated PLB on serine-16 (pPLB Ser[16]), (**E**) phosphorylated PLB on threonine-17 (pPLB Thr[17]), (**F**) ryanodine receptor (RYR), (**G**) calsequestrin (CSQ), (**H**) sodium-calcium exchanger (NCX), and (**I**) L-type Ca^{2+} channel (L Channel) normalized to β-actin (internal control). Quantification of (**J**) SERCA2a, (**K**) pPLB Ser[16], and (**L**) pPLB Thr[17] normalized to total PLB. Data are expressed as mean ± standard deviation. Student's *t*-test for independent samples. * $p < 0.05$ vs. C.

4. Discussion

The purpose of this study was to investigate the contribution of alterations in the expression of Ca^{2+}-handling proteins and SERCA2a activity to myocardial performance during severe food restriction. The nutritional protocol used in this study was sufficient in duration and intensity to cause malnutrition, which is in agreement with previous studies [12,13,17]. The main findings are that malnutrition reduces the myocardial L-type calcium channels protein expression and alters SERCA2a activity, as expressed by the decrement of the mechanical responses to extracellular Ca^{2+} elevation in the absence and presence of cyclopiazonic acid. Overall, these changes in the calcium-handling capacity of the cardiac myocyte are critical in the development of myocardial dysfunction during severe food restriction.

In agreement with previous literature [11–15,17,19,31], the 50% reduction of food intake used in this study promoted drastic changes in the body composition of animals, as visualized by a reduction in all of the variables related to fat and body weight. Malnutrition causes damage to several segments of the body system, including a substantial reduction of hind limb muscles such as the soleus, plantaris, gastrocnemius, and extensor digitorum longus muscles [12,13,15,37,38]. In addition, food restriction leads to severe changes in the cardiorespiratory structures, particularly the weights of the lungs, atria, and ventricles [13,14,16]. These above responses, which were also observed in our study, are consistent with increased protein catabolism, as well as decreased protein synthesis, since the restriction of pellet intake produces protein and energy deficiency [38].

In isolated papillary muscle analysis, malnutrition resulted in decreased contractile (DT and +dT/dt) and relaxation (−dT/dt) capacity, which is in accordance with functional impairments observed in precedent studies [12,13,19,21,22,24,39]. At the molecular level, the quantity and quality of contractile proteins and the availability of energy and Ca^{2+} are crucial factors in the regulation of myocardial performance [40,41]. Previous work from our laboratory showed that severe food restriction causes a reduction of myofibril density and myofilament disorganization [12,17,20,26]. Furthermore, the authors reported myosin isoform remodeling with the down-regulation of alpha- (V_1) and up-regulation of beta-myosin heavy chain (V_3) genes, which present faster and lower ATPase activity, respectively [24,30]. In addition, the constant low availability of substrates during the malnutrition period may lead to mitochondrial injury [17,20], leading to a lack of ATP production, which may impair the capacity of the ATP-dependent proteins.

The primary aim of this study was to evaluate the role of Ca^{2+}-handling proteins in the development of myocardial dysfunction induced by malnutrition. Myocardial Ca^{2+}-handling inadequacies are among the most critical responses to cardiac aggression [12–17,19,29–31]. Previous studies reported alterations in the activity [13,15,16] and content [14,15,31] of Ca^{2+}-handling regulatory components, which is consistent with our current observations. With any cardiac remodeling process at the macro or micro level, the initial intent of adaptations in Ca^{2+} handling is to attenuate the insult imposed on the myocardium in an effort to maintain the functional balance. However, these adaptations are incapable of enduring the aggression over time, contributing to the establishment of the dysfunctional condition [42,43]. In the present study, 90 days of severe food restriction resulted in alterations in SERCA2a activity and, among the analyzed proteins, only a reduction in L-type calcium channels protein expression. During the Ca^{2+} transient, the SERCA2a is an ATP-dependent protein that is responsible for the reuptake of Ca^{2+} into the sarcoplasmic reticulum of the cardiomyocyte, which is crucial for the maintenance of the cytosolic concentration. Although we did not observe a reduction of SERCA2a protein expression, impaired activity was noted in the papillary muscle assay in the presence of cyclopiazonic acid, as noted previously [19]. This finding is likely due to the low-energy capacity in severe food restriction, which limits the ATP required for proper SERCA2a function. Impaired SERCA2a activity may provoke the elevation of diastolic Ca^{2+}, which would lead to the increased importance of the exchanger Na^+/Ca^{2+} (NCX) cytosolic Ca^{2+} extrusion. Since the NCX mechanism has a reduced capacity to remove cytosolic Ca^{2+}, alterations in Ca^{2+} handling persist.

In disagreement with our hypothesis, in which changes in several Ca^{2+}-handling proteins were expected, our study showed only reduction in the expression of the L-type calcium channel protein due to food restriction, which is in accordance with a previous report [15]. Although the reason for this outcome is unclear, we speculate that this may be an adaptive response, serving as a cellular defense mechanism in response to dysfunctional SERCA2a activity. In this regard, myocyte Ca^{2+} influx is decreased in order to avoid Ca^{2+} accumulation, particularly during diastole. However, the reduction of Ca^{2+} influx across the plasma membrane as a result of decreased L-type calcium channel expression over time may lead to impairments in cardiomyocyte force development, as evidenced by reduced values of DT and +dT/dt in animals suffering from malnutrition. Although the functional control of the L-type channel is well described, little is known about the processes involved in the expression of this sarcolemma channel. The L-type channel is composed of three subunits (α_{1C}, $\alpha_2\delta$, and β) [44], with the α_{1C} pore-forming subunit analyzed in this study. The expression of the α_{1C} subunit protein is regulated by β and α adrenergic systems [45], angiotensin II [46], and calcium flux through the L-type channel [47–49]. Based on our results, it is possible that the cytosolic Ca^{2+} "leftover", due to SERCA2a dysfunction, activated the auto-regulation mechanism of L-type channel protein expression on the distal carboxyl-terminus via a process sensitive to Ca^{2+} flux alterations [47–49]. In conditions in which there is undue cytosolic Ca^{2+} accumulation, the hyperactivation of this regulatory mechanism may be related to the decrease in the protein expression of these channels. For better visualization of the food restriction effects over the myocardium, we constructed a figure placing the main molecular alterations and suggested a hypothesis that attributed cardiac dysfunction to malnutrition, as can be seen in Figure 4.

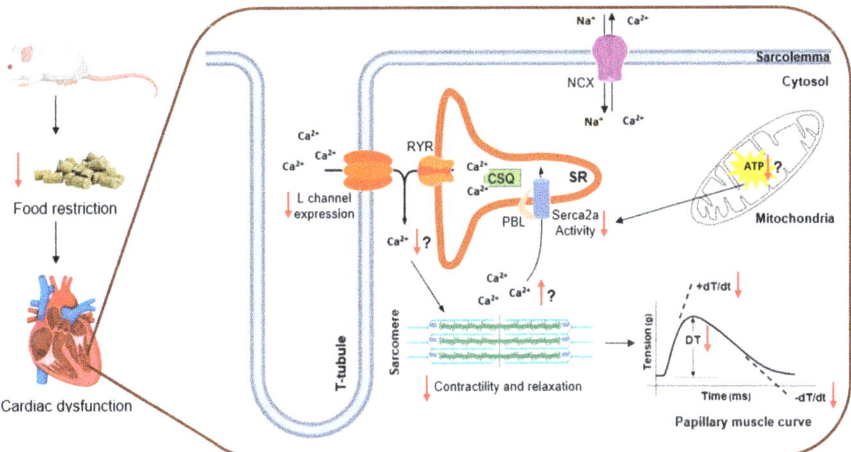

Figure 4. Overview of food restriction effects over myocardium and presumable interactions between molecular changes and myocardial mechanical impairment. Food restriction reduced sarcoplasmic reticulum Ca^{2+}-ATPase (SERCA2a) activity and L-type calcium channel (L channel) expression and impaired myocardial mechanical performance. Hypothetical adenosine triphosphate (ATP) depletion due to malnutrition could reduce the SERCA2a activity, leading to a high amount of residual cytosolic calcium. In turn, residual calcium signals to deplete the release of calcium by the ryanodine receptor (RYR), which is achieved by the reduced calcium entry secondary to the decreased protein expression of the L channel. This sequence of events involving the myocardial calcium handling would be a plausible mechanism by which the food restriction deteriorates the cardiac function. Ca^{2+}, calcium; Na^+, sodium; NCX, sodium-calcium exchanger; SR, sarcoplasmic reticulum; CSQ, calsequestrin; PLB, phospholamban; DT, developed tension; +dT/dt, positive tension derivative; −dT/dt negative tension derivative.

Therefore, our results show that malnutrition, due to 90 days of severe food restriction, leads to cardiac dysfunction. The deterioration of myocardial function may be the consequence of a reduction of L-type calcium channels protein expression and impaired SERCA2a activity. Future studies are required to understand the mechanisms that cause inadequacies in L-type Ca^{2+} calcium channels and

SERCA2a activity during severe food restriction, which ultimately contribute to malnutrition-induced myocardial dysfunction.

Author Contributions: A.F.d.D. and A.C.C. conceived the design of the study. A.F.d.D., A.P.L.-L., A.S.L., D.H.S.d.C. and L.C.d.T. performed the experiments. A.F.d.D., V.L.d.S., S.L.B.d.S., G.A.F.M., P.G.S., D.F.V. analyzed the data. C.R.P. performed the statistical analysis. A.F.d.D., V.L.d.S., S.C.K. and A.C.C. wrote the paper. S.L.B.d.S., G.A.F.M., P.G.S. and D.F.V revised the manuscript. All authors read and approved the final version of the manuscript.

Funding: This research was funded by the São Paulo Research Foundation (FAPESP), grant numbers 2009/52862-0 and 2010/50961-9.

Conflicts of Interest: The authors declare no conflicts of interest.

References

1. Alugoju, P.; Narsimulu, D.; Bhanu, J.U.; Satyanarayana, N.; Periyasamy, L. Role of quercetin and caloric restriction on the biomolecular composition of aged rat cerebral cortex: An FTIR study. *Spectrochim. Acta A Mol. Biomol. Spectrosc.* **2019**, *220*, 117128. [CrossRef] [PubMed]

2. Mattison, J.A.; Roth, G.S.; Beasley, T.M.; Tilmont, E.M.; Handy, A.M.; Herbert, R.L.; Longo, D.L.; Allison, D.B.; Young, J.E.; Bryant, M.; et al. Impact of caloric restriction on health and survival in rhesus monkeys from the NIA study. *Nature* **2012**, *489*, 318–321. [CrossRef] [PubMed]

3. Trepanowski, J.F.; Canale, E.R.; Marshall, E.K.; Kabir, M.M.; Bloomer, R.J. Impact of caloric and dietary restriction regimens on markers of health and longevity in humans and animals: A summary of available findings. *Nutr. J.* **2011**, *10*, 107. [CrossRef] [PubMed]

4. Kenyon, C.J. The genetics of ageing. *Nature* **2010**, *464*, 504–512. [CrossRef] [PubMed]

5. Rebrin, I.; Kamzalov, S.; Sohal, R.S. Effects of age and caloric restriction on glutathione redox state in mice. *Free Radic. Biol. Med.* **2003**, *35*, 626–635. [CrossRef]

6. Weindruch, R.; Sohal, R.S. Caloric Intake and Aging. *N. Engl. J. Med.* **1997**, *337*, 986–994. [CrossRef] [PubMed]

7. Brent, B.; Obonyo, N.; Akech, S.; Shebbe, M.; Mpoya, A.; Mturi, N.; Berkley, J.A.; Tulloh, R.M.R.; Maitland, K. Assessment of Myocardial Function in Kenyan Children with Severe, Acute Malnutrition: The Cardiac Physiology in Malnutrition (CAPMAL) Study. *JAMA Netw. Open* **2019**, *2*, e191054. [CrossRef] [PubMed]

8. Bebars, G.M.; Askalany, H.T. Assessment of left ventricular systolic and diastolic functions in severely malnourished children. *Egypt. Pediatr. Assoc. Gaz.* **2018**. [CrossRef]

9. Silverman, J.A.; Chimalizeni, Y.; Hawes, S.E.; Wolf, E.R.; Batra, M.; Khofi, H.; Molyneux, E.M. The effects of malnutrition on cardiac function in African children. *Arch. Dis. Child.* **2016**, *101*, 166–171. [CrossRef] [PubMed]

10. El-Sayed, H.L.; Nassar, M.F.; Habib, N.M.; Elmasry, O.A.; Gomaa, S.M. Structural and functional affection of the heart in protein energy malnutrition patients on admission and after nutritional recovery. *Eur. J. Clin. Nutr.* **2006**, *60*, 502–510. [CrossRef] [PubMed]

11. Cicogna, A.C.; Padovani, C.R.; Okoshi, K.; Aragon, F.F.; Okoshi, M.P. Myocardial Function during Chronic Food Restriction in Isolated Hypertrophied Cardiac Muscle. *Am. J. Med. Sci.* **2000**, *320*, 244–248. [CrossRef]

12. Okoshi, M.P.; Okoshi, K.; Pai, V.D.; Pai-Silva, M.D.; Matsubara, L.S.; Cicogna, A.C. Mechanical, biochemical, and morphological changes in the heart from chronic food-restricted rats. *Can. J. Physiol. Pharmacol.* **2001**, *79*, 754–760. [CrossRef] [PubMed]

13. Gut, A.L.; Okoshi, M.P.; Padovani, C.R.; Aragon, F.F.; Cicogna, A.C. Myocardial dysfunction induced by food restriction is related to calcium cycling and beta-adrenergic system changes. *Nutr. Res.* **2003**, *23*, 911–919. [CrossRef]

14. Sugizaki, M.M.; Leopoldo, A.P.L.; Conde, S.J.; Campos, D.S.; Damato, R.; Leopoldo, A.S.; Nascimento, A.F.D.; Júnior, S.D.A.O.; Cicogna, A.C. Upregulation of mRNA myocardium calcium handling in rats submitted to exercise and food restriction. *Arq. Bras. Cardiol.* **2011**, *97*, 46–52. [CrossRef]

15. De Tomasi, L.C.; Bruno, A.; Sugizaki, M.M.; Lima-Leopoldo, A.P.; Nascimento, A.F.; Júnior, S.A.; Pinotti, M.F.; Padovani, C.R.; Leopoldo, A.S.; Cicogna, A.C. Food restriction promotes downregulation of myocardial L-type Ca^{2+} channels. *Can. J. Physiol. Pharmacol.* **2009**, *87*, 426–431. [CrossRef] [PubMed]

16. Gut, A.L.; Sugizaki, M.M.; Okoshi, M.P.; Carvalho, R.F.; Pai-Silva, M.D.; Aragon, F.F.; Padovani, C.R.; Okoshi, K.; Cicogna, A.C. Food restriction impairs myocardial inotropic response to calcium and beta-adrenergic stimulation in spontaneously hypertensive rats. *Nutr. Res.* **2008**, *28*, 722–727. [CrossRef] [PubMed]

17. Sugizaki, M.M.; Carvalho, R.F.; Aragon, F.F.; Padovani, C.R.; Okoshi, K.; Okoshi, M.P.; Zanati, S.G.; Pai-Silva, M.D.; Novelli, E.L.B.; Cicogna, A.C. Myocardial Dysfunction Induced by Food Restriction is Related to Morphological Damage in Normotensive Middle-Aged Rats. *J. Biomed. Sci.* **2005**, *12*, 641–649. [CrossRef] [PubMed]

18. Okoshi, K.; Fioretto, J.; Okoshi, M.; Cicogna, A.; Aragon, F.; Matsubara, L.; Matsubara, B. Food restriction induces in vivo ventricular dysfunction in spontaneously hypertensive rats without impairment of in vitro myocardial contractility. *Braz. J. Med. Biol. Res.* **2004**, *37*, 607–613. [CrossRef]

19. Sugizaki, M.M.; Leopoldo, A.S.; Okoshi, M.P.; Bruno, A.; Conde, S.J.; Lima-Leopoldo, A.P.; Padovani, C.R.; Carvalho, R.F.; Nascimento, A.F.D.; De Campos, D.H.S.; et al. Severe food restriction induces myocardial dysfunction related to SERCA2 activity. *Can. J. Physiol. Pharmacol.* **2009**, *87*, 666–673. [CrossRef]

20. Sugizaki, M.M.; Pai-Silva, M.D.; Carvalho, R.F.; Padovani, C.R.; Bruno, A.; Nascimento, A.F.; Aragon, F.F.; Novelli, E.L.B.; Cicogna, A.C. Exercise training increases myocardial inotropic response in food restricted rats. *Int. J. Cardiol.* **2006**, *112*, 191–201. [CrossRef]

21. Fioretto, J.R.; Querioz, S.S.; Padovani, C.R.; Matsubara, L.S.; Okoshi, K.; Matsubara, B.B. Ventricular remodeling and diastolic myocardial dysfunction in rats submitted to protein-calorie malnutrition. *Am. J. Physiol. Heart Circ. Physiol.* **2002**, *282*, H1327–H1333. [CrossRef] [PubMed]

22. Hilderman, T.; McKnight, K.; Dhalla, K.S.; Rupp, H.; Dhalla, N.S. Effects of long-term dietary restriction on cardiovascular function and plasma catecholamines in the rat. *Cardiovasc. Drugs Ther.* **1996**, *10*, 247–250. [CrossRef] [PubMed]

23. Katzeff, H.L.; Powell, S.R.; Ojamaa, K. Alterations in cardiac contractility and gene expression during low-T3 syndrome: Prevention with T3. *Am. J. Physiol. Metab.* **1997**, *273*, E951–E956. [CrossRef] [PubMed]

24. Haddad, F.; Bodell, P.W.; McCue, S.A.; Herrick, R.E.; Baldwin, K.M. Food restriction-induced transformations in cardiac functional and biochemical properties in rats. *J. Appl. Physiol.* **1993**, *74*, 606–612. [CrossRef] [PubMed]

25. Kirsch, A.; Savabi, F. Effect of food restriction on the phosphocreatine energy shuttle components in rat heart. *J. Mol. Cell. Cardiol.* **1992**, *24*, 821–830. [CrossRef]

26. Okoshi, M.P.; Okoshi, K.; Matsubara, L.S.; Pai-Silva, M.D.; Gut, A.L.; Padovani, C.R.; Pai, V.D.; Cicogna, A.C. Myocardial remodeling and dysfunction are induced by chronic food restriction in spontaneously hypertensive rats. *Nutr. Res.* **2006**, *26*, 567–572. [CrossRef]

27. Rossi, M.A.; Zucoloto, S. Ultrastructural changes in nutritional cardiomiopathy of protein-calorie malnourished rats. *Br. J. Exp. Pathol.* **1982**, *63*, 242–253. [PubMed]

28. McKnight, K.A.; Rupp, H.; Beamish, R.E.; Dhalla, N.S. Modification of catecholamine-induced changes in heart function by food restriction in rats. *Cardiovasc. Drugs Ther.* **1996**, *10*, 239–246. [CrossRef] [PubMed]

29. O'Brien, P.J.; Shen, H.; Bissonette, D.; Jeejeebhoy, K.N. Effects of hypocaloric feeding and refeeding on myocardial Ca and ATP cycling in the rat. *Mol. Cell. Biochem.* **1995**, *142*, 151–161. [CrossRef] [PubMed]

30. Klebanov, S.; Herlihy, J.T.; Freeman, G.L. Effect of long-term food restriction on cardiac mechanics. *Am. J. Physiol. Circ. Physiol.* **1997**, *273*, H:2333–H:2342. [CrossRef]

31. A Vizotto, V.; Carvalho, R.F.; Sugizaki, M.M.; Lima, A.P.; Aragon, F.F.; Padovani, C.R.; Castro, A.V.B.; Pai-Silva, M.D.; Nogueira, C.R.; Cicogna, A.C. Down-regulation of the cardiac sarcoplasmic reticulum ryanodine channel in severely food-restricted rats. *Braz. J. Med. Biol. Res.* **2007**, *40*, 27–31. [CrossRef] [PubMed]

32. Ghazalpour, A.; Bennett, B.; Petyuk, V.A.; Orozco, L.; Hagopian, R.; Mungrue, I.N.; Farber, C.R.; Sinsheimer, J.; Kang, H.M.; Furlotte, N.; et al. Comparative Analysis of Proteome and Transcriptome Variation in Mouse. *PLoS Genet.* **2011**, *7*, e1001393. [CrossRef] [PubMed]

33. Foss, E.J.; Radulovic, D.; Shaffer, S.A.; Ruderfer, D.M.; Bedalov, A.; Goodlett, D.R.; Kruglyak, L. Genetic basis of proteome variation in yeast. *Nat. Genet.* **2007**, *39*, 1369–1375. [CrossRef] [PubMed]

34. Leopoldo, A.S.; Lima-Leopoldo, A.P.; Sugizaki, M.M.; Nascimento, A.F.D.; De Campos, D.H.S.; Luvizotto, R.D.A.M.; Castardeli, É.; Alves, C.A.B.; Brum, P.C.; Cicogna, A.C.; et al. Involvement of L-type calcium channel and serca2a in myocardial dysfunction induced by obesity. *J. Cell. Physiol.* **2011**, *226*, 2934–2942. [CrossRef] [PubMed]

35. Campos, D.H.S.; De Leopoldo, A.S.; Lima-Leopoldo, A.P.; Nascimento, A.F.; Do Oliveira-Junior, S.A.; De Silva, D.C.T.; Sugizaki, M.M.; Padovani, C.R.; Cicogna, A.C. Obesity Preserves Myocardial Function During Blockade of the Glycolytic Pathway. *Arq. Bras. Cardiol.* **2014**, *103*, 330–337. [CrossRef] [PubMed]

36. Bradford, M.M. A rapid and sensitive method for the quantitation of microgram quantities of protein utilizing the principle of protein-dye binding. *Anal. Biochem.* **1976**, *72*, 248–254. [CrossRef]

37. Maxwell, L.; Enwemeka, C.; Fernandes, G. Effects of exercise and food restriction on rat skeletal muscles. *Tissue Cell* **1992**, *24*, 491–498. [CrossRef]

38. Katzeff, H.L.; Ojamaa, K.M.; Klein, I. The effects of long-term aerobic exercise and energy restriction on protein synthesis. *Metabolism* **1995**, *44*, 188–192. [CrossRef]

39. Cicogna, A.C.; Padovani, C.R.; Okoshi, K.; Matsubara, L.S.; Aragon, F.F.; Okoshi, M.P. The influence of temporal food restriction on the performance of isolated cardiac muscle. *Nutr. Res.* **2001**, *21*, 639–648. [CrossRef]

40. Hanft, L.M.; Korte, F.S.; McDonald, K.S. Cardiac function and modulation of sarcomeric function by length. *Cardiovasc. Res.* **2008**, *77*, 627–636. [CrossRef]

41. Sequeira, V.; Nijenkamp, L.L.; Regan, J.A.; Van Der Velden, J. The physiological role of cardiac cytoskeleton and its alterations in heart failure. *Biochim. Biophys. Acta (BBA) Biomembr.* **2014**, *1838*, 700–722. [CrossRef] [PubMed]

42. Gorski, P.A.; Ceholski, D.K.; Hajjar, R.J. Altered myocardial calcium cycling and energetics in heart failure —A rational approach for disease treatment. *Cell Metab.* **2015**, *21*, 183–194. [CrossRef] [PubMed]

43. Lou, Q.; Janardhan, A.; Efimov, I.R. Remodeling of Calcium Handling in Human Heart Failure. *Pharm. Biotechnol.* **2012**, *740*, 1145–1174.

44. Fan, I.Q.; Chen, B.; Marsh, J.D. Transcriptional Regulation of L-type Calcium Channel Expression in Cardiac Myocytes. *J. Mol. Cell. Cardiol.* **2000**, *32*, 1841–1849. [CrossRef]

45. Golden, K.L.; Ren, J.; Dean, A.; Marsh, J.D. Norepinephrine regulates the in vivo expression of the L-type calcium channel. *Mol. Cell. Biochem.* **2002**, *236*, 107–114. [CrossRef] [PubMed]

46. Tsai, C.T.; Wang, D.L.; Chen, W.P.; Hwang, J.J.; Hsieh, C.S.; Hsu, K.L.; Tseng, C.D.; Lai, L.P.; Tseng, Y.Z.; Chiang, F.T.; et al. Angiotensin II Increases Expression of α1C Subunit of L-Type Calcium Channel Through a Reactive Oxygen Species and cAMP Response Element–Binding Protein–Dependent Pathway in HL-1 Myocytes. *Circ. Res.* **2007**, *100*, 1476–1485. [CrossRef] [PubMed]

47. Schroder, E.; Magyar, J.; Burgess, D.; Andres, D.; Satin, J. Chronic verapamil treatment remodels ICa,L in mouse ventricle. *Am. J. Physiol. Heart Circ. Physiol.* **2007**, *292*, H1906–H1916. [CrossRef] [PubMed]

48. Schroder, E.; Byse, M.; Satin, J. The L-Type Calcium Channel C-terminus Auto-regulates Transcription. *Circ. Res.* **2009**, *104*, 1373–1381. [CrossRef] [PubMed]

49. Satin, J.; Schroder, E.A.; Crump, S.M. L-type Calcium Channel Auto-Regulation of Transcription. *Cell Calcium* **2011**, *49*, 306–313. [CrossRef]

© 2019 by the authors. Licensee MDPI, Basel, Switzerland. This article is an open access article distributed under the terms and conditions of the Creative Commons Attribution (CC BY) license (http://creativecommons.org/licenses/by/4.0/).

Article

Six Weeks of Calorie Restriction Improves Body Composition and Lipid Profile in Obese and Overweight Former Athletes

Joanna Hołowko [1], Małgorzata Magdalena Michalczyk [2,*], Adam Zając [2], Maja Czerwińska-Rogowska [1], Karina Ryterska [1], Marcin Banaszczak [1], Karolina Jakubczyk [1] and Ewa Stachowska [1]

1 Department of Biochemistry and Human Nutrition, Pomeranian Medical University, 71-460 Szczecin, Poland
2 Department of Sport Nutrition, The Jerzy Kukuczka Academy of Physical Education in Katowice, 40-065 Katowice, Poland
* Correspondence: m.michalczyk@awf.katowice.pl; Tel.: +4-832-207-5342; Fax: +4-832-207-5200

Received: 4 June 2019; Accepted: 25 June 2019; Published: 27 June 2019

Abstract: Objective: The aim of the study was to compare the impact of 6 weeks of reducing daily caloric intake by 20% of total daily energy expenditure (TDEE)-CRI vs. reducing daily caloric intake by 30% of TDEE-CRII on body mass reduction and insulin metabolism in former athletes. Methods: 94 males aged 35.7 ± 5.3 years, height 180.5 ± 4.1 cm, and body mass 96.82 ± 6.2 kg were randomly assigned to the CRI ($n = 49$) or CRII ($n = 45$) group. Thirty-one participants (18 subjects from CRI and 13 from CRII) resigned from the study. The effects of both diets on the body composition variables (body mass—BM; body fat—BF; fat free mass—FFM; muscle mass—MM; total body water—TBW), lipid profile (total lipids—TL; total cholesterol—TCh; HDL cholesterol—HDL; LDL cholesterol—LDL; triglycerides—TG), and glucose control variables (glucose—GL, insulin—I, HOMA-IR, insulin-like growth factor-1—IGF-1, leptin and adiponectin) were measured. Results: After adhering to the CR I diet, significant differences were observed in FFM, MM and TG. After adhering to the CR II diet, significant differences were registered in tCh, TL and LDL. Both diets had a significant influence on leptin and adiponectin concentrations. Significant differences in FFM, MM, and tCh were observed between the CR I and CR II groups. At the end of the dietary intervention, significant differences in BF, FFM, MM and TBW were observed between the CR I and CR II groups. Conclusion: The 6 weeks of CR II diet appeared to be more effective in reducing BF and lipid profile and proved to be especially suitable for subjects with high body fat content and an elevated level of lipoproteins and cholesterol. Both reductive diets were effective in improving the levels of leptin and adiponectin in obese former athletes.

Keywords: calorie restriction diet; body mass reduction; insulin; IGF-1; leptin; adiponectin

1. Introduction

In many sports disciplines, and especially those in which muscular strength and body mass are important for performance, lack of physical activity and diet modification after the end of the career cause a drastic increase in body mass [1]. This often leads former athletes to become overweight or even obese [1–3]. In spite of the fact that this phenomenon appears to be frequent, the problem of excessive weight and obesity has not been studied extensively among former athletes, as their health problems remain outside of the mainstream of scientific interest [1]. Recently, only a few papers have been published regarding this topic [1–5], indicating that the maintenance of normal body mass in former athletes is a serious health issue [6]. Weight gain in this population is a consequence of three parallel processes: a decrease in the basal metabolic rate, the reduction of daily energy expenditure, and the excessive consumption of food [3,4]. The reduction in training intensity leads to a loss of muscle mass, which, in turn, determines the value of the necessary daily caloric intake. Former athletes are often accustomed to excessive calorie intake, which causes gains in body mass [4]. At the same time, a significant reduction in the volume and intensity of daily physical activity is observed, and as a consequence, a decreased resting metabolic rate occurs [7].

Currently, there are no clearly defined standards of nutrition for athletes terminating their sports careers [1,3]. It appears that the metabolic changes caused by intense exercise and nutrition specific for particular sports require a different approach to the diet of both the current and former athletes [6]. It is essential to consider that there are sport disciplines that require enormous energy consumption and expenditure during training and competition, as well as sports that prefer calorie restriction and low body mass [8]. In aesthetic sports disciplines that concentrate on movement coordination and physique (e.g., artistic gymnastics), calorie restriction and body mass control are key factors of peak performance [9]. Thus, a correctly adjusted diet for a standard subject may not be adequate for former athletes, justifying the search for more efficient dietary management of athletes at the end of their sports careers. A logical solution for former athletes with increased body mass and fat content includes substantial calorie restrictions (CRs) [1]. The available studies confirm the positive influence of CR diets on body mass reduction, blood pressure, variables of glucose metabolism, lipid profile, and immune response [1,10–12]. Several studies have confirmed that reducing the amount of calories in the diet by approximately 30% can lead to significant health benefits [13,14].

Besides poor eating habits, especially excess calorie intake, leptin and adiponectin are important variables that influence excessive weight and obesity. Therefore, they have drawn the attention of scientists engaged in metabolism and obesity studies. Leptin and adiponectin independently and inversely influence such phenomena as the insulin resistance of tissues, glucose metabolism, and vessel inflammation [15]. It has been proven that leptin can be associated with the complications resulting from obesity, such as hypertension and cardiovascular diseases [16]. Adiponectin has an inverse effect to leptin. This cytokine has an anti-atherosclerotic effect; it increases insulin sensitivity [17] and has anti-inflammatory properties [18]. The concentration of adiponectin in plasma is inversely proportional to the body mass index (BMI), and the concentration of insulin and triglycerides is directly proportional to the concentration of HDL [19].

Apart from the mentioned hormones, insulin-like growth factor (IGF-1) also plays a key role in glucose and lipids metabolism [20]. IGF-1 administration has been shown to reduce serum glucose levels in healthy individuals but also in insulin-resistant patients [20]. IGF-1 promotes fatty acid transport in muscle and its inhibition causes severe consequences like insulin resistance and even diabetes [20,21]. A deficiency of IGF-1 in adults is associated with impaired muscle mass, bone density, and lipid levels [20]. On the other hand, increased levels of free IGF-1 are observed in obese subjects [22].

Excess body fat is a risk factor for insulin resistance [1,23,24]. There are a number of well-established tests used to measure insulin resistance (IR) [25], including the homeostasis model assessment (HOMA). The HOMA-IR model is a simple, noninvasive method for predicting insulin resistance in middle-aged people with proper glucose tolerance [25]. This mathematical model is based on the reciprocal loop theory between the liver and β-cells of the pancreas, which regulate the concentration of glucose and insulin. The model can be used to evaluate the function of the β-cells of the pancreas as well as the level of insulin resistance [21]. To determine the HOMA index, fasting glucose and insulin values must be available. A higher level of HOMA-IR is observed in overweight, obese subjects but also in former athletes with body mass imbalance [1,23,24].

Overweight and obese subjects are prone to metabolic diseases such as hypercholesterolemia or hyperglycemia [1]. Thus, it is of great importance to monitor the above-mentioned blood variables in these patients under calorie-restriction diets [1]. It is expected that reduced calorie intake would decrease the concentration of blood glucose and insulin, while at the same time improving the lipid profile, decreasing the risk of diabetes [25]. A calorie restriction diet should also regulate the level of adiponectin, which has a favorable effect on the cardiovascular system.

The aim of this study was to determine whether a dietary intervention based on the introduction of two different types of calorie restriction induce significant changes in body mass and body composition, as well as metabolic indicators, such as insulin resistance (HOMA-IR), concentration of leptin, adiponectin, IGF-1, glucose, total cholesterol, triglycerides, and HDL and LDL cholesterol. The first reductive diet decreased total daily energy expenditure (TDEE) by 20% (CR I), while the second one was based on a 30% reduction of TDEE (CR II).

2. Materials and Methods

2.1. Subjects

Ninety-four (94) Caucasian males were initially qualified for the study. All of the qualified individuals were obese former athletes as determined by the body mass index (BMI ≥ 30). The study subjects included former athletes recruited from the following sport disciplines: canoeing, rowing, swimming, athletics, soccer, as well as weightlifting and powerlifting. During their career, which lasted 10 years on average, each subject trained intensively from 6 to 10 times per week. The time period since the end of their careers did not exceed 5 years. Volunteers were randomly assigned to one of the two intervention groups CRI: $n = 49$ and CRII $n = 45$. For randomization, we used the sealed envelopes method. Specifically, during the first visit to the laboratory, each participant drew an envelope with the prescribed diet. The participants who were classified into the CRI group, consumed daily 20% calories less than the total daily energy expenditure (TDEE), whereas the participants assigned to CRII group reduced their daily calorie intake by 30% of TDEE (Table 1). Thirty-one participants resigned from the study (18 subjects from CRI and 13 from CRII) (Figure 1). These individuals were not able maintain the calorie-restricted diets and consumed fast foods, sweets, and alcohol, which were not included in the prescribed diets. Of all the subjects recruited for the study, 31 from CRI and 32 from the CRII group completed the experiment. After conducting the experiment, the study participants were further divided into three subgroups (Figure 1): CRI—1.5–2.5 kg, $n = 13$/2.5–3.0 kg, $n = 10$/above 3.0 kg, $n = 8$; CRII—1.5–2.5 kg, $n = 14$/2.5–3.0 kg, $n = 11$/above 3.0 kg, $n = 7$, depending on the range of body mass reduction.

Table 1. Characteristics of the applied diets.

Nutrients	CR I MEAN ± SD	CR II MEAN ± SD
TEI, kJ	9589.73 ± 150.6	8924.48 ± 217.6
TEI, kcal	2292 ± 36	2133 ± 52
CARBOHYDRATES, %	50 ± 0.4	50 ± 0.2
CARBOHYDRATES, g/kg/body mass	2.7 ± 0.3	2.6 ± 0.5
SIMPLE SUGARS, %	<10	<10
FIBER, g/day	33 ± 0.2	34 ± 0.5
FAT, %	30 ± 0.6	30 ± 0.8
FAT, g/kg body mass	0.8 ± 0.2	0.75 ± 0.3
CHOLESTEROL, mg/day	300 ± 2.1	300 ± 1.7
PROTEIN, %	20 ± 0.7	20 ± 0.06
PROTEIN, g/kg body mass	1.2 ± 0.6	1.1 ± 0.4

Note: TEI—Total Energy Intake.

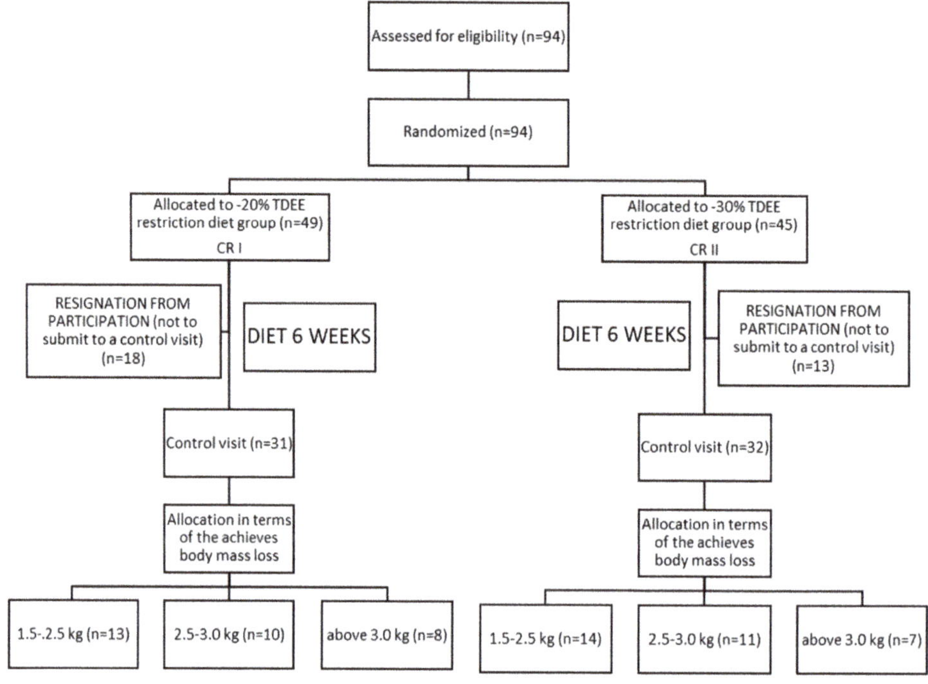

Figure 1. Scheme of the experimental protocol.

During the initial visit, the volunteers were advised by a dietician to maintain their habitual lifestyle and current physical activity. The exclusion criteria were as follows: The intake of any supplements with established antioxidant properties; energy expenditure of physical activity > 3000 kcal/week; hypercholesterolemia (total cholesterol > 8.0 mM or dyslipidemia therapy); diabetes (glucose > 126 mg/dL or diabetes treatment); hypertension (systolic blood pressure >140 mmHg and/or diastolic blood pressure > 90 mmHg or antihypertensive treatment); multiple allergies; celiac disease or other intestinal diseases; any condition that could limit the mobility of the subject, making laboratory visits difficult; life-threatening diseases or conditions which could worsen adherence to the measurements or treatments; vegetarianism or the need for other specific diets; and alcoholism or other drug addiction. Written informed consent was obtained from all participants. The study protocol was approved by the ethics committee of the Pomeranian Medical University in Szczecin, Poland (ethic reference KB-0012/53/11) and conformed to the ethical guidelines of the 1975 Declaration of Helsinki.

2.2. Dietary Intervention

The dietary intervention lasted 6 weeks. The first group of participants received a mix diet with a calorie reduction of 20% of the TDEE-CRI, while the second group was prescribed a similar diet with a calorie reduction of 30% of TDEE-CRII. TDEE was calculated according to the commonly accepted model (TDEE = RMR × AF) [26]. Resting metabolic rate (RMR) was measured during the first visit and during the subsequent control visit by means of the Fitmate apparatus (Pro, COSMED, Rome, Italy). The precise Cosmed Fitmate system was used to determine the RMR energy expenditure. The whole test took approximately 20 min. The device uses the calorimetry method and directly measures the amount of oxygen uptake (with accuracy of ± 0.02%) which makes it possible to measure resting energy expenditure (REE), the resting metabolic rate (RMR) and the basic metabolic rate (BMR). The Canopy measuring cap was used for the measurement. The activity factor (AF) was determined based on the available indicator—2.0-high activity/1.6-medium/1.4-low/and 1.2-sedentary lifestyle [26].

In order to establish the AF, each participant during the first visit was administered a validated International Physical Activity Questionnaire—IPAQ. This questionnaire refers to the work-related PA as well as activities performed at home and its surroundings, any kinds of activities performed in free time (in relation to moderate or intensive physical activity) and time spent sitting. The CRI and CRII diet model provided a total daily intake of 2192 ± 54 kcal and 2133 ± 42 kcal, respectively. The following proportions of carbohydrates, fats and protein were prescribed for both study groups: CRI—3 g/kg/body mass of carbohydrates, 0.8 g/kg/body mass of fats and approximately 1.2 g/kg/body mass of protein. In the CRII diet, the participants consumed 2.9 g/kg/body mass of carbohydrates, 0.74 g/kg/body mass of fats and 1.1 g/kg/body mass of protein. The composition of the diets is presented in Table 1. The applied macronutrient composition was recommended according to the actual norms of nutrition for the Polish population as published by the Food and Nutrition Institute [27]. The meals were prepared in the form of 24-h menus for 7 days of the week. They also consisted of instructions for five meals per day. The particular diet composition was analyzed using DIETETYK 6.0 software (Jumar, Poland). The scheme of the experimental protocol is shown in Figure 1.

2.3. Diet Control

After 3 and 6 weeks of the experiment, the prescribed diets were checked for quantity and quality. The participants were asked to complete a 72 h food diary (2 weekdays and 1 weekend day). To increase the accuracy of the recorded data, each participant received a diet diary booklet that contained menus, pages to record foods, and photographs of food that depicted portion choices for a common food item. The dietician indicated that each study participant should record the food brand and portion size. The amounts consumed were recorded in household units, by volume or by measuring with a ruler. In addition, each subject was interviewed about their dietary patterns in the previous 6 weeks. The dietary records were validated by a nutritionist. Diets were reconstructed from diary entries received from participants in the Diet 5 program and referred to the currently valid standards recommended by the National Food and Nutrition Institute in Warsaw. Nutrient analyses were carried out using the corresponding Polish food table [27]. The participants who did not report to the control visit at the laboratory, as well as those who did not reach the recommended calorie values programed in the experiment due to the excessive consumption of sweetened beverages or alcohol, were eliminated from the study.

2.4. Anthropometric Data

The evaluations of body composition (body mass—BM, kg; body fat BF, %; fat free mass—FFM, %; muscle mass—MM, %; total body water—TBW, %) were carried out twice using electric impedance multi-frequency measurement BIA-1, Akern, Bioresearch SRL, PONASSIEVE, Florence, Italy [28,29]. Based on the obtained values, BMI was calculated using the following formula (BMI = body mass $(kg)/height[m]^2$). The anthropometric data of former athletes before the beginning of the study is summarized in Table 2.

Table 2. The anthropometrics characteristic of the participant at the baseline.

Variable	CRI, $n = 31$ Mean ± SD	CR II, $n = 32$ Mean ± SD
BM, kg	92.3 ± 11.5	89.5 ± 14.0
BMI	28.1 ± 4.2	28.3 ± 4.1
BF, %	30.4 ± 8.0	29.3 ± 8.2
FFM, %	62.84 ± 8.36	58.98 ± 16.34
FFM, kg	67.68 ± 4.77	63.47 ± 17.48
MM, %	49.13 ± 4.59	43.36 ± 11.34
TBW, %	45.24 ± 5.26	42.35 ± 11.57

Note: CR—calorie restriction, BM—body mass; BMI—body mass index, BF—body fat, FFM—free fat mass, MM—muscle mass, TBW—total body water.

2.5. Biochemical Variables and Test Procedures

The following blood metabolic variables were evaluated in all study participants: glucose (GL), total cholesterol (TCh), HDL cholesterol (HDL), LDL cholesterol (LDL), triglycerides (TG), total lipids, insulin-like growth factor-1 (IGF-1), leptin, adiponectin, HOMA-IR, and insulin (I). Fasting blood samples were collected between 08:00 and 10:00 am. at the beginning of the study and after 6 weeks of intervention. After an overnight fast, venous blood samples for lipid analyses were collected into tubes with EDTA anticoagulant, and for glucose estimation in tubes with sodium fluoride and for hormones and IgF-1 in serum tubes. The blood was immediately placed on ice or in a refrigerator, and samples were centrifuged at 2000× g (2500 rpm) for 10 min at 4 °C within 2 h of collection. The plasma was then immediately stored in conditions to minimize artificial oxidation (i.e., with an antioxidant cocktail under an inert atmosphere). Standard blood biochemical analyses such as blood lipid profile and glucose concentration were carried out at the University Hospital Laboratory. The ELISA-automated microparticle enzyme immunoassay test kits for quantitative assessments of leptin, adiponectin and IGF-1 (all from R & D Systems, Minneapolis, MN, USA) were employed. HOMA-IR was calculated from the formula—fasting insulin level × fasting glucose divided by 22.5 [30].

2.6. Statistical Analysis

The Shapiro-Wilk, Levene, and Mauchly's tests were used in order to verify the normality, homogeneity, and sphericity of the sample's data variances, respectively. The verification of the differences between the analyzed values before and after the diet intervention as well as between CRI and CRII diets were verified using ANOVA with repeated measures. Statistical significance was set at $p < 0.05$. Effect sizes (η^2 Eta-squared) were reported for the results where appropriate (Table 3). Parametric effect sizes were defined as large for $\eta^2 > 0.14$, as moderate for 0.06 and small for 0.01 [31,32]. The statistical significance was set at $p < 0.05$. All statistical analyses were performed using Statistica 9.1 and the results were presented as means with standard deviations.

Table 3. Results of Shapiro-Wilk tests for CR I and CR II.

Variables	Shapiro-Wilk Test	
	CR I	**CR II**
BM, kg	0.756	0.767
BMI	0.687	0.707
BF, %	0.686	0.715
FFM, %	0.713	0.743
FFM, kg	0.730	0.760
MM, %	0.749	0.778
TBW, %	0.764	0.791
GL, mmol/L	0.792	0.817
TCh, mg/dL	0.805	0.828
HDL, mg/dL	0.814	0.837
LDL, mg/dL	0.825	0.846
TG, mg/dL	0.835	0.855
TL, mg/dL	0.844	0.863
I, mg/dL	0.851	0.869
HOMA-IR	0.858	0.874
Leptin, ng/mL	0.791	0.813
Adiponectin, ng/mL	0.815	0.833
IGF-1, ng/mL	0.818	0.841

3. Results

Of the 96 randomly chosen participants, 63 completed the 6-week study (Figure 1). No significant differences were noted between CR I and CR II groups for all measured variables before and after the diet intervention, except tCh and LDL (Table 4). After the diet intervention, a statistically significant decrease in TG, TL and leptin, and an increase in adiponectin levels were observed in both the CR I and the CR II group (Table 4). A high value of SD in BM before and after the CRI and CRII diet (Table 4), resulted in the groups being divided into subgroups, depending on the body mass reduction. It was revealed that after 6 weeks of diet in the CRI 1.5–2.5 subgroup, FFM (kg) and MM (%) were significantly reduced (Table 5). There were significant differences in BF, FFM (%), MM, TBW (Table 5), and tCh (Table 5) between CRI and CRII in the 3 kg subgroup. After the 6-week intervention, in the CRI 1.5–2.5 subgroup there was a significant TG decrease (Table 6). In addition, after the same amount of time in the three subgroups of CR II, a significant decrease was observed in the tCh TL and LDL level (Table 6). Both CR I and CR II induced significant increases in adiponectin concentration in all subgroups, while leptin concentration was significantly reduced in all subgroups after the CR I and CR II diet model (Table 6). After the diet intervention, a statistically significant decrease in leptin and an increase in adiponectin levels were observed in all subgroups in both the CR I and CR II diet (Table 6).

Table 4. Differences in body composition, lipid profile and glucose control variables before and after CRI and CRII diet.

	Variable	CR I (n = 31)			CR II (n = 32)			ES CR II vs. CR I
		Before	After	ES	Before	After	ES	
		Mean ± SD	Mean ± SD		Mean ± SD	Mean ± SD		
Body composition	BM, kg	92.3 ± 11.5	90.4 ± 10.7	0.014	89.5 ± 14.0	87.1 ± 12.6	0.031	0.037
	BMI	28.1 ± 4.2	27.8 ± 3.7	0.011	28.3 ± 4.1	27.3 ± 4.0	0.010	0.011
	BF, %	30.4 ± 8.0	29.2 ± 5.8	0.013	30.71 ± 5.9	29.0 ± 7.7	0.027	0.014
	FFM, kg	62.84 ± 8.36	58.95 ± 7.26	0.015	59.98 ± 16.34	58.02 ± 14.82	0.024	0.012
	FFM, %	67.68 ± 4.77	67.62 ± 4.95	0.010	66.39 ± 14.10	63.47 ± 17.48	0.055	0.045
	MM, %	49.13 ± 4.59	48.70 ± 6.07	0.011	48.39 ± 12.62	43.36 ± 11.34	0.058	0.051
	TBW, kg	45.24 ± 5.26	45.18 ± 4.83	0.010	43.95 ± 11.80	42.35 ± 11.57	0.026	0.034
Lipid profile	tCh, mg/dL	197.2 ± 40.3	171.4 ± 53.4 *	0.343 *	198.1 ± 40.5	187.3 ± 37.1 #*	0.241	0.351 #*
	HDL, mg/dL	51.6 ± 13.4	47.0 ± 16.2	0.061	54.7 ± 13.3	56.4 ± 17.9	0.021	0.061
	LDL, mg/dL	107.6 ± 28.3	102.5 ± 64.7	0.072	122.8 ± 42.0	116.3 ± 34.8 #*	0.253	0.347 #*
	TG, mg/dL	102.3 ± 57.3	89.7 ± 54.8 *	0.333	108.5 ± 45.8	83.0 ± 48.7 *	0.249 *	0.057
	TL, mg/dL	599.51 ± 102.7	520.80 ± 105.1 *	0.349 *	630.44 ± 124.4	572.93 ± 97.3 *	0.135 *	0.355
Glucose control	GL, mmol/L	5.24 ± 0.83	4.86 ± 1.27	0.070	5.34 ± 1.53	5.17 ± 1.01	0.013	0.067
	I, mg/dL	10.9 ± 6.9	8.7 ± 5.2	0.034	10.0 ± 8.2	9.06 ± 5.8	0.021	0.028
	HOMA-IR	2.6 ± 2.3	2.4 ± 1.1	0.010	2.7 ± 1.3	2.0 ± 1.4	0.010	0.012
	Leptin, ng/mL	13.78 ± 4.8	9.45 ± 6.9 *	0.148 *	13.27 ± 4.7	9.69 ± 3.9 *	0.144 *	0.019
	Adiponectin, ng/mL	3.33 ± 1.3	7.10 ± 3.2 *	0.166 *	3.83 ± 1.2	6.93 ± 3.3 *	0.147 *	0.022
	IGF-1, ng/mL	43.93 ± 4.6	42.50 ± 14.8	0.041	44.76 ± 10.4	42.93 ± 6.6	0.023	0.011

Note: ES—effect size, BM—body mass; BMI—body mass index, BF—body fat, FFM—free fat mass, MM—muscle mass, TBW—total body water, TL—total lipids, TG—triglycerides, tCh—total cholesterol, HDL—cholesterol HDL, LDL—cholesterol LDL, GL—glucose, I—insulin, IGF-1—insulin-like growth factor, *—statistically significant difference with $p < 0.01$ compared with before and ES for after, #—statistically significant difference with $p < 0.05$ compared with the CRI and ES for CRII; #*—statistically significant difference with $p < 0.05$ compared with the CRI and ES for CRII and after.

Table 5. Results of the CR I and CR II diet on body mass and composition before and after the diet intervention.

	CR I					
Variables	1.5–2.5 kg		2.5–3.0 kg		Over 3 kg	
	Before	After	Before	After	Before	After
BM, Kg	97.95 ± 4.5	92.26 ± 3.1	99.60 ± 6.1	94.75 ± 4.2	105.67 ± 5.2	100.91 ± 8.1
BMI	31.26 ± 1.2	28.14 ± 0.2	30.65 ± 2.13	29.23 ± 1.3	32.19 ± 2.5	30.86 ± 2.1
BF, %	30.68 ± 3.1	29.92 ± 2.1	32.46 ± 5.3	31.85 ± 3.4	37.61 ± 3.4	37.10 ± 5.3
FFM, %	69.32 ± 4.2	68.05 ± 3.64	67.60 ± 4.2	66.54 ± 5.3	63.02 ± 4.2	62.39 ± 5.2
FFM, Kg	59.52 ± 3.1	55.75 ± 2.36 *	60.70 ± 4.1	55.91 ± 4.2	55.94 ± 4.1	53.53 ± 2.5
MM, %	49.97 ± 5.3	46.91 ± 2.77 *	49.56 ± 3.1	46.97 ± 2.7	42.96 ± 2.6	42.67 ± 2.9
TBW, %	50.87 ± 5.3	49.19 ± 3.36	49.61 ± 4.2	48.55 ± 3.1	47.59 ± 3.9	46.84 ± 2.8

Table 5. *Cont.*

Variables	CR II					
	1.5–2.5 kg		2.5–3.0 kg		Over 3 kg	
	Before	After	Before	After	Before	After
BM, Kg	96.95 ± 8.9	93.37 ± 6.3	100.43 ± 4.2	96.27 ± 4.3	98.67 ± 3.8	93.15 ± 2.1
BMI	30.08 ± 1.3	28.99 ± 1.4	30.78 ± 2.1	29.56 ± 1.1	30.40 ± 2.1	28.30 ± 1.4
BF, %	32.89 ± 2.1	31.86 ± 2.4	34.44 ± 2.6	33.75 ± 1.3	33.61 ± 2.1 #	31.21 ± 0.9 #
FFM, %	67.60 ± 3.1	65.84 ± 4.2	65.56 ± 4.2	65.21 ± 3.8	68.79 ± 4.9 #	67.90 ± 5.3
FFM, Kg	58.48 ± 4.1	57.86 ± 3.2	58.53 ± 3.6	55.98 ± 4.1	59.71 ± 3.2	54.86 ± 3.1
MM, %	47.83 ± 2.1	46.92 ± 3.5	46.83 ± 2.6	44.74 ± 2.5	49.86 ± 2.1 #	46.64 ± 2.5
TBW, %	50.25 ± 4.1	48.86 ± 3.4	48.85 ± 3.6	47.82 ± 4.1	50.59 ± 3.2 #	48.56 ± 3.5

Note: BM—body mass; BMI—body mass index, BF—body fat, FFM—free fat mass, MM—muscle mass, TBW—total body water, *—statistically significant difference with $p < 0.05$ compared with baseline, #—statistically significant difference with $p < 0.05$ compared with the CRI.

Table 6. Results of the CR I and CR II diet on lipid profile and glucose control variables before and after the diet intervention in particular subgroups.

	Variables	CR I					
		1.5–2.5 kg		2.5–3.0 kg		Over 3 kg	
		Before	After	Before	After	Before	After
Lipid profile	TL, mg/dL	735.63 ± 33.5	609.65 ± 41.23	775.00 ± 47.3	665.36 ± 63.1	608.86 ± 51.3	528.50 ± 72.1
	TG, mg/dL	131.63 ± 13.37	84.41 ± 14.91 *	127.30 ± 8.2	84.45 ± 13.1	126.00 ± 11.2	82.63 ± 8.2
	tCh, mg/dL	201.56 ± 35.26	162.35 ± 26.13	200.90 ± 26.1	166.55 ± 21.4	181.14 ± 38.5	165.00 ± 27.2
	LDL, mg/dL	132.25 ± 14.81	104.47 ± 12.1	139.90 ± 12.5	110.45 ± 9.2	120.86 ± 17.1	99.87 ± 15.5
	HDL, mg/dL	49.81 ± 2.94	49.12 ± 3.2	46.37 ± 5.3	47.62 ± 8.2	49.43 ± 3.2	50.07 ± 4.1
Glucose control	GL, mmol/L	5.30 ± 1.42	4.78 ± 0.6	5.50 ± 0.8	4.95 ± 1.7	5.51 ± 1.5	5.12 ± 1.2
	I, mg/dL	11.08 ± 2.62	10.14 ± 1.3	13.52 ± 4.1	12.86 ± 3.1	11.79 ± 3.7	10.18 ± 4.2
	HOMA-IR	2.68 ± 0.31	2.43 ± 0.24	3.36 ± 1.3	3.15 ± 1.5	2.82 ± 0.8	2.36 ± 0.7
	Leptin, ng/mL	12.32 ± 1.2	9.76 ± 1.3 *	13.10 ± 1.4	8.45 ± 0.4 *	15.43 ± 2.3	9.19 ± 1.3 *
	Adiponectin, ng/mL	3.51 ± 0.2	6.45 ± 0.5 *	3.50 ± 0.3	7.55 ± 0.5 *	3.12 ± 0.6	7.20 ± 1.2 *
	IGF-1, ng/mL	42.23 ± 0.9	41.90 ± 2.1	43.06 ± 1.1	41.60 ± 0.8	44.21 ± 3.4	43.10 ± 1.5

	Variables	CR II					
		1.5–2.5 kg		2.5–3.0 kg		Over 3 kg	
		Before	After	Before	After	Before	After
Lipid profile	TL, mg/dL	639.71 ± 42.3	559.36 ± 23.2 *	632.70 ± 54.2	540.40 ± 21.1 *	649.14 ± 32.1	548.57 ± 24.5 *
	TG, mg/dL	99.50 ± 4.2	71.36 ± 12.1	109.00 ± 12.3	76.70 ± 21.5	109.00 ± 15.1	73.86 ± 23.6
	tCh, mg/dL	204.54 ± 18.1	183.64 ± 13.1 *	199.40 ± 14.2	172.80 ± 11.2 *	206.71 ± 14.1 #	177.57 ± 12.2 *
	HDL, mg/dL	55.50 ± 8.2	56.63 ± 6.1	50.80 ± 3.1	49.89 ± 4.1	50.42 ± 4.1	50.76 ± 3.4
	LDL, mg/dL	131.43 ± 9.4	112.64 ± 7.3 *	126.90 ± 8.2	107.60 ± 8.2 *	134.5 ± 7.27	112.29 ± 9.4 *
Glucose control	GL, mmol/L	5.07 ± 1.1	5.04 ± 0.8	5.34 ± 0.7	5.12 ± 0.9	5.39 ± 0.8	5.10 ± 1.1
	I, mg/dL	10.77 ± 1.8	8.34 ± 1.3	12.29 ± 1.3	8.72 ± 2.8	14.99 ± 2.1	9.90 ± 3.1
	HOMA-IR	2.52 ± 0.5	1.93 ± 0.9	2.91 ± 0.7	2.13 ± 0.5	3.57 ± 1.4	2.46 ± 1.1
	Leptin, ng/mL	11.86 ± 2.1	8.13 * ± 2.5	12.10 ± 3.1	9.65 * ± 0.3	15.15 ± 1.1	10.30 * ± 1.2
	Adiponectin, ng/mL	4.08 ± 1.2	8.10 * ± 2.6	3.84 ± 3.2	7.20 * ± 0.4	3.60 ± 0.8	6.95 * ± 0.4
	IGF-1, ng/mL	44.35 ± 2.7	43.16 ± 3.1	42.94 ± 2.9	43.08 ± 1.5	42.40 ± 1.9	41.30 ± 0.8

Note: TL—total lipids, TG—triglycerides, tCh—total cholesterol, HDL—cholesterol HDL, LDL—cholesterol LDL, GL—glucose, I—insulin, IGF-1—insulin-like growth factor, *—statistically significant difference with $p < 0.05$ compared with baseline, #—statistically significant difference with $p < 0.05$ compared with the CRI.

4. Discussion

The current literature does not provide scientifically-based nutritional standards for former athletes [33–35]. However, there are studies concerning nutritional methods and observations of the health status of these subjects [5,36–38]. In the present study, from the 96 randomly chosen participants to consume caloric-restriction diets, 63 completed the 6-week study. Thirty-one participants (18 subjects from CR I group and 13 subjects from CR II) resigned from the study. The high number of participants who discontinued the study was caused by many diet restriction such as the need to have regular meals or no snacking. When participants met with the dietitians, they which informed them that despite their initial readiness, they could not resign from their usual diet habits especially from snacking, eating sweets or consuming alcohols. Unfortunately, it is a typical problem, psychological rather than physiological, common for obese and overweight individuals. Although these people are aware of the fact that high body mass and fat content significantly increase the risk of developing many fatal disorders, they cannot commit fully to the dietary restrictions. Studies show that individuals who decide to reduce their body mass and fat content, apart from dietetics, require psychological care [39,40].

In the present study, both CR I (reducing daily caloric intake by 20% of TDEE) and CRII diets (reducing daily caloric intake by 30% of TDEE), when adjusted to the caloric needs of a participant, helped to reduce body mass, thus improving BMI. In former athletes who applied the reducing daily caloric intake by 30% of TDEE and lowered their body mass by 1.5–2.5 kg, 2.5–3.0 kg, and over 3.0 kg, a significant improvement in lipid variables (tCh, LDL and TL) was observed, as insulin and levels were decreased and HOMA-IR was reduced. When comparing the CRI and CRII groups, it appeared that a more drastic reduction of calories observed in the CRII model improved the variables of lipid metabolism to a greater extent. This was especially visible in subjects who reduced their body mass by more than 3 kg. In former athletes who reduced daily caloric intake by 20% of TDEE and achieved a body mass loss of 1.5–2.5 kg, a significant decrease in FFM and MM was observed. This suggests that a calorie restriction diet does not protect the muscle tissue from being catabolized [41]. The lower glucose concentration resulting from this calorie restriction can be explained by a greater uptake of glucose by increased muscle mass.

Losing weight and increasing physical activity independently has a beneficial effect on the glucose metabolism and insulin sensitivity [42]. Improvement in insulin sensitivity with dietary intervention depends on the number of calories reduced. The more restrictive the diet, the greater improvement in insulin sensitivity [42]. Our study participants demonstrated no significant improvement in glucose, I or HOMA- IR, as they consumed a relatively high number of calories daily (2192 ± 51 kcal). Insulin sensitivity improvement can be observed in healthy subjects even in short-lasting but very low-calorie diets, where calorie intake can only be 500 kcal/day [43]. Such a diet model induces a marked decrease in abdominal adipo tissue, reduction in adipocyte size, an increased adipo tissue gene expression of mitochondrial biogenesis markers and non-mitochondrial oxygen consumption pathways, as well as improved whole-body insulin sensitivity [43]. Other authors confirm that calorie restriction helps decrease body mass, reduces insulin, glucose and HOMA-IR, and stabilizes their level, even if the subjects are not physically active [1,10,12]. Research on reducing daily caloric intake by 30% of TDEE has also been carried out on subjects with type 2 diabetes. The results indicate that this relatively mild caloric restriction helps decrease the level of insulin in diabetic patients [44]. Weiss and colleagues [45] in their research performed on 52 men and women aged 45–65 observed a decrease in the level of insulin and glucose and an improvement in insulin sensitivity after the introduction of a diet based on calorie restriction.

The loss of body mass due to calorie restriction significantly affects leptin and adiponectin secretion in fat tissue [46]. Indeed, strong evidence indicates that lowering body mass contributes to the reduction of blood leptin concentration. In the research carried out by Wing et al. [47] on 52 obese women, after the introduction of a calorie-restricted diet, the average decrease of body mass was 8.1 kg and the concentration of leptin was lowered from 30.1 to 20.4 ng/mL [47]. Sartorio et al. [48] in a study that lasted 3 weeks and involved a group of 54 obese patients, achieved a statistically significant reduction in plasma leptin concentration in both men and women (women: from 41.1 ± 3.6 ng/mL to 29.9 ± 3.0 ng/mL; men: from 19.4 ± 2.6 ng/mL to 11.6 ± 1.3 ng/mL) [48]. Adiponectin demonstrates the opposite effect. Its level rises in the blood in proportion to the reduction of body mass [49,50], a phenomenon that was also confirmed in the present study. Adiponectin can influence body mass reduction through the stimulation of the release of free fatty acids and glucose from peripheral tissues [51].

The present study did not show any changes in the blood concentration of IGF-1 in any of the groups. This is a rather unfavorable phenomenon because IGF-1 can promote muscle lipids and glucose metabolism and reduce insulin resistance [20,52]. Fontana et al. [22] in a study performed on men and woman showed that a 2-year calorie restriction diet caused a significant decrease in the concentration of IGF-1.

The former athletes participating in this study had a significantly increased body fat mass, which varied from 28% to 30% in most cases. Similar findings were reported by other authors [1]. Overweight and obesity among former athletes has been observed more often in athletes participating in speed–strength sport disciplines and those with weight categories in comparison to aerobic endurance sports [5,53,54]. This association is strongly related to body mass and fat mass during their sports careers [55–58]. Among NFL and NHL players, shot putters, and wrestlers in the heavyweight division, body mass helps in achieving better sport results. Borchers et al. [59] found that the prevalence of obesity was 21% in former NFL players. Kujala et al. [60] also confirmed that former power-sports athletes had a higher body mass compared with endurance athletes, especially long-distance runners and cyclists. Similarly, Albuquerque et al. [4] and Hyman et al. [54] showed that obesity is common in retired NFL players.

Not all former athletes have problems with body mass. Former endurance-trained athletes have a much smaller likelihood of obesity [53,56]. Marquet et al. [53] confirmed that former cyclists have a significantly lower prevalence of obesity than the general population [53]. This may be caused by previously increased energy expenditure due to significant training loads, as well as high total daily energy expenditure compared with sedentary subjects [61]. In addition, Vogt et al. [62] indicated that elite cyclists had a 30% higher daily energy expenditure during the competitive season compared to preseason training. Other authors also explain this phenomenon by pointing to the very active lifestyle of retired endurance athletes [3,53]. Cyclists or marathoners even after the end of their careers maintain proper body weight and optimal body fat [49]. Phil and Jurimae [4] indicate that more than half of the former athletes in Finland engage in regular leisure-time physical activity or compete in different kinds of sports throughout their adult life, maintaining normal body weight. Similar observations were reported by Sarna et al. [2] in former Estonian athletes.

It can be concluded that calorie-restriction diets are effective in reducing body mass in overweight and obese former athletes. Reducing daily caloric intake by 30% of TDEE is especially suitable for subjects with high body fat content, as well as increased glucose, insulin, lipoprotein and cholesterol levels. The reduction of body mass by 3 kg or more allows for a significant improvement in the lipid profile, leading to favorable endocrine changes. The concentration of leptin is reduced while that of adiponectin likely increases, stimulating the release of free fatty acids and glucose from peripheral tissues. It seems that former athletes should adopt a dietary strategy which allows for the preservation of muscle mass while reducing body fat content. Reducing daily caloric intake by 30% of TDEE can be recommended for overweight and obese former athletes to maintain proper body mass and health.

The present study has several strengths but also some limitations. The former includes a unique, large research group, the use of two calorically-different diet models, and a relatively long 6-week dietary intervention. The main limitation is the adherence of the study participants to the prescribed diet, effective control of dietary intake by the subjects, and the inability to unify the groups by preparing the same variable-calorie meals at the same time of day. These variables likely affected the final results of weight reduction, which are often quite divergent within the same dietary group.

5. Conclusions

For overweight and obese former athletes, a calorie restriction diet based on a 30% reduction of TDEE is more effective with regards to improved body mass, lipid profile, and reduced insulin and HOMA-IR levels compared to a 20% calorie restriction diet. Both the 20% and 30% of TDEE restriction diets reduced the levels of leptin and increased adiponectin concentration.

Availability of Data and Material: The datasets used and/or analyzed during the current study are available from the corresponding author upon reasonable request.

Author Contributions: J.H. and E.S. conceived and designed the research protocol. M.M.M. and A.Z., J.H., and E.S. performed analysis and data interpretation and wrote and edited the manuscript. M.C.-R., K.R., M.B., and K.J. collected the data, performed statistical analysis, provided the lab equipment for analysis and performed blood analysis. All authors read and approved the final manuscript.

Funding: The research project was supported by grant NCN KB-0012/53/11.

Acknowledgments: The authors would like to thank Paweł Cięszczyk, Paweł Wysokiński, Dominika Maciejewska, Krzysztof Ficek, Krzysztof Wilk, Jakub Piotrowski, and Anna Lubkowska for providing valuable comments while preparing the manuscript.

Conflicts of Interest: The authors declare no conflict of interest.

Abbreviations

CRI	Caloric restriction diet I
CRII	Caloric restriction diet II
BM	Body mass
BMI	Body mass index
BF	Body fat
FFM	Fat free mass
MM	Muscle mass
TBW	Total body water
GL	Glucose
TCh	Total cholesterol
HDL	HDL cholesterol
LDL	LDL cholesterol
TG	Triglycerides
TL	Total lipids
IGF-1	Insulin-like growth factor-1
I	Insulin
HOMA-IR	Homeostasis model assessment- insulin resistance
CRs	Calorie restrictions
TDEE	Total daily energy expenditure
AF	Activity factor
RMR	Resting metabolic rate
ANOVA	Analysis of variance
HOMA-IR	Homeostasis model assessment- insulin resistance
CRs	Calorie restrictions
TDEE	Total daily energy expenditure
AF	Activity factor
RMR	Resting metabolic rate
ANOVA	Analysis of variance

References

1. Maciejewska, D.; Michalczyk, M.; Czerwińska-Rogowska, M.; Banaszczak, M.; Ryterska, K.; Jakubczyk, K.; Piotrwski, J.; Hołowko, J.; Drozd, A.; Wysokińki, P.; et al. Seeking Optimal Nutrition for Healthy Body Mass Reduction among Former Athletes. *J. Hum. Kinet.* **2017**, *60*, 63–75. [CrossRef] [PubMed]

2. Sarna, S.; Sahi, T.; Koskenvuo, M.; Kaprio, J. Increased life expectancy of world class male athletes. *Med. Sci. Sports Exerc.* **1993**, *25*, 237–244. [CrossRef] [PubMed]

3. O'Kane, J.W.; Teitz, C.C.; Fontana, S.M.; Lind, B.K. Prevalence of obesity in adult population of former college rowers. *J. Am. Board Fam. Pract.* **2002**, *15*, 451–456. [PubMed]

4. Pihl, E.; Jurimae, T. Relationships between body weight change and cardiovascular disease risk factors in male former athletes. *Int. J. Obes.* **2001**, *25*, 1057–1062. [CrossRef] [PubMed]

5.   Albuquerque, F.N.; Kuniyoshi, F.H.S.; Calvin, A.D.; Sierra-Johnson, J.; Romero-Corral, A.; Lopez-Jimenez, F.; George, C.F.; Rapoport, D.M.; Vogel, R.A.; Khandheria, B.; et al. Sleep-disordered breathing, Hypertension and Obesity in Retired National Football League Players. *J. Am. Coll. Cardiol.* **2010**, *56*, 1432–1433. [CrossRef] [PubMed]

6.   Batista, C.; Soares, J.M. Are former elite athletes more protected against metabolic syndrome? *J. Cardiol.* **2013**, *61*, 440–445. [CrossRef] [PubMed]

7.   King, N.A.; Caudwell, P.; Hopkins, M.; Byrne, N.M.; Colley, R.; Hills, A.P.; Stubbs, J.R.; Blundell, J.E. Metabolic and Behavioral Compensatory Responses to Exercise Interventions: Barriers to Weight Loss. *Obesity* **2007**, *15*, 1373–1383. [CrossRef]

8.   Bescós, R.; Rodríguez, F.A.; Iglesias, X.; Knechtle, B.; Benítez, A.; Marina, M.; Padullés, J.M.; Torrado, P.; Vazquez, J.; Rosemann, T. Nutritional behavior of cyclists during a 24-hour team relay race: A field study report. *J. Int. Soc. Sports Nutr.* **2012**, *9*, 3. [CrossRef]

9.   Hulmi, J.J.; Isola, V.; Suonpää, M.; Järvinen, N.J.; Kokkonen, M.; Wennerström, A.; Nyman, K.; Perola, M.; Ahtiainen, J.P.; Häkkinen, K. The Effects of Intensive Weight Reduction on Body Composition and Serum Hormones in Female Fitness Competitors. *Front. Physiol.* **2016**, *7*, 689. [CrossRef]

10.   Holloszy, J.O.; Fontana, L. Caloric Restriction in Humans. *Exp. Gerontol.* **2007**, *42*, 709–712. [CrossRef]

11.   Fontana, L.; Meyer, T.E.; Klein, S.; Holloszy, J.O. Long-term calorie restriction is highly effective in reducing the risk for atherosclerosis in humans. *Proc. Natl. Acad. Sci. USA* **2004**, *101*, 6659–6663. [CrossRef] [PubMed]

12.   Larson-Meyer, D.E.; Heilbronn, L.K.; Redman, L.M.; Newcomer, B.R.; Frisard, M.I.; Anton, S.; Smith, S.R.; Alfonso, A.; Ravussin, E.; the Pennington CALERIE Team. Effect of calorie restriction with or without exercise on insulin sensitivity, beta-cell function, fat cell size, and actopic lipid in overweight subjects. *Diabetes Care* **2006**, *29*, 1337–1344. [CrossRef]

13.   Wang, Y. Molecular Links between Caloric Restriction and Sir2/SIRT1 Activation. *Diabetes Metab. J.* **2014**, *38*, 321–329. [CrossRef] [PubMed]

14.   Sharples, A.P.; Hughes, D.C.; Deane, C.S.; Saini, A.; Selman, C.; Stewart, C.E. Longevity and skeletal muscle mass: the role of IGF signalling, the sirtuins, dietary restriction and protein intake. *Aging Cell* **2015**, *14*, 511–523. [CrossRef]

15.   Nguyen, P.A.H.; A Heggermont, W.; Vanhaverbeke, M.; Dubois, C.; Vydt, T.; Vörös, G.; Van Der Schueren, B.; Overbergh, L.; Mathieu, C.; Desmet, W.; et al. Leptin-adiponectin ratio in pre-diabetic patients undergoing percutaneous coronary intervention. *Acta Cardiol.* **2015**, *70*, 640–646. [CrossRef] [PubMed]

16.   Hall, M.E.; Harmancey, R.; Stec, D.E. Lean heart: Role of leptin in cardiac hypertrophy and metabolism. *World J. Cardiol.* **2015**, *7*, 511–524. [CrossRef] [PubMed]

17.   Dyck, D.J. Adipokines as regulators of muscle metabolism and insulin sensitivity. *Appl. Physiol. Nutr. Metab.* **2009**, *34*, 396–402. [CrossRef]

18.   Komai, N.; Morita, Y.; Sakuta, T.; Kuwabara, A.; Kashihara, N. Anti-tumor necrosis factor therapy increases serum adiponectin levels with the improvement of endothelial dysfunction in patients with rheumatoid arthritis. *Mod. Rheumatol.* **2007**, *17*, 385–390. [CrossRef]

19.   Chan, K.C.; Chou, H.H.; Wu, D.J.; Wu, Y.L.; Huang, C.N. Diabetes mellitus has an additional effect on coronary artery disease. *Jpn. Heart J.* **2004**, *45*, 921–927. [CrossRef]

20.   Aguirre, G.A.; De Ita, J.R.; De La Garza, R.G.; Castilla-Cortazar, I. Insulin-like growth factor-1 deficiency and metabolic syndrome. *J. Transl. Med.* **2016**, *14*, 2486. [CrossRef]

21.   Frystyk, J.; Vestbo, E.; Skjærbaek, C.; Mogensen, C.E.; Ørskov, H. Free insulin-like growth factors in human obesity. *Metabolism* **1995**, *44*, 37–44. [CrossRef]

22.   Fontana, L.; Villareal, D.T.; Das, S.K.; Smith, S.R.; Meydani, S.N.; Pittas, A.G.; Klein, S.; Bhapkar, M.; Rochon, J.; Ravussin, E.; et al. Effects of 2-year calorie restriction on circulating levels of IGF-1, IGF-binding proteins and cortisol in nonobese men and women: A randomized clinical trial. *Aging Cell* **2016**, *15*, 22–27. [CrossRef] [PubMed]

23.   Wallace, T.M.; Matthews, D.R. The assessment of insulin resistance in man. *Diabet. Med.* **2002**, *19*, 527–534. [CrossRef] [PubMed]

24.   Emami, M.; Behforouz, A.; Jarahi, L.; Zarifian, A.; Rashidlamir, A.; Rashed, M.M.; Khaleghzade, H.; Ghaneifar, Z.; Safarian, M.; Azimi-Nezhad, M.; et al. The risk of developing obesity, insulin resistance, and metabolic syndrome in former power-sports athletes—Does sports career termination increase the risk. *Indian J. Endocrinol. Metab.* **2018**, *22*, 515–519. [CrossRef] [PubMed]

25. Johnson, M.L.; Distelmaier, K.; Lanza, I.R.; Irving, B.A.; Robinson, M.M.; Konopka, A.R.; Shulman, G.I.; Nair, K.S. Mechanism by Which Caloric Restriction Improves Insulin Sensitivity in Sedentary Obese Adults. *Diabetes* **2016**, *65*, 74–84. [CrossRef] [PubMed]

26. Jarosz, M. *Nutrition Norms for the Polish Population*; Institute of Nutrition: Warsaw, Poland, 2017; pp. 21–39. (In Polish)

27. Kunachowicz, H.; Nadolna, I.; Iwanow, K.; Przygoda, B. *Nutritional Value of Products and Traditional Dishes*; PZWL: Warszawa, Poland, 2005. (In Polish)

28. Fosbøl, M.; Zerahan, B. Contemporary methods of body composition measurement. *Clin. Physiol. Funct. Imaging* **2014**, *35*. [CrossRef] [PubMed]

29. Kyle, U.G.; Bosaeus, I.; De Lorenzo, A.D.; Deurenberg, P.; Elia, M.; Gómez, J.M.; Heitmann, B.L.; Kent-Smith, L.; Melchior, J.-C.; Pirlich, M.; et al. Bioelectrical impedance analysis—part I: Review of principles and methods. *Clin. Nutr.* **2004**, *23*, 1430–1453. [CrossRef] [PubMed]

30. Matthews, D.R.; Hosker, J.P.; Rudenski, A.S.; Naylor, B.A.; Treacher, D.F.; Turner, R.C. Homeostasis model assessment: Insulin resistance and beta-cell function from fasting plasma glucose and insulin concentrations in man. *Diabetologia* **1985**, *28*, 412–419. [CrossRef] [PubMed]

31. Maszczyk, A.; Gołaś, A.; Pietraszewski, P.; Roczniok, R.; Zajac, A.; Stanula, A. Application of Neural and Regression Models in Sports Results Prediction. *Procedia Soc. Behav. Sci.* **2014**, *117*, 482–487. [CrossRef]

32. Maszczyk, A.; Golas, A.; Czuba, M.; Krol, H.; Wilk, M.; Stastny, P.; Goodwin, J.; Kostrzewa, M.; Zajac, A. Emg analysis and modelling of flat bench press using artificial neural networks. *S. Afr. J. Res. Sport Phys. Educ. Recreat.* **2016**, *38*, 91–103.

33. Jäger, R.; Kerksick, C.M.; Campbell, B.I.; Cribb, B.I.; Wells, S.D.; Skwiat, T.M.; Purpura, M.; Ziegenfuss, T.; Ferrando, A.A.; Arent, S.M.; et al. International Society of Sports Nutrition Position Stand: Protein and exercise. *J. Int. Soc. Sports Nutr.* **2017**, *14*, 20. [CrossRef] [PubMed]

34. Michalczyk, M.M.; Czuba, M.; Zydek, G.; Zając, A.; Langfort, J. Dietary Recommendations for Cyclists during Altitude Training. *Nutrients* **2016**, *8*, 377. [CrossRef] [PubMed]

35. Helms, E.R.; Aragon, A.; Fitschen, P.J. Evidence-based recommendations for natural bodybuilding contest preparation: nutrition and supplementation. *J. Int. Soc. Sports Nutr.* **2014**, *11*, 20. [CrossRef] [PubMed]

36. Fogelholm, M.; Kaprio, J.; Sarna, S. Healthy lifestyles of former Finnish word class athletes. *Med. Sci. Sports Exerc.* **1994**, *26*, 224–229. [CrossRef]

37. Saarni, S.E.; Rissanen, A.; Sarna, S.; Koskenvuo, M.; Kaprio, J. Weight cycling of athletes and subsequent weight gain in middleage. *Int. J. Obes.* **2006**, *30*, 1639–1644. [CrossRef]

38. Sunman, M.L.; Hoerr, S.L.; Sprague, H.; Olson, H.W.; Quinn, T.J. Lifestyle variables as predictors of survival in former college men. *Nutr. Res.* **1991**, *11*, 141–148. [CrossRef]

39. Martin, A.; Booth, J.N.; Laird, Y.; Sproule, J.; Reilly, J.J.; Saunders, D.H. Physical activity, diet and other behavioural interventions for improving cognition and school achievement in children and adolescents with obesity or overweight. *Cochrane Database Syst. Rev.* **2018**, *3*, CD009728. [CrossRef]

40. Lau, D.C.; Douketis, J.D.; Morrison, K.M.; Hramiak, I.M.; Sharma, A.M.; Ur, E. Obesity Canada Clinical Practice Guidelines Expert Panel. 2006 Canadian clinical practice guidelines on the management and prevention of obesity in adults and children [summary]. *CMAJ* **2007**, *176*, 1–13. [CrossRef]

41. Racette, S.B.; Rochon, J.; Uhrich, M.L.; Villareal, D.T.; Das, S.K.; Fontana, L.; Bhapkar, M.; Martin, C.K.; Redman, L.M.; Fuss, P.J.; et al. Effects of Two Years of Calorie Restriction on Aerobic Capacity and Muscle Strength. *Med. Sci. Sports Exerc.* **2017**, *49*, 2240–2249. [CrossRef]

42. Lin, C.Y.; Chen, P.C.; Kuo, H.K.; Lin, L.Y.; Lin, J.W.; Hwang, J.J. Effects of obesity, physical activity, and cardiorespiratory fitness on blood pressure, inflammation, and insulin resistance in the National Health and Nutrition Survey 1999–2002. *Nutr. Metab. Cardiovasc. Dis.* **2009**, *20*, 713–719. [CrossRef]

43. Vink, R.G.; Roumans, N.J.; Čajlaković, M.; Cleutjens, J.P.M.; Boekschoten, M.V.; Fazelzadeh, P.; Vogel, M.A.A.; Blaak, E.E.; Mariman, E.C.; Van Baak, M.A.; et al. Diet-induced weight loss decreases adipose tissue oxygen tension with parallel changes in adipose tissue phenotype and insulin sensitivity in overweight humans. *Int. J. Obes.* **2017**, *41*, 722–728. [CrossRef] [PubMed]

44. Meehan, C.A.; Cochran, E.; Mattingly, M.; Gorden, P.; Brown, R.J.; Stergios, P. Mild Caloric Restriction Decreases Insulin Requirements in Patients with Type 2 Diabetes and Severe Insulin Resistance. *Medicine* **2015**, *94*, e1160. [CrossRef] [PubMed]

45. Weiss, E.P.; Albert, S.G.; Reeds, D.N.; Kress, K.S.; Ezekiel, U.R.; McDaniel, J.L.; Patterson, B.W.; Klein, S.; Villareal, D.T. Calorie Restriction and Matched Weight Loss from Exercise: Independent and Additive Effects on Glucoregulation and the Incretin System in Overweight Women and Men. *Diabetes Care* **2015**, *38*, 1253–1262. [CrossRef] [PubMed]

46. MacLean, P.S.; Higgins, J.A.; Giles, E.D.; Sherk, V.D.; Jackman, M.R. The role for adipose tissue in weight regain after weight loss. *Obes. Rev.* **2015**, *16*, 45–54. [CrossRef] [PubMed]

47. Wing, R.R.; Sinha, M.K.; Considine, R.V.; Caro, J.F.; Lang, W. Relationship Between Weight Loss Maintenance and Changes in Serum Leptin Levels. *Horm. Metab. Res.* **1996**, *28*, 698–703. [CrossRef] [PubMed]

48. Sartorio, A.; Agosti, F.; Resnik, M.; Lafortuna, C.L. Effects of a 3-week integrated body weight reduction program on leptin levels and body composition in severe obese subjects. *J. Endocrinol. Investig.* **2003**, *26*, 250–256. [CrossRef] [PubMed]

49. Whitehead, J.P.; Richards, A.A.; Hickman, I.J.; Macdonald, G.A.; Prins, J.B. Adiponectin—A key adipokine in the metabolic syndrome. *Diabetes Obes. Metab.* **2006**, *8*, 264–280. [CrossRef] [PubMed]

50. Paniagua, J.A. Nutrition, insulin resistance and dysfunctional adipose tissue determine the different components of metabolic syndrome. *World J. Diabetes* **2016**, *7*, 483–514. [CrossRef]

51. Yamauchi, T.; Kamon, J.; Minokoshi, Y.; Ito, Y.; Waki, H.; Uchida, S.; Yamashita, S.; Noda, M.; Kita, S.; Ueki, K.; et al. Adiponectin stimulates glucose utilization and fatty-acid oxidation by activating AMP-activated protein kinase. *Nat. Med.* **2002**, *8*, 1288–1295. [CrossRef]

52. Frysak, Z.; Schovanek, J.; Iacobone, M.; Karasek, D. Insulin-like Growth Factors in a clinical setting: Review of IGF-I. *Biomed. Pap.* **2015**, *159*, 347–351. [CrossRef]

53. Marquet, L.-A.; Brown, M.; Tafflet, M.; Nassif, H.; Mouraby, R.; Bourhaleb, S.; Toussaint, J.-F.; Desgorces, F.-D. No effect of weight cycling on the post-career BMI of weight class elite athletes. *BMC Public Health* **2013**, *13*, 510. [CrossRef] [PubMed]

54. Hyman, M.H.; Dang, D.L.; Liu, Y. Differences in Obesity Measures and Selected Comorbidities in Former National Football League Professional Athletes. *J. Occup. Environ. Med.* **2012**, *54*, 816–819. [CrossRef] [PubMed]

55. Franchini, E.; Brito, C.J.; Artioli, G.G. Weight loss in combat sports: physiological, psychological and performance effects. *J. Int. Soc. Sports Nutr.* **2012**, *9*, 52. [CrossRef] [PubMed]

56. Abbey, E.L.; Wright, C.J.; Kirkpatrick, C.M. Nutrition practices and knowledge among NCAA Division III football players. *J. Int. Soc. Sports Nutr.* **2017**, *14*, 13. [CrossRef] [PubMed]

57. Michalczyk, M.M.; Zajac, A.; Mikolajec, K.; Zydek, G.; Langfort, J. No Modification inBlood Lipoprotein Concentration but Changes in Body Composition after 4 Weeks of Low Carbohydrate Diet (LCD) Followed by 7 Days of Carbohydrate Loading in Basketball Players. *J. Hum. Kinet.* **2018**, *65*, 125–137. [CrossRef] [PubMed]

58. Michalczyk, M.M.; Chycki, J.; Zajac, A.; Maszczyk, A.; Zydek, G.; Langfort, J. Anaerobic Performance after a Low-Carbohydrate Diet (LCD) Followed by 7 Days of Carbohydrate Loading in Male Basketball Players. *Nutrients* **2019**, *11*, 778. [CrossRef] [PubMed]

59. Borchers, J.R.; Clem, K.L.; Habash, D.L.; Nagaraja, H.N.; Stokley, L.M.; Best, T.M. Metabolic syndrome and insulin resistance in Division 1 collegiate football players. *Med. Sci. Sports Exerc.* **2009**, *41*, 2105–2110. [CrossRef]

60. Kujala, U.M.; Kaprio, J.; Taimela, S.; Sarna, S. Prevalence of diabetes, hypertension, and ischemic heart Obesity in College Rowers 455 disease in former elite athletes. *Metabolism* **1994**, *43*, 1255–1260. [CrossRef]

61. Melanson, E.L. The effect of exercise on non-exercise physical activity and sedentary behavior in adults. *Obes. Rev.* **2017**, *18*, 40–49. [CrossRef]

62. Vogt, S.; Schumacher, Y.O.; Roecker, K.; Dickhuth, H.H.; Schoberer, U.; Schmid, A.; Heinrich, L. Power Output during the Tour de France. *Int. J. Sports Med.* **2007**, *28*, 756–761. [CrossRef]

© 2019 by the authors. Licensee MDPI, Basel, Switzerland. This article is an open access article distributed under the terms and conditions of the Creative Commons Attribution (CC BY) license (http://creativecommons.org/licenses/by/4.0/).

Review

Mechanisms of Lifespan Regulation by Calorie Restriction and Intermittent Fasting in Model Organisms

Dae-Sung Hwangbo [1,*,†], **Hye-Yeon Lee** [2,†], **Leen Suleiman Abozaid** [1] **and Kyung-Jin Min** [2,*]

[1] Department of Biology, University of Louisville, Louisville, KY 40292, USA; leen.abozaid@louisville.edu

[2] Department of Biological Sciences, Inha University, Incheon 22212, Korea; 319127@inha.ac.kr

* Correspondence: ds.hwangbo@louisville.edu (D.-S.H.); minkj@inha.ac.kr (K.-J.M.);
 Tel.: +1-502-852-5934 (D.-S.H.); +82-32-860-8193 (K.-J.M.)

† Equal contributions were made by these authors.

Received: 31 March 2020; Accepted: 22 April 2020; Published: 24 April 2020

Abstract: Genetic and pharmacological interventions have successfully extended healthspan and lifespan in animals, but their genetic interventions are not appropriate options for human applications and pharmacological intervention needs more solid clinical evidence. Consequently, dietary manipulations are the only practical and probable strategies to promote health and longevity in humans. Caloric restriction (CR), reduction of calorie intake to a level that does not compromise overall health, has been considered as being one of the most promising dietary interventions to extend lifespan in humans. Although it is straightforward, continuous reduction of calorie or food intake is not easy to practice in real lives of humans. Recently, fasting-related interventions such as intermittent fasting (IF) and time-restricted feeding (TRF) have emerged as alternatives of CR. Here, we review the history of CR and fasting-related strategies in animal models, discuss the molecular mechanisms underlying these interventions, and propose future directions that can fill the missing gaps in the current understanding of these dietary interventions. CR and fasting appear to extend lifespan by both partially overlapping common mechanisms such as the target of rapamycin (TOR) pathway and circadian clock, and distinct independent mechanisms that remain to be discovered. We propose that a systems approach combining global transcriptomic, metabolomic, and proteomic analyses followed by genetic perturbation studies targeting multiple candidate pathways will allow us to better understand how CR and fasting interact with each other to promote longevity.

Keywords: aging; lifespan; longevity; calorie restriction; fasting

---

1. Introduction

1.1. Opening Sentences

Almost all organisms, except for a few species including perennial plants, lobsters, quahog, rockfish, and Testudinidae, undergo a series of biological processes referred to as "aging" and "senescence." [1]. Biological aging is generally defined as "a series phenomenon of functional, structural, and biochemical changes that occur throughout cells and organs, disrupting homeostasis in the body and ultimately leading to death" [2]. Prior to the early twentieth century, studies on human aging were not considered important because humans lived for a relatively short period of about 35 to 45 years. Since that time, technology and human medicine have greatly advanced, the human lifespan has increased, and research into human longevity and healthy living has increased. One of the breakthroughs of the research is that the aging process can be retarded by dietary manipulations.

1.2. History of Dietary Manipulations for Health and Longevity

In the early 1900s, there was some evidence that dietary manipulations affect health and longevity of organisms. Reduction of food intake decreased the occurrence of cancers in rodents [3], and increased the lifespan in aged female rats [4] and fruit flies [5]. The basic concept of caloric restriction (CR) was founded in the late 1930s. Ingle et al. reported that the reduction of food intake increased the lifespan of planktonic cladoceran, *Daphnia longispina* [6], and McCay et al. showed that restricted diet extended the lifespan of rats two fold compared to rats on a normal diet [7]. Since the late 1930s, the term CR has become more widely used, and, in the 1940s, many researchers reported that CR retarded or prevented the onset of age-related diseases such as kidney disease, tumors, and leukemia [8–12]. From the 1950s to the 1980s, the longevity effect of CR was also reported in other species. CR decreased the mortality rate in *Tokophrya infusionum* (Protozoan) [13], *Philodina acuticornis* (rotifera) [14], *Lebistes reticulates* (fish) [15], *Caenorhabditis elegans* (nematode) [16], *Rattus norvegicus* (rat) [17,18], and *Mus musculus* (mouse) [19,20]. In addition to limiting the feeding amount, controlling the feeding period (e.g., intermittent feeding) was also researched during these decades [18,20–23]. In the 1980s, several sources of evidence started to indicate that the dietary composition was the controlling determinant for the longevity effect of CR, and the term dietary restriction (DR) began to be widely used. Several studies have shown that reduced calorie intake by alteration of nutrient content, such as fat, carbohydrates, or amino acids, can have different effects on longevity in model animals [24–26]. In the 1990s, results of studies into the effects of CR in rhesus monkey (*Macaca mulatta*), non-human primates (NHP) were published by three groups—the National Institute on Aging (NIA) [27], the Wisconsin National Primate Research Center (WNPRC) [28], and the University of Maryland [29].

In the 2000s, the term intermittent feeding underwent a slight change and became intermittent fasting (IF). IF is a dietary manipulation that cycles between periods of ad libitum feeding and periods of fasting, including alternate-day fasting (ADF) and periodic fasting (PF) [30]. Although the effects of IF on health and longevity have not been elucidated as clearly as those of CR, there is evidence indicating a positive effect of IF on aging [31,32]. Recently, the concept of IF merged with that of the circadian rhythm and a new diet regimen, time restricted feeding (TRF), has emerged. TRF is a slight variation of IF interventions in which food intake is limited to 12 h each day without a change in the total calorie intake of the normal diet [31–35]. TRF has been reported to reduce the incidence of aging-related diseases and delay aging without an actual reduction in food intake.

1.3. Key Determinant of Lifespan Regulation through Diet Manipulation

CR regards the daily caloric intake per se as a key determinant in lifespan regulation. For example, a reduction of calorie intake without a reduction of protein intake increased the lifespan of rats [25], and lifespan was not altered in rats fed isocaloric diets in which either fat or mineral components had been reduced [26,36]. These studies indicated that the total calories are a key determinant in regulating the lifespan of rats. However, recent evidence had indicated that the amount of calorie intake might not be a key determinant of lifespan regulation by CR. The lifespans of rats and fruit flies have been increased by nutritional changes or protein reduction while providing the same calorie intake [37–41]. Moreover, the results of several studies have suggested that amino acids are key modulators of lifespan in organisms [42,43]. Furthermore, reducing only one type of amino acid, methionine, is sufficient enough to increase the lifespan of yeast, nematodes, fruit flies, and rodents [44–47]. Beneficial effects of TRF on health and longevity indicated that there might be a third determinant in lifespan extension, other than total calories or nutrient composition, since TRF exerts its effect without exhibiting notable changes in total calories or nutrient composition [31]. A more thorough investigation into the key determinant(s) of nutrient restriction effect is necessary.

2. Animal Models and Protocols of Dietary Manipulation

2.1. Yeast (Saccharomyces Cerevisiae)

Yeast aging is classified into two different types as replicative and chronological aging [48]. Replicative aging is defined by the number of daughter cells produced by a mother cell, while chronological aging is defined by the time in which a nondividing cell can maintain viability. Although two yeast aging paradigms have been used in aging studies, replicative aging is more widely used in CR-related aging studies. Generally, CR in yeast is performed by reducing the glucose level in growth medium, which commonly contains 2% peptone, 1% yeast extract and 2% glucose. The concentrations of glucose are reduced to ~0.5–0.005% for CR [49]. In these settings, replicative lifespan of budding yeast was extended by about 10 times in the low-dose glucose medium compared to the lifespan of control [50–53]. Yeast is also cultured in water in order to undergo fasting [54].

2.2. Nematode (Caenorhabditis Elegans)

C. elegans has several advantages in aging studies—a relatively short lifespan/reproductive cycle, a translucent body, it is easy to culture, has a small genome, and there are many available mutants [55]. DR is mainly performed in nematodes by controlling the concentration of the bacteria such as *Escherichia coli* in the media that they feed [54,56]. In the worms, genetic perturbations that mimic DR were also introduced by inhibiting specific nutrient transporters [57] and reducing pharyngeal pumping [58]. For IF, worms are placed every other day in medium with and without bacteria [59,60]. This IF regimen (alternate 2 days eating/ 2 days fasting) successfully extended lifespan in the worms [59,60]. Furthermore, chronic fasting also increased the lifespan of worms compared to normal diet-fed worms [61,62].

2.3. Fruit Fly (Drosophila Melanogaster)

The fruit fly, *D. melanogaster,* is another invertebrate model organism widely used for aging and dietary intervention studies [63]. Similar to *C. elegans*, the fruit fly also has many advantages such as a relatively short lifespan and high productivity. However, compared to *C. elegans*, the fruit fly has more complicated and diverse tissues such as the heart and kidney that are functionally homologous to mammals [63]. Gene manipulation and editing tools are also readily available to study the genes of interest in a time- and tissue-controlled manner [63]. Furthermore, their simple food composition allows for easy manipulation of the food component in experiments. Although the composition of the food medium is diverse among laboratories, the most general method for DR supplementation in the fruit fly is dilution of the food ingredients including yeast as a protein source, sugar, or fat from an ad libitum medium. Food reduction or diluted food has also been consistently shown to extend the lifespan in fruit flies [40,64,65]. Furthermore, limiting amino acids such as methionine or limiting protein sources were sufficient to increase the lifespan of fruit flies [40,41,46,66]. A relatively diverse fasting study design can be carried out in fruit flies. In the case of ADF, food is provided every two days and fasting is performed for 24 h. Recent studies have found that a 2-day fed:5-day fasted IF regime [67] and a TRF regime with daily access to food during the day and water access during the night [68] can be implemented in fruit flies. In the IF regime's case, flies were treated for IF for the first 30 days of adulthood and then switched to an ad libitum diet due to high mortality by fasting in older flies [67]. In this regimen, IF increased the lifespan of fruit flies [67]. However, a 3 h or 6 h starvation during the day was not enough to extend the lifespan [65]. Additionally, TRF did not increase the median lifespan of fruit flies, although TRF improved the muscle performance and attenuated age-related cardiac dysfunction [31,68].

2.4. Rodents

Although research results showing longevity manipulation by dietary modulation in nematodes and fruit flies are thought-provoking and motivating, the complexities of human physiology block the direct application of such results in humans. In this regard, rodents can fill some of the gaps between them and humans because, compared to fruit flies, nematodes, and yeast, rodents have a closer phylogenic relationship to humans and greater similarities in their physiological features and process. Many studies have shown beneficial effects of CR/DR on aging in rodents. For example, CR/DR reduced the incidence of age-related diseases such as cancer, neurodegenerative diseases, and cardiovascular diseases and prolonged lifespan by 30% in rats and 15% in mice [24,69–71]. Rodents, including mice and rats, were the first experimental model systems used to investigate the effect of CR on lifespan [7]. Generally, to conduct CR in rodents, the total consumed volume of food is thoroughly controlled so that 20–50% of calories are reduced compared to ad libitum food administration. [72,73]. In addition to this traditional CR administration, trials modulating macromolecule composition such as proteins or carbohydrates were also attempted. Similarly, reducing the concentration of specific amino acids such as methionine or tryptophan is another form of dietary modulation and was shown to extend lifespan [42,47,71,74–78]. To assess the effects of fasting regimen in rodents, IF can be conducted so that rodents are provided with only water or minimal nutrients for less than 24 h followed by a normal diet period of 48 h, whereas PF can be conducted so that rodents are fasted for approximately 48 h, returned to normal feeding and then fasted again at least one week later [79]. To conduct TRF, food access can be regulated by transferring mice daily between cages with ad libitum food and cages with water only [80,81]. In rodent models, the effects of IF on lifespan are not yet conclusive. IF with every other day fasting or fasting for one day every three to four days extended the lifespan of rodents [82–85]. However, a study showed that IF introduced at 10 months of age had no effect on mean lifespan in C57BL/6J mice or decreased the lifespan in A/J mice [83]. Unlike IF, multiple studies showed that TRF inhibits several chronic diseases and tumor progression and increases lifespan in rodents [86–88].

2.5. Non-Human Primates

The use of NHP in dietary studies provides unique evidence that cannot be obtained by studying a lower-order model animal. Although the results of NHP studies have high reliability in human applications, NHP studies can encounter several technical, financial and ethical difficulties. Three independent groups, the NIA, the WNPRC, and the University of Maryland have investigated, or are currently investigating, the beneficial effects of CR on NHP by using the rhesus monkey model. A research group at the University of Maryland have focused on the effects of short-term CR on obesity and diabetes [89,90], while the NIA and WNPRC have been investigating the effects of CR in rhesus monkeys throughout their entire lifetime. Although the rhesus monkeys in the CR groups were provided with about 70% food compared to ad libitum groups in both the NIA and WNPRC studies, there is a key difference between them in terms of dietary composition [91–93]. The NIA provided unpurified natural ingredient-based food, while the WNPRC provided a purified diet to monkeys [91–93]. Although the exact information of food ingredients is not available in natural ingredient-based food, it provides phytochemicals and minerals which might have beneficial effects on health and lifespan. On the other hand, a purified diet has an advantage in that nutrient composition of the diet is more defined, allowing the manipulation of specific components of the diet. In addition, the NIA provided approximate ad libitum intake considering their age and bodyweight for the maturing control monkeys without overfeeding, but the WNPRC established the ad libitum reference for each individual and implemented CR based on individual standards [91–93]. Lifelong CR in rhesus monkeys led to lifespan extension at the WNPRC [91], but there was no lifespan extension effect by CR at NIA [92]. The NIA used the food that was lower in calories and fat, and higher in protein and fiber compared to food used by the WNPRC. These dietary manipulations conducted at the NIA led to a longer lifespan of the control old-onset groups from the median lifespan of rhesus monkey. The median lifespan of rhesus monkey was similar to what was previously reported as the 90th percentile of this species (~35 years old). In addition, juvenile/adult males without CR in the NIA showed similar median lifespans compared to the lifespan of monkey with CR in the WNPRC. Thus, it suggests that the difference in diet between the control and the CR group was insufficient to change lifespan.

However, since the NIA uses rhesus macaques of various ages, sex, and different genetic backgrounds (Indian and Chinese), it showed results that can compare the effect of CR according to the differences in age/sex/genetic background. Although the results of the effect of CR on the lifespan of rhesus monkey were different, both groups present health benefits of CR such as loss of weight and fat, and reduced risk of cancer and cardiovascular disorder. Thus, if all variables were controlled, it was suggested that CR can robustly increase lifespan in monkeys and also suggest applications in humans [93].

3. Dietary Manipulations for Human Application

Many studies have shown that dietary manipulation can retard the aging process through some well conserved mechanisms in diverse organisms from yeast to NHP. The determination of conserved mechanisms that produce beneficial effects of dietary manipulation in humans would require additional investigation, due to the limited number of studies examining the effects of CR/IF in humans. However, several epidemiological and cross-sectional studies using centenarians and individuals who volunteered CR practice indicate the beneficial effect of CR in humans. Epidemiological data can be gathered from people who follow food restrictions due to religious guidelines. For example, Muslims ingest no food or water for approximately 15 h between sunrise and sunset for a month during Ramadan every year. Thus, this long-term food restriction during Ramadan could be considered a human IF model.

Some studies have shown that Ramadan fasting has the effect of promoting human health [94]. A Comprehensive Assessment of the Long-term Effects of Reducing Intake of Energy (CALERIE) research program was designed to systematically investigate sustained CR effects in healthy volunteer humans over a two-year period [95]. The CALERIE program produced several results that demonstrate the beneficial effect of CR on aging and health in humans, including observation of an increase in metabolism and a decrease in oxidative stress [96,97]; however, the study did not indicate the presence of beneficial effects of CR on age-related bone and muscle impairment [98,99]. Additionally, some studies have shown that IF can improve metabolic health and physiological function in humans. IF reduced fat mass, lean mass, and body weight in healthy humans and obese patients [33,100–104]. Similarly, IF improved lipid and glucose metabolism, reduced inflammatory response, lowered blood pressure, and improved cardiovascular health [102,105–109]. Several studies have shown that IF is an effective intervention, especially for people who are overweight or diabetic. IF reduced overall fat mass and decreased insulin resistance [103,110–113]. Some researchers also conducted the studies to evaluate the effects of TRF on human health, and demonstrated that TRF improved insulin sensitivity, blood pressure, oxidative stress, and quality of life in overweight or diabetic adults [35,114,115]. Results of the studies also showed that TRF improved cardiovascular function and other indicators of healthspan (e.g., walking distance and heart rate) in healthy middle-aged and older adults [116] although weight loss observed with other IF methods were not accompanied by TRF. These results suggest that IF including TRF may be a promising manipulation to extend the healthspan of humans.

4. Molecular Mechanisms of CR and IF

The ultimate goal for animal studies on CR/IF is to identify the conserved molecular mechanisms that can extend the healthspan of humans. Healthspan, the period of life that is free from disease, is measured by examining declines of functional health parameters and disease states. Because healthspan is a multifactorial complex phenotype that is significantly affected by genotypes (G) and environmental factors (E) as well as complicated interactions between them (G × E), measuring healthspan often gets complicated [117]. Furthermore, delayed functional aging in one parameter is not always necessarily linked to the extension of healthspan in different health parameters [117]. In fact, by depending on the types of health parameters and experimental approaches, different healthspan results were observed from the studies that used the same long-lived mutant animals [117]. Unlike healthspan, lifespan is unequivocally recorded by simply following the mortality of individual organisms. Lifespan extension in animal models is strongly correlated with a decrease in morbidity and an increase in health. Therefore, although we believe that results of health-related parameters from animal CR/IF studies are likely to be translatable to human healthspan, we will focus on the mechanisms of lifespan extension in animal models in this manuscript.

Although not complete, studies for the last two decades on CR have provided a great amount of details about the mechanisms of CR. Recent advances in OMICs and bioinformatic techniques followed by organism level genetic perturbation analyses significantly extended our knowledge on the molecular mechanisms that mediate lifespan extension by CR. A current understanding is that CR works through the key nutrient and stress-responsive metabolic signaling pathways including IIS/FOXO, TOR, AMPK, Sirtuins, NRF2, and autophagy. While these pathways regulate CR independently, cross-talks among these pathways as well as upstream master networks such as circadian clock were also suggested to regulate lifespan extension by CR. Although the number of reports on IF is less than CR, recent studies clearly demonstrated that IF also extends lifespan in both vertebrate and invertebrate model organisms [60,67,79,83,118,119]. Notably, increased survival by nutrient deprivation was also observed in prokaryotic *E.coli* cells, emphasizing that fasting-related lifespan extension is evolutionarily conserved [79]. However, there is still a lack of comprehensive understanding for the mechanisms responsible for lifespan extension by IF. As nutrient-dependent interventions, CR and IF were suggested to share a common strategy: the reduction of caloric intake and nutrients that limit longevity. In fact, CR and IF also result in common metabolic and physiological changes in multiple tissues and organs (Figure 1) [32]. For example, ketone bodies, insulin sensitivity, and adiponectin are increased while insulin, IGF-1, and leptin are decreased. Overall inflammatory response and oxidative stress are reduced by both regimens [32]. They also cause similar behavioral changes such as increased hunger response and cognitive response [32]. Accordingly, it is widely accepted that common molecular mechanisms may mediate the lifespan extension by CR and IF. A proposed model for the mechanisms underlying the lifespan extension by CR and IF relatively follow the notion that both CR and IF alter the activity of common key metabolic pathways, namely, TOR, IIS, and sirtuin pathways (Figure 1) [120]. However, there must be independent mechanisms as well due to one major difference between CR and IF in that IF aims to extend lifespan without an overall reduction in caloric intake by taking advantage of the molecular pathways that respond to fasting [30,32,121].

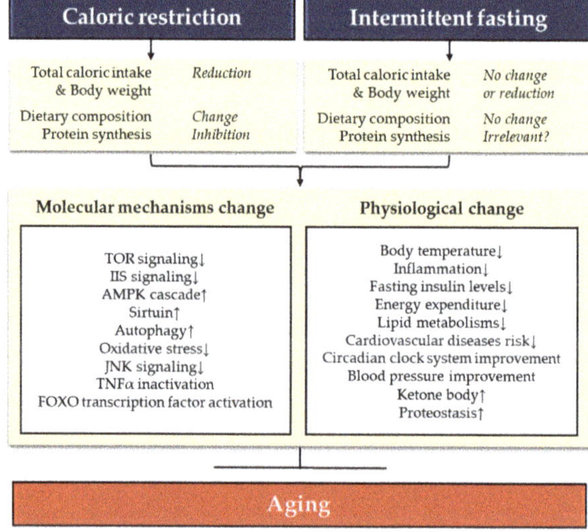

Figure 1. Possible anti-aging mechanisms of caloric restriction (CR) and intermittent fasting (IF). Different dietary interventions by CR and IF result in similar molecular and physiological changes that promote longevity in model organisms. Patterns of individual dietary, metabolic, molecular, and physiological parameters can be different depending on the types of CR and IF as well as the animal models. See the main text for details.

Chronic CR that results in the extension of healthspan and lifespan usually involves a body weight loss in animal models [119]. Body weight loss is also often observed in animals under IF [119]. This is an important issue in both practical and mechanistic perspectives. Although a modest body weight loss may be beneficial for overall health, a severe loss of body weight may counteract beneficial effects on other health parameters. Mechanistically, it is possible that CR and IF result in extension of healthspan and lifespan at the cost of body weight reduction. In this sense, it is interesting to note that a loss of body weight can be decoupled from other beneficial effects by IF [30,114]. This raises an important question of whether fasting by itself may induce some, if not all, extension of healthspan and lifespan at least by IF. Although a weight loss was observed in the participants of the CALERIE trial (also seen in Section 3), the weight loss was mild and within the normal range of health while improving other health parameters [95–97]. Therefore, although further investigations are required for the reciprocal relationship between body weight and the efficacy of CR/IF, we favor the idea that that body weight reduction by CR and IF are side effects that are not the mechanistic determinant for the benefits of CR and IF.

CR and IF significantly reorganize genomic, metabolomic, and proteomic landscapes in local tissues as well as in the global organism level in an age, sex, and strain-dependent manner. However, these molecular changes in gene expression, metabolites, and proteomes do not necessarily represent whether those changes are causal factors for CR- and IF-mediated lifespan extension. Genetic perturbation studies in animal models must be followed in order to link them to lifespan regulation by CR and IF. Therefore, in this review, we will primarily focus on the molecular pathways that were genetically tested for CR and IF effects on lifespan, leaving out much of correlative studies describing the physiological and metabolic traits affected by CR and IF. Because genetic perturbation studies and OMICs data for IF are significantly less than those of CR, we will first discuss molecular mechanisms of CR followed by whether those mechanisms overlap with IF.

4.1. AMPK-TOR Signaling

In eukaryotes, the target of the Rapamycin (TOR) pathway plays a central role in nutrient and energy sensing to control cellular and organismal growth [122–124]. The TOR pathway regulates growth and metabolism by promoting protein synthesis in response to nutritional availability including dietary amino acids [124]. A number of genetic studies showed that suppression or downregulation of the TOR pathway extend lifespan in multiple model organisms including the yeast *S. cerevisiae* [54,125–129], the worm *C. elegans* [60,130–142], the fly *D. melanogaster* [143,144], and the mouse *M. musculus* [145–148]. As CR downregulates the TOR signaling cascade, it has long been suggested that CR may extend lifespan by at least partially suppressing the TOR pathway at the cost of reduced growth. In fact, mutant animals for the components of the TOR pathway were often shown to fail or decrease in lifespan extension by CR [54,125–129,136,141,143,144], indicating that the TOR pathway antagonizes the full benefit of CR-mediated lifespan extension. As a key amino acid sensing pathway, this may explain that restriction of protein alone, specifically by single amino acids methionine and tryptophan in the diet, were sufficient to extend lifespan.

In addition to amino acids, the TOR pathway is also regulated by cell energy status through AMP-dependent protein kinase (AMPK), a conserved energy sensor in eukaryotes [149,150]. Increased AMP:ATP ratio by energy depletion such as CR activates AMPK, which in turn inhibits the TOR pathway [149]. Thus, CR activates AMPK while suppressing the TOR cascade subsequently. Unlike the TOR pathway where it extends lifespan when suppressed, AMPK extends lifespan in model organisms when activated [136,151–153]. Importantly, similar to the TOR pathway, genetic perturbation studies also showed that AMPK mediates lifespan extension by CR. For example, lifespan extension by CR in worms was suppressed in the mutant worms for *aak-2*, one of the catalytic subunits of AMPK [136]. However, it is interesting to note that another type of CR in worms (i.e., feeding diluted bacteria in liquid culture) did not require AMPK signaling to extend lifespan [154]. Although this discrepancy needs further investigation particularly into their methods including the nutritional value in each type of the CR protocols, it is possible that non-overlapping mechanisms between CR and IF may be responsible. In other words, fasting-related mechanism independent of CR may contribute to this difference. In this sense, it is interesting to note that mild nutritional stress through feeding 2-deoxy-D-glucose (2-DG) or food deprivation, which mimic fasting, extended lifespan in worms through AMPK signaling [155,156]. An indication for this explanation can be drawn from mammalian studies. While acute starvation readily activates AMPK, activation of AMPK depends on the duration and type of CR [157]. In some cases, extended CR failed to activate AMPK [157]. Thus, it is possible that the AMPK-TOR dependent lifespan extension could partially be due to the mechanisms induced by fasting, parts of which may be independent of CR. Supporting this hypothesis, it is noteworthy that Honjoh et al. showed that lifespan extension by IF (by every-other-day feeding) was dependent on RHEB, a small GTPase protein that activates the TOR pathway by directing binding to the TOR Kinase [158], at least in worms [60]. As they also showed that RHEB-dependent IF-mediated lifespan extension was partially due to IIS/FOXO signaling, their results support the idea that tightly regulated networks between IIS/FOXO and TOR signaling cascade may mediate both DR and IF-dependent lifespan extension.

4.2. IIS-FOXO Signaling

In mammals, growth hormone (GH) secreted from the pituitary gland promotes somatic growth by activating a cascade of downstream hormonal signaling such as Insulin/Insulin-like growth factor-1 signaling (IIS) [120,159]. Activated IIS signaling cascade by GH mediates the translocation of its main downstream targets, forkhead box protein O (FOXO) transcription factors, to the cytoplasm from the nucleus [160]. In the absence or reduction of GH/IIS signals, the FOXO transcription factors translocate into the nucleus and promote the expression of their target genes involved in cell death, cell cycle arrest, DNA repair, stress resistance, and detoxification [160], all of which are attributed to promote longevity by switching organismal metabolic status from somatic growth to maintenance [161]. Although there is no system equivalent to GH in lower organisms such as yeast, worms, and flies [120], a number of observations reported for the last two decades strongly support the idea that downregulation of IIS and activation of FOXO transcription factors extend lifespan in these animal models (reviewed in [120,159,162,163]). In fact, of the > 40 genetic mutations that have been reported to extend lifespan in the mouse and the rat models, approximately one third of them are involved in GH and IIS [164]. Because CR reduces GH and IIS [164], it is generally accepted that CR extends lifespan by limiting GH/IIS signaling and subsequently expressing pro-longevity genes by activating FOXO transcription factors [165]. To date, there are mixed results reported for the question of whether the IIS-FOXO signaling cascade is responsible for CR-mediated lifespan extension. For example, Bonkowski et al. reported that dwarf mice with targeted disruption of the GH receptor failed to extend overall, median, or average lifespan by CR (food reduction by 30% compared to ad libitum) [69], suggesting that CR extends lifespan by downregulation of IIS. Alternatively, in another study, CR (30% CR) further extended the lifespan of the long-lived dwarf mice with GH production that was selectively suppressed in the pituitary gland, spleen, and thymus [166], suggesting that lifespan extension by GH suppression may occur through an independent mechanism of CR. Alternatively, these results also imply that GH signaling in other tissues such as the liver and testis should be also suppressed for a full benefit of lifespan extension by CR [166], raising an important question regarding the tissues critical for CR-mediated lifespan extension. Interestingly, these data show a clear dissociation of lifespan extension by GH suppression from its dwarfism (small body size caused by GH suppression), opening an important possibility that CR may extend lifespan without the cost of growth reduction. Similar to the dwarf mice mutant for the GH receptor [69], CR failed to extend lifespan of both heterozygous and homozygous mutant mice for FOXO3 [167], showing that IIS-FOXO signaling is indeed required for the full benefit of CR-mediated lifespan extension. More complicated observations were reported in lower organisms. In flies, multiple studies suggest that although IIS-FOXO signaling modulates longevity response to CR, it appears not to be the main player of CR [168–170]. In worms, it is still inconclusive whether IIS-FOXO is required for CR-mediated lifespan extension because mutant worms for DAF-16, the sole ortholog of FOXO transcription factors, showed a different longevity response depending on the types of CR [154]. While a relatively considerable amount of research has been done on the relationship between IIS-FOXO and CR-dependent lifespan extension, no direct genetic studies testing whether IIS-FOXO mediates IF-dependent lifespan have been reported. However, functional studies characterizing the reciprocal effect between IF and IIS-FOXO signaling suggests that IIS-FOXO may be at least partially responsible for IF-dependent lifespan extension. For example, in mammals, key metabolic and physiological changes attributed to lifespan extension by CR include increased insulin sensitivity, stress resistance, and immune function with reduced inflammation. Recent studies demonstrated that IF also shows these beneficial changes, displaying a promising prospect that IF may also increase lifespan through IIS-FOXO signaling.

4.3. Sirtuins

Sirtuins, silent information regulator 2 (sir2) proteins, are protein deacetylases that require NAD$^+$ as a cofactor for the deacetylation reaction [171]. Because NAD$^+$ and its reduced form NADH are involved in many important cellular metabolic pathways, sirtuins function as metabolic sensors that represent the metabolic state of the cell. As NAD$^+$ accumulates under nutritional stress and activates sirtuins [172], it was suggested that activation of sirtuins may extend lifespan, possibly through the mechanisms that extend lifespan by CR and/or IF. In fact, it was shown that genetic overexpression of sirtuins extended lifespan in multiple model organisms including yeast [173], worms [50,174–182], flies [178,183–185], and mice [186,187]. Similarly, pharmacological activation of sirtuins by feeding resveratrol extended lifespan in some of these animals [178,188]. Furthermore, it was also shown that the sirtuin family genes were required for the lifespan extension by CR in these animal models [50,178,183–185]. For example, when SIR2 was deleted, CR by glucose dilution failed to extend lifespan in yeast [50]. However, it is interesting to note that, while a milder CR (0.5% glucose) in yeast required SIR2 for lifespan extension [50], a severe form (0.05% glucose) of CR extended lifespan independent of SIR2 [189]. It would be important to test whether this severe form of CR extend lifespan by the mechanisms related to fasting. In this case, it would also be critical to identify the threshold concentration of glucose that differentiates fasting from CR. Characterizing global changes in transcriptome and metabolome between these *sir2*-dependent mild CR and *sir2*-independent severe CR (aka fasting) would be also critical to better understand the relationship between CR and fasting. In flies, increased lifespan by *sir2* overexpression was not further extended by CR [183]. On the other hand, CR failed to extend the lifespan of null mutant flies for sir2 [183]. It was also shown that genetic knockdown of *sir2* in fat body suppressed the lifespan extension by CR [185]. These reports support the idea that *sir2* plays a critical role in CR-dependent lifespan extension. In worms, whether *sir-2.1* (the ortholog of *sir2* in yeast and flies) is necessary for CR-mediated lifespan extension or not was dependent on the type of CR-treatment [154]. It would be interesting to test whether the type of CR that does not require *sir-2.1* extends lifespan by activating the pathway that extends lifespan by fasting. Despite all of these observations that support the idea that sirtuins are important mediators of CR, there are conflicting claims about the role of sirtuins in pro-longevity and CR-mediated lifespan extension in lower eukaryotic organisms [189–191]. This discrepancy may be due to differences in dosage of sirtuins, tissue septicity, and CR administration protocols [189–192]. For example, lifespan extension by overexpression of sirtuins depends on the levels of sirtuins [184,185,192,193]. When *sir2* was expressed over 45 fold, it resulted in a shortened lifespan while a modest overexpression up to 11 fold increased lifespan [193]. Therefore, the impact of sirtuins on aging, CR-mediated, and possibly IF-mediated lifespan extension needs to be thoroughly studied [189–192]. In mice, knockout mutants for SIRT1, one of the seven mammalian sirtuins homologous to invertebrate sirtuins [194], failed to extend lifespan under CR [195,196], confirming that sirtuins' role in CR-mediated lifespan extension is conserved across species. In addition, similar to the lower organisms, multiple studies demonstrated that activation of sirtuins extended lifespan in mice [186,187]. Overall, if some degree of variability in published data is tolerated [189–191], it can be concluded that the sirtuin pathway is key for CR-mediated lifespan extension in both invertebrate and vertebrate model organisms. However, despite the observations that NAD$^+$ levels are increased by fasting and that sirtuins are involved in the benefits of fasting in physiological and pathological level [32,197], whether SIRT1 or the other mammalian sirtuins (SIRT1-6) play a role in IF-mediated lifespan extension is poorly understood. There is no lifespan data yet shown in animal models that specifically tested for the involvement of sirtuins in IF-mediated lifespan extension. It was recently revealed that fasting induced *dSirt4* (a *Drosophila* sirtuin family member localized to mitochondria) and over-expression of *dSirt4* extended lifespan [198]. It would be of great interest to test whether *dSirt4* mediates the CR- and IF-dependent lifespan extensions. Furthermore, considering the fact that the levels of sirtuins can result in opposite results in lifespan [193], it would also be important to profile the expression levels of sirtuins by different types of CR and IF.

4.4. Circadian Clock

Circadian (~24 h) clocks control a wide range of rhythmic metabolic, physiological, and behavioral parameters by communicating timing information via rhythmic transcription of output genes [199]. The misalignment of these internal clocks with 24 h environmental cycles are known to adversely impact metabolism, aging, and age-related disease [200,201]. Because the circadian clock orchestrates daily metabolism in response to cellular needs and nutritional availability, it was proposed to mediate the beneficial effect of CR [191,202]. A series of recent observations suggested that the circadian clock may play a master role in CR-dependent lifespan extension [203,204]. For example, it was shown that CR for two months in early life was sufficient enough to increase the amplitude of core clocks in the mouse liver [204,205]. As loss of rhythmic expression of clock-controlled genes (CCGs) is implicated as a cause of aging, these results suggest that CR may promote longevity by strengthening the rhythmic regulation of metabolism and physiology. In this regard, it is remarkable that CR failed to extend lifespan of knockout mice for Bmal1, one of the core circadian clock transcription factors [206], indicating that a functional circadian clock system is indeed necessary for CR-dependent lifespan extension in mice. Similar to mice, in flies, Katewa et al. reported that CR also increased the amplitude of core clock genes [203]. They also showed that genetic perturbation that increases clock function also resulted in lifespan extension in a diet-dependent manner [203]. Furthermore, they showed that homozygous mutants for *timeless*, a core clock gene in flies, failed to extend lifespan under CR to the level of wild type [203], indicating that circadian clock is also determinant of CR-dependent lifespan extension in flies. However, whether circadian clock is required for CR-mediated lifespan in flies needs cautious analysis as inconsistent results were reported, possibly due to uncontrolled environmental factors such as intestinal microbiome among the fly population [203,207,208]. With these observations in mice and flies, one important question is how exactly the circadian clock mediates the beneficial effect of CR. It is noticeable that transcriptional and post-transcriptional regulation of most known CR effectors such as GH/IGF-1, FOXO, TOR, AMPK, sirtuins, and NRF2 are directly or indirectly under the control of the circadian clock [32,202]. This raises the possibility for the circadian clock to play a master role in CR-mediated lifespan extension by simultaneously controlling these CR pathways. For example, in mice, cellular production of NAD$^+$, a key co-factor of sirtuins that promotes CR-dependent lifespan extension, is under the circadian clock. During fasting at night, the NAD$^+$ level is increased, which, in turn, activates sirtuins [32]. Similarly, nutritional input from feeding during the day increases ATP:AMP ratio and amino acid availability, thereby increasing the IIS and TOR pathways while suppressing the AMPK cascade. This process facilitates anabolic reactions and may promote aging. On the other hand, metabolism is switched to catabolic reactions by decreased ATP:AMP ratio and amino acid availability during fasting at night. Consequently, fasting at night suppresses the IIS and TOR pathways while activating the AMPK cascade and FOXO transcription factors, which subsequently give rise to anti-aging effects. Therefore, the circadian clock system may promote longevity by relaying the anti-aging signals induced by CR and IF.

One outstanding question is whether it is the total caloric/diet intake, rhythmic oscillation between feeding and fasting, or fasting itself (time and duration of fasting) that determines the beneficial effect of CR and IF. At least in mice, recent studies provided evidence that supports fasting as the key factor for CR- and IF-mediated lifespan extension. A systemic monitoring of food consumption behavior revealed that mice given the CR diet tended to limit their feeding time to a narrow temporal window, self-imposing and mimicking TRF [209]. Thus, mice under CR experienced a longer fasting time than when under AL diet [209], suggesting the possibility that it was not the calorie but the timing of food consumption or duration of fasting that confers longer lifespan in CR. Another study unequivocally demonstrated that mice under TRF extended lifespan even when they were under AL diet [88]. This study proved that controlling time-of-feeding can override the anti-longevity effect of caloric intake and is sufficient for lifespan extension [88]. This may explain why lifespan was not extended in mice when they were allowed to eat a hypo-caloric diet all day, although their overall caloric intake was comparable to that of CR [42]. Because these studies show that eating pattern (i.e., circadian fasting time and duration) rather than nutritional value (i.e., calorie and composition) determines lifespan, lifespan extension by CR and IF could occur at least partially through non-overlapping independent molecular mechanisms. Therefore, these observations strongly argue that molecular mechanisms responsible for lifespan extension by CR utilize some of the metabolic changes that occur during fasting. In this sense, lifespan extension by restricting specific nutrients such as methionine may also be due to changes in eating patterns that mimic TRF and IF as in Mitchell et al. [88]. With the evidence that restriction of caloric intake as well as specific nutrients such as methionine are sufficient to extend lifespan, these studies also indicate that there are both common and independent mechanisms underlying CR- and IF- mediated lifespan extension. Unlike CR studies in mice, where they have to fast once they consumed all the food that is given to them, CR in invertebrate models such as flies and worms allows them to have constant access to food. In fact, although there are daily rhythms in feeding behavior, flies do feed continuously over 24 h [210,211], removing the possibility that CR-mediated lifespan extension in flies is through the mechanisms by which IF extends lifespan. Furthermore, a genome-wide expression analysis revealed that global expression changes by CR and TRF differ from each other [212]. Importantly, this study also showed that the gene expression signature of TRF is also different from an extended starvation, raising the possibility that the molecular changes responsible for IF-mediated lifespan extension are different from that of CR, but also may not be from extremely severe fasting conditions. Gill et al. also reported that TRF ameliorates age-dependent heart failure by a mechanism independent of starvation and CR [212]. They showed that global transcriptional response to TRF is very different from that of starvation and CR [212]. Instead, they discovered that the circadian clock and clock-controlled TCP-1 ring complex chaperonin mediate the TRF effect. It will be of great interest to test whether TRF promotes longevity in flies, in which case these pathways might also mediate lifespan extension by TRF. Discovering the contribution of circadian clock to the benefits of TRF in Gill et al.'s study is not unexpected, considering the role of the circadian clock system to regulate daily metabolism and physiology in response to rhythmic environmental signals including the light:dark cycle and food consumption. Despite all of this compelling evidence, contribution of circadian clock to CR in worms and yeast is less understood due to their lack of a homologous system of a circadian clock pathway. However, they contain oscillatory metabolic fluctuations and behavior which need to undergo further studies for whether their CR response can be also modified by a circadian oscillatory mechanism [213–215].

5. Conclusions and Future Directions

5.1. Coordinated Regulation between IIS, TOR, AMPK, Sirtuins, and Circadian Clock

The ultimate goal of animal studies for CR and IF is to uncover evolutionarily conserved molecular mechanisms for the beneficial effect of CR and IF, and to eventually apply them to humans. Despite recent progress in our understanding of CR and IF, there are multiple challenges to overcome in order to achieve this goal. One such challenge is that there still lacks a comprehensive understanding of coordinated regulation among the key molecular pathways known and suggested to mediate CR and IF, namely, IIS, FOXO, TOR, AMPK, sirtuins, and the circadian clock. Molecular characterization of these pathways showed that they are tightly linked to and intertwined with each other in response to cellular nutritional state. However, the majority of animal studies performed so far on these pathways for the impact of CR and IF have been limited to testing and identifying single genes and pathways. Considering the impact of these pathways on systemic metabolism and physiology in many different tissues and organs, it is unlikely that a single gene or pathway is solely responsible for the lifespan extension by CR and IF. One way to solve this issue is to target multiple genes and pathways simultaneously [154,216]. For example, Hou et al. postulated that perturbation of multiple pathways would result in an additive or synergic effect in lifespan extension compared to the lifespan extensions by any single gene perturbation [217]. Using *C. elegans* as a model organism, they took advantage of the temporally resolved global transcriptome analysis followed by a systems biology approach. From this approach, they discovered that a combination of downregulation of IIS, downregulation of TOR, and upregulation of AMPK strongly resembled the transcriptomic change induced by CR [217]. Further genetic testing confirmed that lifespan was maximized when all of these perturbations were combined. More importantly, they also discovered that CR failed to further extend lifespan in these animals [42], showing that a simultaneous targeting of multiple candidate pathways may increase the power to detect hidden mechanisms for CR and IF.

5.2. Limits of Animal Studies for CR and IF

The amount of food that animals consume (meal size) and the time/duration of food consumption (meal timing) that animals take are key factors to interpret CR and IF results in animal models. Unlike rodent models where food is readily provided and removed from experimental animals, these parameters (i.e., meal size and meal timing/duration) are hardly controlled in the lower organisms widely used for CR and IF studies such as yeast, worms, and flies. Regardless of the method of choice for CR and IF, these animals basically feed ad libitum when they are provided food. A bigger challenge is that it is not practically easy to measure the amount of food they consumed, which is an important confounding factor to interpreting CR and IF data. An unignorable number of different, often contradictory, results from different strains and/or laboratory on CR and IF may be at least partially due to these factors. Importantly, these limits also put roadblocks on the translation of animal studies for CR and IF into human applications. In addition to these practical limits, the interspecies differences in physiology, metabolism, reproduction, and behavior between model organisms and humans serve as additional confounding factors for human translatability. For example, rodents have much higher metabolic rates than humans [218], yet similar fasting and feeding protocols are often used for IF. In addition to these intrinsic differences between model organisms and humans, intraspecies variations (differences in the population of the same species; also seen Section 5.3) often add to the complexity of human translation of animal studies. In flies, although some beneficial effects were observed by TRF (12 h of fasting during the dark phase of the day) on cardiac function and other metabolic and behavioral parameters such as body weight and sleep [212], an increased mortality was observed by 12 h of fasting in some young (<2 weeks) wild types flies (D.S. Hwangbo, unpublished data). On the other hand, some other wild type flies were strongly resistant to an extended period of fasting (up to 5 days), at least when they were young, during the IF regime of 2 day feeding:5 day fasting [67]. We speculate that, due to the confounding factors arising from the interspecies and intraspecies differences, the degree of beneficial effect of CR and IF on healthspan and lifespan in humans might not be equivalent to that of animal models [4,219]. Therefore, for the best working CR and IF protocols for human translations, we propose that multifactorial models should be developed to accommodate these confounding factors that interfere with the interpretation of animal results to human applications.

5.3. Individual Variations

From a practical perspective, IF is often thought of as a milder form of CR and generally considered to be easier for human implication. Beyond the evolutionary difference in metabolism and physiology between animals and human, potential interactions between genetic variations among human populations and the candidate mechanisms for CR and IF should not be overlooked. Human lifespan is affected by multiple genetic and non-genetic factors including population origin and interactions between the nuclear/mitochondrial genome and microbiomes [220]. It was suggested that only about 10–25% of human lifespan variation is explained by genetic factors [159], emphasizing the importance of the interactions between genetic background and environmental factors [221]. In animal models, some physiological and metabolic traits, especially lifespan, are strongly affected by genetic backgrounds and variations as well as non-genetic factors such as symbiotic microbiome and water balance [222]. When a collection of recombinant inbred mouse strains were tested for lifespan under ad libitum diet and CR (40% reduction compared to ad libitum diet) diet, a wide range of lifespan responses were observed in both ad libitum and CR diets [223,224]. For example, the mean lifespan of female mice on ad libitum diet varied from 407 to 1208 days. Strikingly, their lifespans on CR diet varied to a greater degree from 113 to 1225 days. Importantly, not only did CR fail in lifespan extension in some lines, but it even shortened lifespan in some lines too [223]. Similarly, a strong variation in lifespan response to diets was observed when a collection of nearly 200 genetically distinct lines of Drosophila (DGRP: Drosophila Genetic Reference Panel) tested for lifespan in ad libitum (5% Yeast) and CR (0.5% Yeast) [225]. In both cases, lifespan response also significantly varied between males and females [223,225], generating a further layer of complication in understanding the mechanisms of CR. A simple interpretation of these animal studies would suggest that a certain type of CR and IF may not be beneficial, but they can be even deleterious depending on genetic variations and sex [32]. Therefore, for human applications of CR and IF, we suggest that individualized genomics and medicine should be established first to take full advantage of CR and IF.

Author Contributions: D.-S.H., H.-Y.L., L.S.A. and K.-J.M. wrote the manuscript. All authors have read and agreed to the published version of the manuscript.

Funding: This work was supported by an Inha University Research Grant.

Conflicts of Interest: The authors declare no conflict of interest.

References

1. Jones, O.R.; Vaupel, J.W. Senescence is not inevitable. *Biogerontology* **2017**, *18*, 965–971. [CrossRef]
2. McDonald, R.B. *Biology of Aging*; Garland Science: New York, NY, USA, 2013.
3. Rous, P. The Influence of Diet on Transplanted and Spontaneous Mouse Tumors. *J. Exp. Med.* **1914**, *20*, 433–451. [CrossRef] [PubMed]
4. Osborne, T.B.; Mendel, L.B.; Ferry, E.L. The Effect of Retardation of Growth Upon the Breeding Period and Duration of Life of Rats. *Science* **1917**, *45*, 294–295. [CrossRef] [PubMed]
5. Loeb, J.; Northrop, J.H. What Determines the Duration of Life in Metazoa? *Proc. Natl. Acad. Sci. USA* **1917**, *3*, 382–386. [CrossRef] [PubMed]
6. Ingle, L.; Wood, T.R.; Banta, A.M. A study of longevity, growth, reproduction and heart rate in Daphnia longispina by limitations in quantity of food. *J. Exp. Zool.* **1937**, *76*, 325–352. [CrossRef]
7. McCay, C.M.; Crowell, M.F.; Maynard, L.A. The effect of retarded growth upon the length of life span and upon the ultimate body size. *J. Nutr.* **1935**, *10*, 63–79. [CrossRef]
8. Tannenbaum, A. The Initiation and Growth of Tumors: Introduction. I. Effects of Underfeeding. *Cancer Res.* **1940**, *38*, 335–350. [CrossRef]
9. Saxton, J.A., Jr.; Kimball, G.C. Relation of nephrosis and other diseases of albino rats to age and to modifications of diet. *Arch. Pathol.* **1941**, *32*, 951–965.
10. McCay, C.M.; Sperling, G.; Barnes, L.L. Growth, ageing, chronic diseases, and life span in rats. *Arch. Biochem.* **1943**, *2*, 469–479.

11. Saxton, J.A.; Boon, M.C.; Furth, J. Observations on the inhibition of development of spontaneous leukemia in mice by underfeeding. *Cancer Res.* **1944**, *4*, 401–409.

12. White, F.R.; White, J.; Mider, G.B.; Kelly, M.G.; Heston, W.E.; David, P.W. Effect of caloricrestriction on mammary tumor formation in strain C3H mice and on the response to painting with methylcholanthrene. *J. Natl. Cancer Inst.* **1944**, *5*, 43–48. [CrossRef]

13. Rudzinska, M.A. The influence of amount of food on the reproduction rate and longevity of a suctorian (*Tokophrya infusionum*). *Science* **1951**, *113*, 10–11. [CrossRef] [PubMed]

14. Fanestil, D.D.; Barrows, C.H. Aging in the rotifer. *J. Gerontol.* **1965**, *20*, 462–469. [PubMed]

15. Comfort, A. Effect of Delayed and Resumed Growth on the Longevity of a Fish (*Lebistes reticulatus*, Peters) in Captivity. *Gerontologia* **1963**, *8*, 150–155. [CrossRef] [PubMed]

16. Klass, M.R. Aging in the nematode *Caenorhabditis elegans*: Major biological and environmental factors influencing life span. *Mech. Ageing Dev.* **1977**, *6*, 413–429. [CrossRef]

17. Ross, M.H. Length of Life and Nutrition in the Rat. *J. Nutr.* **1961**, *75*, 197–210. [CrossRef]

18. Beauchene, R.E.; Bales, C.W.; Smith, C.A.; Tucker, S.M.; Mason, R.L. The effect of feed restriction on body composition and longevity of rats. *Physiologist* **1979**, *22*, 8.

19. Weindruch, R.; Walford, R. Dietary restriction in mice beginning at 1 year of age: Effect on life-span and spontaneous cancer incidence. *Science* **1982**, *215*, 1415–1418. [CrossRef]

20. Cheney, K.E.; Liu, R.K.; Smith, G.S.; Leung, R.E.; Mickey, M.R.; Walford, R.L. Survival and disease patterns in C57BL/6J mice subjected to undernutrition. *Exp. Gerontol* **1980**, *15*, 237–258. [CrossRef]

21. Leveille, G.A. The long-term effects of meal-eating on lipogenesis, enzyme activity, and longevity in the rat. *J. Nutr.* **1972**, *102*, 549–556. [CrossRef]

22. Gerbase-DeLima, M.; Liu, R.K.; Cheney, K.E.; Mickey, R.; Walford, R.L. Immune function and survival in a long-lived mouse strain subjected to undernutrition. *Gerontologia* **1975**, *21*, 184–202. [CrossRef] [PubMed]

23. Goodrick, C.L.; Ingram, D.K.; Reynolds, M.A.; Freeman, J.R.; Cider, N.L. Effects of intermittent feeding upon growth and life span in rats. *Gerontology* **1982**, *28*, 233–241. [CrossRef] [PubMed]

24. Davis, T.A.; Bales, C.W.; Beauchene, R.E. Differential effects of dietary caloric and protein restriction in the aging rat. *Exp. Gerontol.* **1983**, *18*, 427–435. [CrossRef]

25. Masoro, E.J.; Iwasaki, K.; Gleiser, C.A.; McMahan, C.A.; Seo, E.J.; Yu, B.P. Dietary modulation of the progression of nephropathy in aging rats: An evaluation of the importance of protein. *Am. J. Clin. Nutr.* **1989**, *49*, 1217–1227. [CrossRef]

26. Iwasaki, K.; Gleiser, C.A.; Masoro, E.J.; McMahan, C.A.; Seo, E.J.; Yu, B.P. Influence of the restriction of individual dietary components on longevity and age-related disease of Fischer rats: The fat component and the mineral component. *J. Gerontol.* **1988**, *43*, B13–B21. [CrossRef] [PubMed]

27. Ingram, D.K.; Cutler, R.G.; Weindruch, R.; Renquist, D.M.; Knapka, J.J.; April, M.; Belcher, C.T.; Clark, M.A.; Hatcherson, C.D.; Marriott, B.M.; et al. Dietary restriction and aging: The initiation of a primate study. *J. Gerontol.* **1990**, *45*, B148–B163. [CrossRef] [PubMed]

28. Kemnitz, J.W.; Weindruch, R.; Roecker, E.B.; Crawford, K.; Kaufman, P.L.; Ershler, W.B. Dietary restriction of adult male rhesus monkeys: Design, methodology, and preliminary findings from the first year of study. *J. Gerontol.* **1993**, *48*, B17–B26. [CrossRef]

29. Hansen, B.C.; Bodkin, N.L. Primary prevention of diabetes mellitus by prevention of obesity in monkeys. *Diabetes* **1993**, *42*, 1809–1814. [CrossRef]

30. Anson, R.M.; Guo, Z.; de Cabo, R.; Iyun, T.; Rios, M.; Hagepanos, A.; Ingram, D.K.; Lane, M.A.; Mattson, M.P. Intermittent fasting dissociates beneficial effects of dietary restriction on glucose metabolism and neuronal resistance to injury from calorie intake. *Proc. Natl. Acad. Sci. USA* **2003**, *100*, 6216–6220. [CrossRef]

31. Longo, V.D.; Panda, S. Fasting, Circadian Rhythms, and Time-Restricted Feeding in Healthy Lifespan. *Cell Metab.* **2016**, *23*, 1048–1059. [CrossRef]

32. Di Francesco, A.; Di Germanio, C.; Bernier, M.; de Cabo, R. A time to fast. *Science* **2018**, *362*, 770–775. [CrossRef] [PubMed]

33. Carlson, O.; Martin, B.; Stote, K.S.; Golden, E.; Maudsley, S.; Najjar, S.S.; Ferrucci, L.; Ingram, D.K.; Longo, D.L.; Rumpler, W.V.; et al. Impact of reduced meal frequency without caloric restriction on glucose regulation in healthy, normal-weight middle-aged men and women. *Metabolism* **2007**, *56*, 1729–1734. [CrossRef] [PubMed]

34. Stote, K.S.; Baer, D.J.; Spears, K.; Paul, D.R.; Harris, G.K.; Rumpler, W.V.; Strycula, P.; Najjar, S.S.; Ferrucci, L.; Ingram, D.K.; et al. A controlled trial of reduced meal frequency without caloric restriction in healthy, normal-weight, middle-aged adults. *Am. J. Clin. Nutr.* **2007**, *85*, 981–988. [CrossRef] [PubMed]

35. Anton, S.D.; Lee, S.A.; Donahoo, W.T.; McLaren, C.; Manini, T.; Leeuwenburgh, C.; Pahor, M. The Effects of Time Restricted Feeding on Overweight, Older Adults: A Pilot Study. *Nutrients* **2019**, *11*, 1500. [CrossRef] [PubMed]

36. Masoro, E.J.; Katz, M.S.; McMahan, C.A. Evidence for the glycation hypothesis of aging from the food-restricted rodent model. *J. Gerontol.* **1989**, *44*, B20–B22. [CrossRef] [PubMed]

37. Dalderup, L.M.; Visser, W. Influence of extra sucrose in the daily food on the life-span of Wistar albino rats. *Nature* **1969**, *222*, 1050–1052. [CrossRef]

38. Iwasaki, K.; Gleiser, C.A.; Masoro, E.J.; McMahan, C.A.; Seo, E.J.; Yu, B.P. The influence of dietary protein source on longevity and age-related disease processes of Fischer rats. *J. Gerontol.* **1988**, *43*, B5–B12. [CrossRef]

39. Mair, W.; Piper, M.D.W.; Partridge, L. Calories do not explain extension of life span by dietary restriction in Drosophila. *PLoS Biol.* **2005**, *3*, e223. [CrossRef]

40. Bruce, K.D.; Hoxha, S.; Carvalho, G.B.; Yamada, R.; Wang, H.D.; Karayan, P.; He, S.; Brummel, T.; Kapahi, P.; Ja, W.W. High carbohydrate-low protein consumption maximizes Drosophila lifespan. *Exp. Gerontol.* **2013**, *48*, 1129–1135. [CrossRef]

41. Min, K.J.; Tatar, M. Restriction of amino acids extends lifespan in Drosophila melanogaster. *Mech. Ageing Dev.* **2006**, *127*, 643–646. [CrossRef]

42. Solon-Biet, S.M.; McMahon, A.C.; Ballard, J.W.O.; Ruohonen, K.; Wu, L.E.; Cogger, V.C.; Warren, A.; Huang, X.; Pichaud, N.; Melvin, R.G.; et al. The Ratio of Macronutrients, Not Caloric Intake, Dictates Cardiometabolic Health, Aging, and Longevity in Ad Libitum-Fed Mice. *Cell Metab.* **2014**, *19*, 418–430. [CrossRef] [PubMed]

43. Grandison, R.C.; Piper, M.D.; Partridge, L. Amino-acid imbalance explains extension of lifespan by dietary restriction in Drosophila. *Nature* **2009**, *462*, 1061–1064. [CrossRef]

44. Zimmerman, J.A.; Malloy, V.; Krajcik, R.; Orentreich, N. Nutritional control of aging. *Exp. Gerontol.* **2003**, *38*, 47–52. [CrossRef]

45. Johnson, J.E.; Johnson, F.B. Methionine restriction activates the retrograde response and confers both stress tolerance and lifespan extension to yeast, mouse and human cells. *PLoS ONE* **2014**, *9*, e97729. [CrossRef] [PubMed]

46. Lee, B.C.; Kaya, A.; Ma, S.; Kim, G.; Gerashchenko, M.V.; Yim, S.H.; Hu, Z.; Harshman, L.G.; Gladyshev, V.N. Methionine restriction extends lifespan of Drosophila melanogaster under conditions of low amino-acid status. *Nat. Commun.* **2014**, *5*, 3592. [CrossRef] [PubMed]

47. Perrone, C.E.; Malloy, V.L.; Orentreich, D.S.; Orentreich, N. Metabolic adaptations to methionine restriction that benefit health and lifespan in rodents. *Exp. Gerontol.* **2013**, *48*, 654–660. [CrossRef]

48. Kaeberlein, M. Longevity and aging in the budding yeast. In *Handbook of Models for Human Aging*; Conn, P.M., Ed.; Elvesier Press: Boston, UK, 2006; pp. 109–120.

49. Murakami, C.J.; Burtner, C.R.; Kennedy, B.K.; Kaeberlein, M. A method for high-throughput quantitative analysis of yeast chronological life span. *J. Gerontol. A Biol. Sci. Med. Sci.* **2008**, *63*, 113–121. [CrossRef]

50. Lin, S.J.; Defossez, P.A.; Guarente, L. Requirement of NAD and SIR2 for life-span extension by calorie restriction in Saccharomyces cerevisiae. *Science* **2000**, *289*, 2126–2128. [CrossRef]

51. Lin, S.J.; Kaeberlein, M.; Andalis, A.A.; Sturtz, L.A.; Defossez, P.A.; Culotta, V.C.; Fink, G.R.; Guarente, L. Calorie restriction extends Saccharomyces cerevisiae lifespan by increasing respiration. *Nature* **2002**, *418*, 344–348. [CrossRef]

52. Jiang, J.C.; Jaruga, E.; Repnevskaya, M.V.; Jazwinski, S.M. An intervention resembling caloric restriction prolongs life span and retards aging in yeast. *FASEB J.* **2000**, *14*, 2135–2137. [CrossRef]

53. Wu, Z.; Liu, S.Q.; Huang, D. Dietary restriction depends on nutrient composition to extend chronological lifespan in budding yeast Saccharomyces cerevisiae. *PLoS ONE* **2013**, *8*, e64448. [CrossRef] [PubMed]

54. Wei, M.; Fabrizio, P.; Hu, J.; Ge, H.; Cheng, C.; Li, L.; Longo, V.D. Life Span Extension by Calorie Restriction Depends on Rim15 and Transcription Factors Downstream of Ras/PKA, Tor, and Sch9. *PLoS Genet.* **2008**, *4*, e13. [CrossRef] [PubMed]

55. Cassada, R.C.; Russell, R.L. The dauerlarva, a post-embryonic developmental variant of the nematode Caenorhabditis elegans. *Dev. Biol.* **1975**, *46*, 326–342. [CrossRef]

56. Sutphin, G.L.; Kaeberlein, M. Dietary restriction by bacterial deprivation increases life span in wild-derived nematodes. *Exp. Gerontol.* **2008**, *43*, 130–135. [CrossRef] [PubMed]

57. Houthoofd, K.; Johnson, T.E.; Vanfleteren, J.R. Dietary Restriction in the Nematode Caenorhabditis elegans. *J. Gerontol. Ser. A* **2005**, *60*, 1125–1131. [CrossRef]

58. Houthoofd, K.; Gems, D.; Johnson, T.E.; Vanfleteren, J.R. Dietary restriction in the nematode Caenorhabditis elegans. *Interdiscip. Top. Gerontol.* **2007**, *35*, 98–114. [CrossRef]

59. Uno, M.; Honjoh, S.; Matsuda, M.; Hoshikawa, H.; Kishimoto, S.; Yamamoto, T.; Ebisuya, M.; Yamamoto, T.; Matsumoto, K.; Nishida, E. A fasting-responsive signaling pathway that extends life span in C. elegans. *Cell Rep.* **2013**, *3*, 79–91. [CrossRef]

60. Honjoh, S.; Yamamoto, T.; Uno, M.; Nishida, E. Signalling through RHEB-1 mediates intermittent fasting-induced longevity in C. elegans. *Nature* **2009**, *457*, 726–730. [CrossRef]

61. Kaeberlein, T.L.; Smith, E.D.; Tsuchiya, M.; Welton, K.L.; Thomas, J.H.; Fields, S.; Kennedy, B.K.; Kaeberlein, M. Lifespan extension in Caenorhabditis elegans by complete removal of food. *Aging Cell* **2006**, *5*, 487–494. [CrossRef]

62. Lee, G.D.; Wilson, M.A.; Zhu, M.; Wolkow, C.A.; de Cabo, R.; Ingram, D.K.; Zou, S. Dietary deprivation extends lifespan in Caenorhabditis elegans. *Aging Cell* **2006**, *5*, 515–524. [CrossRef]

63. Piper, M.D.W.; Partridge, L. Drosophila as a model for ageing. *Biochim. Biophys. Acta Mol. Basis Dis.* **2018**, *1864*, 2707–2717. [CrossRef] [PubMed]

64. Piper, M.D.; Partridge, L. Dietary restriction in Drosophila: Delayed aging or experimental artefact? *PLoS Genet.* **2007**, *3*, e57. [CrossRef]

65. Grandison, R.C.; Wong, R.; Bass, T.M.; Partridge, L.; Piper, M.D. Effect of a standardised dietary restriction protocol on multiple laboratory strains of Drosophila melanogaster. *PLoS ONE* **2009**, *4*, e4067. [CrossRef]

66. Partridge, L.; Piper, M.D.; Mair, W. Dietary restriction in Drosophila. *Mech. Ageing Dev.* **2005**, *126*, 938–950. [CrossRef] [PubMed]

67. Catterson, J.H.; Khericha, M.; Dyson, M.C.; Vincent, A.J.; Callard, R.; Haveron, S.M.; Rajasingam, A.; Ahmad, M.; Partridge, L. Short-Term, Intermittent Fasting Induces Long-Lasting Gut Health and TOR-Independent Lifespan Extension. *Curr. Biol.* **2018**, *28*, 1714–1724.e4. [CrossRef] [PubMed]

68. Villanueva, J.E.; Livelo, C.; Trujillo, A.S.; Chandran, S.; Woodworth, B.; Andrade, L.; Le, H.D.; Manor, U.; Panda, S.; Melkani, G.C. Time-restricted feeding restores muscle function in Drosophila models of obesity and circadian-rhythm disruption. *Nat. Commun.* **2019**, *10*, 2700. [CrossRef] [PubMed]

69. Bonkowski, M.S.; Rocha, J.S.; Masternak, M.M.; Al Regaiey, K.A.; Bartke, A. Targeted disruption of growth hormone receptor interferes with the beneficial actions of calorie restriction. *Proc. Natl. Acad. Sci. USA* **2006**, *103*, 7901–7905. [CrossRef]

70. Swindell, W.R. Dietary restriction in rats and mice: A meta-analysis and review of the evidence for genotype-dependent effects on lifespan. *Ageing Res. Rev.* **2012**, *11*, 254–270. [CrossRef]

71. Speakman, J.R.; Mitchell, S.E.; Mazidi, M. Calories or protein? The effect of dietary restriction on lifespan in rodents is explained by calories alone. *Exp. Gerontol.* **2016**, *86*, 28–38. [CrossRef]

72. Vaughan, K.L.; Kaiser, T.; Peaden, R.; Anson, R.M.; de Cabo, R.; Mattison, J.A. Caloric Restriction Study Design Limitations in Rodent and Nonhuman Primate Studies. *J. Gerontol. Ser. A* **2018**, *73*, 48–53. [CrossRef]

73. Fontana, L.; Nehme, J.; Demaria, M. Caloric restriction and cellular senescence. *Mech. Ageing Dev.* **2018**, *176*, 19–23. [CrossRef] [PubMed]

74. De Marte, M.L.; Enesco, H.E. Influence of low tryptophan diet on survival and organ growth in mice. *Mech. Ageing Dev.* **1986**, *36*, 161–171. [CrossRef]

75. Ooka, H.; Segall, P.E.; Timiras, P.S. Histology and survival in age-delayed low-tryptophan-fed rats. *Mech. Ageing Dev.* **1988**, *43*, 79–98. [CrossRef]

76. Richie, J.P., Jr.; Leutzinger, Y.; Parthasarathy, S.; Malloy, V.; Orentreich, N.; Zimmerman, J.A. Methionine restriction increases blood glutathione and longevity in F344 rats. *FASEB J.* **1994**, *8*, 1302–1307. [CrossRef]

77. Miller, R.A.; Buehner, G.; Chang, Y.; Harper, J.M.; Sigler, R.; Smith-Wheelock, M. Methionine-deficient diet extends mouse lifespan, slows immune and lens aging, alters glucose, T4, IGF-I and insulin levels, and increases hepatocyte MIF levels and stress resistance. *Aging Cell* **2005**, *4*, 119–125. [CrossRef]

78. Sun, L.; Sadighi Akha, A.A.; Miller, R.A.; Harper, J.M. Life-span extension in mice by preweaning food restriction and by methionine restriction in middle age. *J. Gerontol. A Biol. Sci. Med. Sci.* **2009**, *64*, 711–722. [CrossRef]

79. Longo, V.D.; Mattson, M.P. Fasting: Molecular Mechanisms and Clinical Applications. *Cell Metab.* **2014**, *19*, 181–192. [CrossRef]

80. Hatori, M.; Vollmers, C.; Zarrinpar, A.; DiTacchio, L.; A.Bushong, E.; Gill, S.; Leblanc, M.; Chaix, A.; Joens, M.; Fitzpatrick, J.A.J.; et al. Time-restricted feeding without reducing caloric intake prevents metabolic diseases in mice fed a high-fat diet. *Cell Metab.* **2012**, *15*, 848–860. [CrossRef]

81. Wang, H.-B.; Loh, D.H.; Whittaker, D.S.; Cutler, T.; Howland, D.; Colwell, C.S. Time-restricted feeding improves circadian dysfunction as well as motor symptoms in the Q175 mouse model of Huntington's disease. *eNeuro* **2018**, *5*. [CrossRef]

82. Mattson, M.P. Energy intake, meal frequency, and health: A neurobiological perspective. *Annu. Rev. Nutr.* **2005**, *25*, 237–260. [CrossRef]

83. Goodrick, C.L.; Ingram, D.K.; Reynolds, M.A.; Freeman, J.R.; Cider, N. Effects of intermittent feeding upon body weight and lifespan in inbred mice: Interaction of genotype and age. *Mech. Ageing Dev.* **1990**, *55*, 69–87. [CrossRef]

84. Carlson, A.J.; Hoelzel, F. Apparent prolongation of the life span of rats by intermittent fasting. *J. Nutr.* **1946**, *31*, 363–375. [CrossRef] [PubMed]

85. Kendrick, D.C. The effects of infantile stimulation and intermittent fasting and feeding on life span in the black-hooded rat. *Dev. Psychobiol.* **1973**, *6*, 225–234. [CrossRef] [PubMed]

86. Chaix, A.; Zarrinpar, A.; Miu, P.; Panda, S. Time-restricted feeding is a preventative and therapeutic intervention against diverse nutritional challenges. *Cell Metab.* **2014**, *20*, 991–1005. [CrossRef] [PubMed]

87. Wu, M.W.; Li, X.M.; Xian, L.J.; Levi, F. Effects of meal timing on tumor progression in mice. *Life Sci.* **2004**, *75*, 1181–1193. [CrossRef]

88. Mitchell, S.J.; Bernier, M.; Mattison, J.A.; Aon, M.A.; Kaiser, T.A.; Anson, R.M.; Ikeno, Y.; Anderson, R.M.; Ingram, D.K.; de Cabo, R. Daily Fasting Improves Health and Survival in Male Mice Independent of Diet Composition and Calories. *Cell Metab.* **2019**, *29*, 221–228.e3. [CrossRef]

89. Bodkin, N.L.; Alexander, T.M.; Ortmeyer, H.K.; Johnson, E.; Hansen, B.C. Mortality and morbidity in laboratory-maintained Rhesus monkeys and effects of long-term dietary restriction. *J. Gerontol. A Biol. Sci. Med. Sci.* **2003**, *58*, 212–219. [CrossRef]

90. Bodkin, N.L.; Ortmeyer, H.K.; Hansen, B.C. Long-term dietary restriction in older-aged rhesus monkeys: Effects on insulin resistance. *J. Gerontol. A Biol. Sci. Med. Sci.* **1995**, *50*, B142–B147. [CrossRef]

91. Colman, R.J.; Anderson, R.M.; Johnson, S.C.; Kastman, E.K.; Kosmatka, K.J.; Beasley, T.M.; Allison, D.B.; Cruzen, C.; Simmons, H.A.; Kemnitz, J.W.; et al. Caloric restriction delays disease onset and mortality in rhesus monkeys. *Science* **2009**, *325*, 201–204. [CrossRef]

92. Mattison, J.A.; Roth, G.S.; Beasley, T.M.; Tilmont, E.M.; Handy, A.M.; Herbert, R.L.; Longo, D.L.; Allison, D.B.; Young, J.E.; Bryant, M.; et al. Impact of caloric restriction on health and survival in rhesus monkeys from the NIA study. *Nature* **2012**, *489*, 318–321. [CrossRef]

93. Mattison, J.A.; Colman, R.J.; Beasley, T.M.; Allison, D.B.; Kemnitz, J.W.; Roth, G.S.; Ingram, D.K.; Weindruch, R.; de Cabo, R.; Anderson, R.M. Caloric restriction improves health and survival of rhesus monkeys. *Nat. Commun.* **2017**, *8*, 14063. [CrossRef] [PubMed]

94. Rouhani, M.; Azadbakht, L. Is Ramadan fasting related to health outcomes? A review on the related evidence. *J. Res. Med. Sci.* **2014**, *19*, 987–992.

95. Rickman, A.D.; Williamson, D.A.; Martin, C.K.; Gilhooly, C.H.; Stein, R.I.; Bales, C.W.; Roberts, S.; Das, S.K. The CALERIE Study: Design and methods of an innovative 25% caloric restriction intervention. *Contemp. Clin. Trials* **2011**, *32*, 874–881. [CrossRef] [PubMed]

96. Meydani, M.; Das, S.; Band, M.; Epstein, S.; Roberts, S. The effect of caloric restriction and glycemic load on measures of oxidative stress and antioxidants in humans: Results from the CALERIE Trial of Human Caloric Restriction. *J. Nutr. Health Aging* **2011**, *15*, 456–460. [CrossRef] [PubMed]

97. Martin, C.K.; Das, S.K.; Lindblad, L.; Racette, S.B.; McCrory, M.A.; Weiss, E.P.; Delany, J.P.; Kraus, W.E.; Team, C.S. Effect of calorie restriction on the free-living physical activity levels of nonobese humans: Results of three randomized trials. *J. Appl. Physiol. (1985)* **2011**, *110*, 956–963. [CrossRef]

98. Villareal, D.T.; Fontana, L.; Weiss, E.P.; Racette, S.B.; Steger-May, K.; Schechtman, K.B.; Klein, S.; Holloszy, J.O. Bone mineral density response to caloric restriction-induced weight loss or exercise-induced weight loss: A randomized controlled trial. *Arch. Intern. Med.* **2006**, *166*, 2502–2510. [CrossRef]

99. Weiss, E.P.; Racette, S.B.; Villareal, D.T.; Fontana, L.; Steger-May, K.; Schechtman, K.B.; Klein, S.; Ehsani, A.A.; Holloszy, J.O.; Washington University School of Medicine, C.G. Lower extremity muscle size and strength and aerobic capacity decrease with caloric restriction but not with exercise-induced weight loss. *J. Appl. Physiol. (1985)* **2007**, *102*, 634–640. [CrossRef]

100. Heilbronn, L.K.; Smith, S.R.; Martin, C.K.; Anton, S.D.; Ravussin, E. Alternate-day fasting in nonobese subjects: Effects on body weight, body composition, and energy metabolism. *Am. J. Clin. Nutr.* **2005**, *81*, 69–73. [CrossRef]

101. Hutchison, A.T.; Liu, B.; Wood, R.E.; Vincent, A.D.; Thompson, C.H.; O'Callaghan, N.J.; Wittert, G.A.; Heilbronn, L.K. Effects of Intermittent Versus Continuous Energy Intakes on Insulin Sensitivity and Metabolic Risk in Women with Overweight. *Obesity* **2019**, *27*, 50–58. [CrossRef]

102. Halberg, N.; Henriksen, M.; Soderhamn, N.; Stallknecht, B.; Ploug, T.; Schjerling, P.; Dela, F. Effect of intermittent fasting and refeeding on insulin action in healthy men. *J. Appl. Physiol. (1985)* **2005**, *99*, 2128–2136. [CrossRef]

103. Klempel, M.C.; Kroeger, C.M.; Bhutani, S.; Trepanowski, J.F.; Varady, K.A. Intermittent fasting combined with calorie restriction is effective for weight loss and cardio-protection in obese women. *Nutr. J.* **2012**, *11*, 98. [CrossRef] [PubMed]

104. Tinsley, G.M.; Forsse, J.S.; Butler, N.K.; Paoli, A.; Bane, A.A.; La Bounty, P.M.; Morgan, G.B.; Grandjean, P.W. Time-restricted feeding in young men performing resistance training: A randomized controlled trial. *Eur. J. Sport Sci.* **2017**, *17*, 200–207. [CrossRef] [PubMed]

105. Hoddy, K.K.; Kroeger, C.M.; Trepanowski, J.F.; Barnosky, A.; Bhutani, S.; Varady, K.A. Meal timing during alternate day fasting: Impact on body weight and cardiovascular disease risk in obese adults. *Obesity* **2014**, *22*, 2524–2531. [CrossRef] [PubMed]

106. Harvie, M.N.; Pegington, M.; Mattson, M.P.; Frystyk, J.; Dillon, B.; Evans, G.; Cuzick, J.; Jebb, S.A.; Martin, B.; Cutler, R.G.; et al. The effects of intermittent or continuous energy restriction on weight loss and metabolic disease risk markers: A randomized trial in young overweight women. *Int. J. Obes.* **2011**, *35*, 714–727. [CrossRef]

107. Azevedo, F.R.; Ikeoka, D.; Caramelli, B. Effects of intermittent fasting on metabolism in men. *Rev. Assoc. Med. Bras. (1992)* **2013**, *59*, 167–173. [CrossRef]

108. Catenacci, V.A.; Pan, Z.; Ostendorf, D.; Brannon, S.; Gozansky, W.S.; Mattson, M.P.; Martin, B.; MacLean, P.S.; Melanson, E.L.; Troy Donahoo, W. A randomized pilot study comparing zero-calorie alternate-day fasting to daily caloric restriction in adults with obesity. *Obesity* **2016**, *24*, 1874–1883. [CrossRef]

109. Stekovic, S.; Hofer, S.J.; Tripolt, N.; Aon, M.A.; Royer, P.; Pein, L.; Stadler, J.T.; Pendl, T.; Prietl, B.; Url, J.; et al. Alternate Day Fasting Improves Physiological and Molecular Markers of Aging in Healthy, Non-obese Humans. *Cell Metab.* **2019**, *30*, 462–476.e6. [CrossRef]

110. Varady, K.A.; Bhutani, S.; Church, E.C.; Klempel, M.C. Short-term modified alternate-day fasting: A novel dietary strategy for weight loss and cardioprotection in obese adults. *Am. J. Clin. Nutr.* **2009**, *90*, 1138–1143. [CrossRef]

111. Johnson, J.B.; Summer, W.; Cutler, R.G.; Martin, B.; Hyun, D.H.; Dixit, V.D.; Pearson, M.; Nassar, M.; Telljohann, R.; Maudsley, S.; et al. Alternate day calorie restriction improves clinical findings and reduces markers of oxidative stress and inflammation in overweight adults with moderate asthma. *Free Radic. Biol. Med.* **2007**, *42*, 665–674. [CrossRef]

112. Varady, K.A.; Bhutani, S.; Klempel, M.C.; Kroeger, C.M.; Trepanowski, J.F.; Haus, J.M.; Hoddy, K.K.; Calvo, Y. Alternate day fasting for weight loss in normal weight and overweight subjects: A randomized controlled trial. *Nutr. J.* **2013**, *12*, 146. [CrossRef]

113. Neeland, I.J.; Turer, A.T.; Ayers, C.R.; Powell-Wiley, T.M.; Vega, G.L.; Farzaneh-Far, R.; Grundy, S.M.; Khera, A.; McGuire, D.K.; de Lemos, J.A. Dysfunctional adiposity and the risk of prediabetes and type 2 diabetes in obese adults. *JAMA* **2012**, *308*, 1150–1159. [CrossRef] [PubMed]

114. Sutton, E.F.; Beyl, R.; Early, K.S.; Cefalu, W.T.; Ravussin, E.; Peterson, C.M. Early Time-Restricted Feeding Improves Insulin Sensitivity, Blood Pressure, and Oxidative Stress Even without Weight Loss in Men with Prediabetes. *Cell Metab.* **2018**, *27*, 1212–1221.e3. [CrossRef] [PubMed]

115. Ravussin, E.; Beyl, R.A.; Poggiogalle, E.; Hsia, D.S.; Peterson, C.M. Early Time-Restricted Feeding Reduces Appetite and Increases Fat Oxidation But Does Not Affect Energy Expenditure in Humans. *Obesity* **2019**, *27*, 1244–1254. [CrossRef] [PubMed]

116. Martens, C.R.; Rossman, M.J.; Mazzo, M.R.; Jankowski, L.R.; Nagy, E.E.; Denman, B.A.; Richey, J.J.; Johnson, S.A.; Ziemba, B.P.; Wang, Y.; et al. Short-term time-restricted feeding is safe and feasible in non-obese healthy midlife and older adults. *GeroScience* **2020**. [CrossRef]

117. Hansen, M.; Kennedy, B.K. Does Longer Lifespan Mean Longer Healthspan? *Trends Cell Biol.* **2016**, *26*, 565–568. [CrossRef]

118. de Cabo, R.; Mattson, M.P. Effects of Intermittent Fasting on Health, Aging, and Disease. *N. Engl. J. Med.* **2019**, *381*, 2541–2551. [CrossRef] [PubMed]

119. Fontana, L.; Partridge, L. Promoting health and longevity through diet: From model organisms to humans. *Cell* **2015**, *161*, 106–118. [CrossRef]

120. Pan, H.; Finkel, T. Key proteins and pathways that regulate lifespan. *J. Biol. Chem.* **2017**, *292*, 6452–6460. [CrossRef]

121. Anson, R.M.; Jones, B.; de Cabod, R. The diet restriction paradigm: A brief review of the effects of every-other-day feeding. *AGE* **2005**, *27*, 17–25. [CrossRef]

122. Nicklin, P.; Bergman, P.; Zhang, B.; Triantafellow, E.; Wang, H.; Nyfeler, B.; Yang, H.; Hild, M.; Kung, C.; Wilson, C.; et al. Bidirectional transport of amino acids regulates mTOR and autophagy. *Cell* **2009**, *136*, 521–534. [CrossRef]

123. Beauchamp, E.M.; Platanias, L.C. The evolution of the TOR pathway and its role in cancer. *Oncogene* **2013**, *32*, 3923–3932. [CrossRef] [PubMed]

124. Kapahi, P.; Chen, D.; Rogers, A.N.; Katewa, S.D.; Li, P.W.; Thomas, E.L.; Kockel, L. With TOR, less is more: A key role for the conserved nutrient-sensing TOR pathway in aging. *Cell Metab.* **2010**, *11*, 453–465. [CrossRef] [PubMed]

125. Kaeberlein, M.; Powers, R.W.; Steffen, K.K.; Westman, E.A.; Hu, D.; Dang, N.; Kerr, E.O.; Kirkland, K.T.; Fields, S.; Kennedy, B.K. Regulation of Yeast Replicative Life Span by TOR and Sch9 in Response to Nutrients. *Science* **2005**, *310*, 1193. [CrossRef]

126. Bonawitz, N.D.; Chatenay-Lapointe, M.; Pan, Y.; Shadel, G.S. Reduced TOR Signaling Extends Chronological Life Span via Increased Respiration and Upregulation of Mitochondrial Gene Expression. *Cell Metab.* **2007**, *5*, 265–277. [CrossRef] [PubMed]

127. Medvedik, O.; Lamming, D.W.; Kim, K.D.; Sinclair, D.A. MSN2 and MSN4 Link Calorie Restriction and TOR to Sirtuin-Mediated Lifespan Extension in Saccharomyces cerevisiae. *PLoS Biol.* **2007**, *5*, e261. [CrossRef]

128. Steffen, K.K.; MacKay, V.L.; Kerr, E.O.; Tsuchiya, M.; Hu, D.; Fox, L.A.; Dang, N.; Johnston, E.D.; Oakes, J.A.; Tchao, B.N.; et al. Yeast Life Span Extension by Depletion of 60S Ribosomal Subunits Is Mediated by Gcn4. *Cell* **2008**, *133*, 292–302. [CrossRef]

129. Wei, M.; Fabrizio, P.; Madia, F.; Hu, J.; Ge, H.; Li, L.M.; Longo, V.D. Tor1/Sch9-Regulated Carbon Source Substitution Is as Effective as Calorie Restriction in Life Span Extension. *PLoS Genet.* **2009**, *5*, e1000467. [CrossRef]

130. Vellai, T.; Takacs-Vellai, K.; Zhang, Y.; Kovacs, A.L.; Orosz, L.; Müller, F. Influence of TOR kinase on lifespan in C. elegans. *Nature* **2003**, *426*, 620. [CrossRef]

131. Jia, K.; Chen, D.; Riddle, D.L. The TOR pathway interacts with the insulin signaling pathway to regulate *C. elegans* larval development, metabolism and life span. *Development* **2004**, *131*, 3897–3906. [CrossRef]

132. Pan, K.Z.; Palter, J.E.; Rogers, A.N.; Olsen, A.; Chen, D.; Lithgow, G.J.; Kapahi, P. Inhibition of mRNA translation extends lifespan in Caenorhabditis elegans. *Aging Cell* **2007**, *6*, 111–119. [CrossRef]

133. Hansen, M.; Taubert, S.; Crawford, D.; Libina, N.; Lee, S.-J.; Kenyon, C. Lifespan extension by conditions that inhibit translation in Caenorhabditis elegans. *Aging Cell* **2007**, *6*, 95–110. [CrossRef] [PubMed]

134. Syntichaki, P.; Troulinaki, K.; Tavernarakis, N. eIF4E function in somatic cells modulates ageing in Caenorhabditis elegans. *Nature* **2007**, *445*, 922–926. [CrossRef] [PubMed]

135. Curran, S.P.; Ruvkun, G. Lifespan Regulation by Evolutionarily Conserved Genes Essential for Viability. *PLoS Genet.* **2007**, *3*, e56. [CrossRef] [PubMed]

136. Greer, E.L.; Dowlatshahi, D.; Banko, M.R.; Villen, J.; Hoang, K.; Blanchard, D.; Gygi, S.P.; Brunet, A. An AMPK-FOXO Pathway Mediates Longevity Induced by a Novel Method of Dietary Restriction in C. elegans. *Curr. Biol.* **2007**, *17*, 1646–1656. [CrossRef] [PubMed]

137. Chen, D.; Pan, K.Z.; Palter, J.E.; Kapahi, P. Longevity determined by developmental arrest genes in Caenorhabditis elegans. *Aging Cell* **2007**, *6*, 525–533. [CrossRef]

138. Hansen, M.; Chandra, A.; Mitic, L.L.; Onken, B.; Driscoll, M.; Kenyon, C. A Role for Autophagy in the Extension of Lifespan by Dietary Restriction in C. elegans. *PLoS Genet.* **2008**, *4*, e24. [CrossRef]

139. Sheaffer, K.L.; Updike, D.L.; Mango, S.E. The Target of Rapamycin Pathway Antagonizes pha-4/FoxA to Control Development and Aging. *Curr. Biol.* **2008**, *18*, 1355–1364. [CrossRef]

140. Tóth, M.L.; Sigmond, T.; Borsos, É.; Barna, J.; Erdélyi, P.; Takács-Vellai, K.; Orosz, L.; Kovács, A.L.; Csikós, G.; Sass, M.; et al. Longevity pathways converge on autophagy genes to regulate life span in Caenorhabditis elegans. *Autophagy* **2008**, *4*, 330–338. [CrossRef]

141. Chen, D.; Thomas, E.L.; Kapahi, P. HIF-1 Modulates Dietary Restriction-Mediated Lifespan Extension via IRE-1 in Caenorhabditis elegans. *PLoS Genet.* **2009**, *5*, e1000486. [CrossRef]

142. Bell, R.; Hubbard, A.; Chettier, R.; Chen, D.; Miller, J.P.; Kapahi, P.; Tarnopolsky, M.; Sahasrabuhde, S.; Melov, S.; Hughes, R.E. A Human Protein Interaction Network Shows Conservation of Aging Processes between Human and Invertebrate Species. *PLoS Genet.* **2009**, *5*, e1000414. [CrossRef]

143. Kapahi, P.; Zid, B.M.; Harper, T.; Koslover, D.; Sapin, V.; Benzer, S. Regulation of Lifespan in Drosophila by Modulation of Genes in the TOR Signaling Pathway. *Curr. Biol.* **2004**, *14*, 885–890. [CrossRef] [PubMed]

144. Zid, B.M.; Rogers, A.N.; Katewa, S.D.; Vargas, M.A.; Kolipinski, M.C.; Lu, T.A.; Benzer, S.; Kapahi, P. 4E-BP Extends Lifespan upon Dietary Restriction by Enhancing Mitochondrial Activity in Drosophila. *Cell* **2009**, *139*, 149–160. [CrossRef] [PubMed]

145. Harrison, D.E.; Strong, R.; Sharp, Z.D.; Nelson, J.F.; Astle, C.M.; Flurkey, K.; Nadon, N.L.; Wilkinson, J.E.; Frenkel, K.; Carter, C.S.; et al. Rapamycin fed late in life extends lifespan in genetically heterogeneous mice. *Nature* **2009**, *460*, 392–395. [CrossRef] [PubMed]

146. Selman, C.; Tullet, J.M.A.; Wieser, D.; Irvine, E.; Lingard, S.J.; Choudhury, A.I.; Claret, M.; Al-Qassab, H.; Carmignac, D.; Ramadani, F.; et al. Ribosomal Protein S6 Kinase 1 Signaling Regulates Mammalian Life Span. *Science* **2009**, *326*, 140–144. [CrossRef] [PubMed]

147. Zhang, H.-M.; Diaz, V.; Walsh, M.E.; Zhang, Y. Moderate lifelong overexpression of tuberous sclerosis complex 1 (TSC1) improves health and survival in mice. *Sci. Rep.* **2017**, *7*, 834. [CrossRef]

148. Wu, J.J.; Liu, J.; Chen, E.B.; Wang, J.J.; Cao, L.; Narayan, N.; Fergusson, M.M.; Rovira, I.I.; Allen, M.; Springer, D.A.; et al. Increased Mammalian Lifespan and a Segmental and Tissue-Specific Slowing of Aging after Genetic Reduction of mTOR Expression. *Cell Rep.* **2013**, *4*, 913–920. [CrossRef]

149. Hardie, D.G. AMP-activated/SNF1 protein kinases: Conserved guardians of cellular energy. *Nat. Rev. Mol. Cell Biol.* **2007**, *8*, 774–785. [CrossRef]

150. Hardie, D.G.; Ross, F.A.; Hawley, S.A. AMPK: A nutrient and energy sensor that maintains energy homeostasis. *Nat. Rev. Mol. Cell Biol.* **2012**, *13*, 251–262. [CrossRef]

151. Apfeld, J.; O'Connor, G.; McDonagh, T.; DiStefano, P.S.; Curtis, R. The AMP-activated protein kinase AAK-2 links energy levels and insulin-like signals to lifespan in C. elegans. *Genes Dev.* **2004**, *18*, 3004–3009. [CrossRef]

152. Weir, H.J.; Yao, P.; Huynh, F.K.; Escoubas, C.C.; Goncalves, R.L.; Burkewitz, K.; Laboy, R.; Hirschey, M.D.; Mair, W.B. Dietary Restriction and AMPK Increase Lifespan via Mitochondrial Network and Peroxisome Remodeling. *Cell Metab.* **2017**, *26*, 884–896.e5. [CrossRef]

153. Mair, W.; Morantte, I.; Rodrigues, A.P.C.; Manning, G.; Montminy, M.; Shaw, R.J.; Dillin, A. Lifespan extension induced by AMPK and calcineurin is mediated by CRTC-1 and CREB. *Nature* **2011**, *470*, 404–408. [CrossRef] [PubMed]

154. Greer, E.L.; Brunet, A. Different dietary restriction regimens extend lifespan by both independent and overlapping genetic pathways in C. elegans. *Aging Cell* **2009**, *8*, 113–127. [CrossRef]

155. Fukuyama, M.; Sakuma, K.; Park, R.; Kasuga, H.; Nagaya, R.; Atsumi, Y.; Shimomura, Y.; Takahashi, S.; Kajiho, H.; Rougvie, A.; et al. elegans AMPKs promote survival and arrest germline development during nutrient stress. *Biol. Open* **2012**, *1*, 929–936. [CrossRef] [PubMed]

156. Schulz, T.J.; Zarse, K.; Voigt, A.; Urban, N.; Birringer, M.; Ristow, M. Glucose Restriction Extends Caenorhabditis elegans Life Span by Inducing Mitochondrial Respiration and Increasing Oxidative Stress. *Cell Metab.* **2007**, *6*, 280–293. [CrossRef] [PubMed]

157. Burkewitz, K.; Zhang, Y.; Mair, W.B. AMPK at the Nexus of Energetics and Aging. *Cell Metab.* **2014**, *20*, 10–25. [CrossRef] [PubMed]

158. Avruch, J.; Hara, K.; Lin, Y.; Liu, M.; Long, X.; Ortiz-Vega, S.; Yonezawa, K. Insulin and amino-acid regulation of mTOR signaling and kinase activity through the Rheb GTPase. *Oncogene* **2006**, *25*, 6361–6372. [CrossRef]

159. Singh, P.P.; Demmitt, B.A.; Nath, R.D.; Brunet, A. The Genetics of Aging: A Vertebrate Perspective. *Cell* **2019**, *177*, 200–220. [CrossRef]

160. Greer, E.L.; Brunet, A. FOXO transcription factors at the interface between longevity and tumor suppression. *Oncogene* **2005**, *24*, 7410–7425. [CrossRef]

161. Martins, R.; Lithgow, G.J.; Link, W. Long live FOXO: Unraveling the role of FOXO proteins in aging and longevity. *Aging Cell* **2016**, *15*, 196–207. [CrossRef]

162. Altintas, O.; Park, S.; Lee, S.-J.V. The role of insulin/IGF-1 signaling in the longevity of model invertebrates, C. elegans and D. melanogaster. *BMB Rep.* **2016**, *49*, 81–92. [CrossRef]

163. van Heemst, D. Insulin, IGF-1 and longevity. *Aging Dis.* **2010**, *1*, 147–157. [PubMed]

164. Hoshino, S.; Kobayashi, M.; Higami, Y. Mechanisms of the anti-aging and prolongevity effects of caloric restriction: Evidence from studies of genetically modified animals. *Aging* **2018**, *10*, 2243–2251. [CrossRef] [PubMed]

165. Jiang, Y.; Yan, F.; Feng, Z.; Lazarovici, P.; Zheng, W. Signaling Network of Forkhead Family of Transcription Factors (FOXO) in Dietary Restriction. *Cells* **2019**, *9*, 100. [CrossRef] [PubMed]

166. Shimokawa, I.; Higami, Y.; Tsuchiya, T.; Otani, H.; Komatsu, T.; Chiba, T.; Yamaza, H. Life span extension by reduction of the growth hormone-insulin-like growth factor-1 axis: Relation to caloric restriction. *FASEB J.* **2003**, *17*, 1108–1109. [CrossRef]

167. Shimokawa, I.; Komatsu, T.; Hayashi, N.; Kim, S.-E.; Kawata, T.; Park, S.; Hayashi, H.; Yamaza, H.; Chiba, T.; Mori, R. The life-extending effect of dietary restriction requires Foxo3 in mice. *Aging Cell* **2015**, *14*, 707–709. [CrossRef]

168. Clancy, D.J.; Gems, D.; Hafen, E.; Leevers, S.J.; Partridge, L. Dietary Restriction in Long-Lived Dwarf Flies. *Science* **2002**, *296*, 319. [CrossRef]

169. Min, K.-J.; Yamamoto, R.; Buch, S.; Pankratz, M.; Tatar, M. Drosophila lifespan control by dietary restriction independent of insulin-like signaling. *Aging Cell* **2008**, *7*, 199–206. [CrossRef]

170. Giannakou, M.E.; Goss, M.; Partridge, L. Role of dFOXO in lifespan extension by dietary restriction in Drosophila melanogaster: Not required, but its activity modulates the response. *Aging Cell* **2008**, *7*, 187–198. [CrossRef]

171. North, B.J.; Verdin, E. Sirtuins: Sir2-related NAD-dependent protein deacetylases. *Genome Biol.* **2004**, *5*, 224. [CrossRef]

172. López-Lluch, G.; Navas, P. Calorie restriction as an intervention in ageing. *J. Physiol.* **2016**, *594*, 2043–2060. [CrossRef]

173. Kaeberlein, M.; McVey, M.; Guarente, L. The SIR2/3/4 complex and SIR2 alone promote longevity in Saccharomyces cerevisiae by two different mechanisms. *Genes Dev.* **1999**, *13*, 2570–2580. [CrossRef] [PubMed]

174. Tissenbaum, H.A.; Guarente, L. Increased dosage of a sir-2 gene extends lifespan in Caenorhabditis elegans. *Nature* **2001**, *410*, 227–230. [CrossRef] [PubMed]

175. Viswanathan, M.; Kim, S.K.; Berdichevsky, A.; Guarente, L. A Role for SIR-2.1 Regulation of ER Stress Response Genes in Determining C. elegans Life Span. *Dev. Cell* **2005**, *9*, 605–615. [CrossRef] [PubMed]

176. Berdichevsky, A.; Viswanathan, M.; Horvitz, H.R.; Guarente, L. C. elegans SIR-2.1 Interacts with 14-3-3 Proteins to Activate DAF-16 and Extend Life Span. *Cell* **2006**, *125*, 1165–1177. [CrossRef] [PubMed]

177. Rizki, G.; Iwata, T.N.; Li, J.; Riedel, C.G.; Picard, C.L.; Jan, M.; Murphy, C.T.; Lee, S.S. The evolutionarily conserved longevity determinants HCF-1 and SIR-2.1/SIRT1 collaborate to regulate DAF-16/FOXO. *PLoS Genet.* **2011**, *7*, e1002235. [CrossRef]

178. Wood, J.G.; Rogina, B.; Lavu, S.; Howitz, K.; Helfand, S.L.; Tatar, M.; Sinclair, D. Sirtuin activators mimic caloric restriction and delay ageing in metazoans. *Nature* **2004**, *430*, 686–689. [CrossRef]

179. Viswanathan, M.; Guarente, L. Regulation of Caenorhabditis elegans lifespan by sir-2.1 transgenes. *Nature* **2011**, *477*, E1–E2. [CrossRef]

180. Ludewig, A.H.; Izrayelit, Y.; Park, D.; Malik, R.U.; Zimmermann, A.; Mahanti, P.; Fox, B.W.; Bethke, A.; Doering, F.; Riddle, D.L.; et al. Pheromone sensing regulates *Caenorhabditis elegans* lifespan and stress resistance via the deacetylase SIR-2.1. *Proc. Natl. Acad. Sci. USA* **2013**, *110*, 5522–5527. [CrossRef]

181. Mouchiroud, L.; Houtkooper, R.H.; Moullan, N.; Katsyuba, E.; Ryu, D.; Cantó, C.; Mottis, A.; Jo, Y.-S.; Viswanathan, M.; Schoonjans, K.; et al. The NAD+/Sirtuin Pathway Modulates Longevity through Activation of Mitochondrial UPR and FOXO Signaling. *Cell* **2013**, *154*, 430–441. [CrossRef]

182. Schmeisser, K.; Mansfeld, J.; Kuhlow, D.; Weimer, S.; Priebe, S.; Heiland, I.; Birringer, M.; Groth, M.; Segref, A.; Kanfi, Y.; et al. Role of sirtuins in lifespan regulation is linked to methylation of nicotinamide. *Nat. Chem. Biol.* **2013**, *9*, 693–700. [CrossRef]

183. Rogina, B.; Helfand, S.L. Sir2 mediates longevity in the fly through a pathway related to calorie restriction. *Proc. Natl. Acad. Sci. USA* **2004**, *101*, 15998–16003. [CrossRef] [PubMed]

184. Bauer, J.H.; Morris, S.N.S.; Chang, C.; Flatt, T.; Wood, J.G.; Helfand, S.L. dSir2 and Dmp53 interact to mediate aspects of CR-dependent lifespan extension in D. melanogaster. *Aging* **2009**, *1*, 38–48. [CrossRef] [PubMed]

185. Banerjee, K.K.; Ayyub, C.; Ali, S.Z.; Mandot, V.; Prasad, N.G.; Kolthur-Seetharam, U. dSir2 in the Adult Fat Body, but Not in Muscles, Regulates Life Span in a Diet-Dependent Manner. *Cell Rep.* **2012**, *2*, 1485–1491. [CrossRef] [PubMed]

186. Kanfi, Y.; Naiman, S.; Amir, G.; Peshti, V.; Zinman, G.; Nahum, L.; Bar-Joseph, Z.; Cohen, H.Y. The sirtuin SIRT6 regulates lifespan in male mice. *Nature* **2012**, *483*, 218–221. [CrossRef] [PubMed]

187. Satoh, A.; Brace, C.S.; Rensing, N.; Cliften, P.; Wozniak, D.F.; Herzog, E.D.; Yamada, K.A.; Imai, S.-I. Sirt1 Extends Life Span and Delays Aging in Mice through the Regulation of Nk2 Homeobox 1 in the DMH and LH. *Cell Metab.* **2013**, *18*, 416–430. [CrossRef]

188. Howitz, K.T.; Bitterman, K.J.; Cohen, H.Y.; Lamming, D.W.; Lavu, S.; Wood, J.G.; Zipkin, R.E.; Chung, P.; Kisielewski, A.; Zhang, L.-L.; et al. Small molecule activators of sirtuins extend Saccharomyces cerevisiae lifespan. *Nature* **2003**, *425*, 191–196. [CrossRef]

189. Kaeberlein, M.; Kirkland, K.T.; Fields, S.; Kennedy, B.K. Sir2-Independent Life Span Extension by Calorie Restriction in Yeast. *PLoS Biol.* **2004**, *2*, e296. [CrossRef]

190. Burnett, C.; Valentini, S.; Cabreiro, F.; Goss, M.; Somogyvári, M.; Piper, M.D.; Hoddinott, M.; Sutphin, G.L.; Leko, V.; McElwee, J.J.; et al. Absence of effects of Sir2 overexpression on lifespan in C. elegans and Drosophila. *Nature* **2011**, *477*, 482–485. [CrossRef]

191. Guarente, L. Calorie restriction and sirtuins revisited. *Genes Dev.* **2013**, *27*, 2072–2085. [CrossRef]

192. Lee, S.-H.; Lee, J.-H.; Lee, H.-Y.; Min, K.-J. Sirtuin signaling in cellular senescence and aging. *BMB Rep.* **2019**, *52*, 24–34. [CrossRef]

193. Whitaker, R.; Faulkner, S.; Miyokawa, R.; Burhenn, L.; Henriksen, M.; Wood, J.G.; Helfand, S.L. Increased expression of Drosophila Sir2 extends life span in a dose-dependent manner. *Aging* **2013**, *5*, 682–691. [CrossRef] [PubMed]

194. Frye, R.A. Phylogenetic Classification of Prokaryotic and Eukaryotic Sir2-like Proteins. *Biochem. Biophys. Res. Commun.* **2000**, *273*, 793–798. [CrossRef] [PubMed]

195. Mercken, E.M.; Hu, J.; Krzysik-Walker, S.; Wei, M.; Li, Y.; McBurney, M.W.; de Cabo, R.; Longo, V.D. SIRT1 but not its increased expression is essential for lifespan extension in caloric-restricted mice. *Aging Cell* **2014**, *13*, 193–196. [CrossRef] [PubMed]

196. Boily, G.; Seifert, E.L.; Bevilacqua, L.; He, X.H.; Sabourin, G.; Estey, C.; Moffat, C.; Crawford, S.; Saliba, S.; Jardine, K.; et al. SirT1 Regulates Energy Metabolism and Response to Caloric Restriction in Mice. *PLoS ONE* **2008**, *3*, e1759. [CrossRef]

197. Zhu, Y.; Yan, Y.; Gius, D.R.; Vassilopoulos, A. Metabolic regulation of Sirtuins upon fasting and the implication for cancer. *Curr. Opin. Oncol.* **2013**, *25*, 630–636. [CrossRef] [PubMed]

198. Wood, J.G.; Schwer, B.; Wickremesinghe, P.C.; Hartnett, D.A.; Burhenn, L.; Garcia, M.; Li, M.; Verdin, E.; Helfand, S.L. Sirt4 is a mitochondrial regulator of metabolism and lifespan in Drosophila melanogaster. *Proc. Natl. Acad. Sci. USA* **2018**, *115*, 1564–1569. [CrossRef]

199. Andreani, T.S.; Itoh, T.Q.; Yildirim, E.; Hwangbo, D.-S.; Allada, R. Genetics of Circadian Rhythms. *Sleep Med. Clin.* **2015**, *10*, 413–421. [CrossRef]

200. Bass, J. Circadian topology of metabolism. *Nature* **2012**, *491*, 348–356. [CrossRef]

201. Manoogian, E.N.C.; Panda, S. Circadian rhythms, time-restricted feeding, and healthy aging. *Ageing Res. Rev.* **2017**, *39*, 59–67. [CrossRef]

202. Chaudhari, A.; Gupta, R.; Makwana, K.; Kondratov, R. Circadian clocks, diets and aging. *Nutr. Healthy Aging* **2017**, *4*, 101–112. [CrossRef]

203. Katewa, S.D.; Akagi, K.; Bose, N.; Rakshit, K.; Camarella, T.; Zheng, X.; Hall, D.; Davis, S.; Nelson, C.S.; Brem, R.B.; et al. Peripheral Circadian Clocks Mediate Dietary Restriction-Dependent Changes in Lifespan and Fat Metabolism in Drosophila. *Cell Metab.* **2016**, *23*, 143–154. [CrossRef] [PubMed]

204. Patel, S.A.; Velingkaar, N.; Makwana, K.; Chaudhari, A.; Kondratov, R. Calorie restriction regulates circadian clock gene expression through BMAL1 dependent and independent mechanisms. *Sci. Rep.* **2016**, *6*, 25970. [CrossRef]

205. Sato, S.; Solanas, G.; Peixoto, F.O.; Bee, L.; Symeonidi, A.; Schmidt, M.S.; Brenner, C.; Masri, S.; Benitah, S.A.; Sassone-Corsi, P. Circadian Reprogramming in the Liver Identifies Metabolic Pathways of Aging. *Cell* **2017**, *170*, 664–677.e11. [CrossRef]

206. Patel, S.A.; Chaudhari, A.; Gupta, R.; Velingkaar, N.; Kondratov, R.V. Circadian clocks govern calorie restriction-mediated life span extension through BMAL1- and IGF-1-dependent mechanisms. *FASEB J.* **2016**, *30*, 1634–1642. [CrossRef] [PubMed]

207. Ulgherait, M.; Chen, A.; Oliva, M.K.; Kim, H.X.; Canman, J.C.; Ja, W.W.; Shirasu-Hiza, M. Dietary Restriction Extends the Lifespan of Circadian Mutants tim and per. *Cell Metab.* **2016**, *24*, 763–764. [CrossRef] [PubMed]

208. Solovev, I.; Shegoleva, E.; Fedintsev, A.; Shaposhnikov, M.; Moskalev, A. Circadian clock genes' overexpression in Drosophila alters diet impact on lifespan. *Biogerontology* **2019**, *20*, 159–170. [CrossRef]

209. Acosta-Rodriguez, V.A.; de Groot, M.H.M.; Rijo-Ferreira, F.; Green, C.B.; Takahashi, J.S. Mice under Caloric Restriction Self-Impose a Temporal Restriction of Food Intake as Revealed by an Automated Feeder System. *Cell Metab.* **2017**, *26*, 267–277.e262. [CrossRef]

210. Xu, K.; Zheng, X.; Sehgal, A. Regulation of feeding and metabolism by neuronal and peripheral clocks in Drosophila. *Cell Metab.* **2008**, *8*, 289–300. [CrossRef]

211. Ro, J.; Harvanek, Z.M.; Pletcher, S.D. FLIC: High-Throughput, Continuous Analysis of Feeding Behaviors in Drosophila. *PLoS ONE* **2014**, *9*, e101107. [CrossRef]

212. Gill, S.; Le, H.D.; Melkani, G.C.; Panda, S. Time-restricted feeding attenuates age-related cardiac decline in Drosophila. *Science* **2015**, *347*, 1265–1269. [CrossRef]

213. Eelderink-Chen, Z.; Mazzotta, G.; Sturre, M.; Bosman, J.; Roenneberg, T.; Merrow, M. A circadian clock in *Saccharomyces cerevisiae*. *Proc. Natl. Acad. Sci. USA* **2010**, *107*, 2043–2047. [CrossRef] [PubMed]

214. Tu, B.P.; Kudlicki, A.; Rowicka, M.; McKnight, S.L. Logic of the Yeast Metabolic Cycle: Temporal Compartmentalization of Cellular Processes. *Science* **2005**, *310*, 1152–1158. [CrossRef] [PubMed]

215. Goya, M.E.; Romanowski, A.; Caldart, C.S.; Bénard, C.Y.; Golombek, D.A. Circadian rhythms identified in *Caenorhabditis elegans* by in vivo long-term monitoring of a bioluminescent reporter. *Proc. Natl. Acad. Sci. USA* **2016**, *113*, E7837–E7845. [CrossRef] [PubMed]

216. Yeo, R.; Brunet, A. Deconstructing Dietary Restriction: A Case for Systems Approaches in Aging. *Cell Metab.* **2016**, *23*, 395–396. [CrossRef] [PubMed]

217. Hou, L.; Wang, D.; Chen, D.; Liu, Y.; Zhang, Y.; Cheng, H.; Xu, C.; Sun, N.; McDermott, J.; Mair, W.B.; et al. A Systems Approach to Reverse Engineer Lifespan Extension by Dietary Restriction. *Cell Metab.* **2016**, *23*, 529–540. [CrossRef] [PubMed]

218. Hulbert, A.J.; Pamplona, R.; Buffenstein, R.; Buttemer, W.A. Life and Death: Metabolic Rate, Membrane Composition, and Life Span of Animals. *Physiol. Rev.* **2007**, *87*, 1175–1213. [CrossRef]

219. Phelan, J.P.; Rose, M.R. Why dietary restriction substantially increases longevity in animal models but won't in humans. *Ageing Res. Rev.* **2005**, *4*, 339–350. [CrossRef]

220. Giuliani, C.; Garagnani, P.; Franceschi, C. Genetics of Human Longevity Within an Eco-Evolutionary Nature-Nurture Framework. *Circ. Res.* **2018**, *123*, 745–772. [CrossRef]

221. Passarino, G.; De Rango, F.; Montesanto, A. Human longevity: Genetics or Lifestyle? It takes two to tango. *Immun. Ageing* **2016**, *13*, 12. [CrossRef]

222. Lucanic, M.; Plummer, W.T.; Chen, E.; Harke, J.; Foulger, A.C.; Onken, B.; Coleman-Hulbert, A.L.; Dumas, K.J.; Guo, S.; Johnson, E.; et al. Impact of genetic background and experimental reproducibility on identifying chemical compounds with robust longevity effects. *Nat. Commun.* **2017**, *8*, 14256. [CrossRef]

223. Liao, C.-Y.; Rikke, B.A.; Johnson, T.E.; Diaz, V.; Nelson, J.F. Genetic variation in the murine lifespan response to dietary restriction: From life extension to life shortening. *Aging Cell* **2010**, *9*, 92–95. [CrossRef] [PubMed]

224. Liao, C.-Y.; Johnson, T.E.; Nelson, J.F. Genetic variation in responses to dietary restriction–an unbiased tool for hypothesis testing. *Exp. Gerontol.* **2013**, *48*, 1025–1029. [CrossRef] [PubMed]
225. Wilson, K.; Beck, J.; Nelson, C.; Hilsabeck, T.; Promislow, D.; Brem, R.; Kapahi, P. Genome-Wide Analyses for Lifespan and Healthspan in D. Melanogaster Reveal Decima as a Regulator of Insulin-Like Peptide Production. *SSRN Electron. J.* **2019**. [CrossRef]

© 2020 by the authors. Licensee MDPI, Basel, Switzerland. This article is an open access article distributed under the terms and conditions of the Creative Commons Attribution (CC BY) license (http://creativecommons.org/licenses/by/4.0/).

Article

Intermittent Fasting for Twelve Weeks Leads to Increases in Fat Mass and Hyperinsulinemia in Young Female Wistar Rats

Ana Cláudia Munhoz [1,*], Eloisa Aparecida Vilas-Boas [1], Ana Carolina Panveloski-Costa [1], Jaqueline Santos Moreira Leite [1], Camila Ferraz Lucena [1], Patrícia Riva [1], Henriette Emilio [2] and Angelo R. Carpinelli [1]

[1] Department of Physiology and Biophysics, Institute of Biomedical Sciences, University of Sao Paulo, 1524 Professor Lineu Prestes avenue, Butanta, São Paulo 05508-900, Brazil; elovilasboas@usp.br (E.A.V.-B.); anakpan@gmail.com (A.C.P.-C.); jaqueline.leite@usp.br (J.S.M.L.); cflucena07@gmail.com (C.F.L.); patricia.riva@gmail.com (P.R.); angelo@icb.usp.br (A.R.C.)

[2] Department of General Biology, Ponta Grossa State University, 4748 General Carlos Cavalcanti avenue, Uvaranas, Parana, PR 84030-900, Brazil; henry.emilio@gmail.com

* Correspondence: anamunoz@icb.usp.br; Tel.: +55-11-98-755-9099

Received: 29 February 2020; Accepted: 6 April 2020; Published: 9 April 2020

Abstract: Fasting is known to cause physiological changes in the endocrine pancreas, including decreased insulin secretion and increased reactive oxygen species (ROS) production. However, there is no consensus about the long-term effects of intermittent fasting (IF), which can involve up to 24 hours of fasting interspersed with normal feeding days. In the present study, we analyzed the effects of alternate-day IF for 12 weeks in a developing and healthy organism. Female 30-day-old Wistar rats were randomly divided into two groups: control, with free access to standard rodent chow; and IF, subjected to 24-hour fasts intercalated with 24-hours of free access to the same chow. Alternate-day IF decreased weight gain and food intake. Surprisingly, IF also elevated plasma insulin concentrations, both at baseline and after glucose administration collected during oGTT. After 12 weeks of dietary intervention, pancreatic islets displayed increased ROS production and apoptosis. Despite their lower body weight, IF animals had increased fat reserves and decreased muscle mass. Taken together, these findings suggest that alternate-day IF promote β -cell dysfunction, especially in developing animals. More long-term research is necessary to define the best IF protocol to reduce side effects.

Keywords: intermittent fasting; fat mass; insulin secretion; pancreatic islet

1. Introduction

The increasing prevalence of obesity around the globe is known to be linked to unhealthy eating patterns and a sedentary lifestyle. Treatments in obesity involve hypocaloric diets associated with physical exercises, producing an overall energy deficit [1,2] and leading to weight loss. One of the many diets that results in weight loss in both humans and laboratory animal models is intermittent fasting (IF). The most common IF protocols adopted by people in attempt to lose weight involve daily fasting for up to 16 hours; or fasting periods of up to 24 hours interspersed with normal feeding days [3,4]. Despite demonstrated weight loss [5–7], more studies are needed to evaluate whether alternate-day IF diets promote health benefits or could cause undesired effects in the long run.

Several studies about IF in both volunteers and animal models have uncovered beneficial effects, such as improved insulin sensitivity and glucose homeostasis [8–10]; improved performance and metabolic efficiency during exercise [11], increased alertness [12], and increased life expectancy [13–15];

a reduction in blood pressure and heart rate [16–20]; a reduction in inflammation; and protection against neurodegeneration [17–19,21–30].

On the other hand, many studies have reported adverse outcomes as a consequence of IF. Diabetic individuals performing IF may exhibit hypoglycaemia, ketoacidosis, dehydration, hypotension, and thrombosis [10,31]. A study in middle-aged men showed a significant increase in blood pressure and total cholesterol [32]. Increased plasma concentrations of cortisol at night have also been reported in several studies following daily fasting for one month, suggesting altered circadian rhythms [33–40]. IF may lead to the worsening of glucose tolerance in non-obese women [41] and cause decreased energy expenditure in young women [42]. Munsters et al. compared plasma insulin and glucose concentrations over only three days of reduced meal frequency and found that the intermittent fasting model produced higher peaks and more abrupt declines in insulin and glucose concentrations, indicating a biological environment prepared for long-term insulin resistance and diabetes [43].

Various physiological parameters are known to be altered in the endocrine pancreas during acute fasting, including insulin syntheses, glucose-stimulated insulin secretion (GSIS), glucose utilization, pancreatic islet metabolism, and pancreatic β-cell sensitivity to glucose [44–50]. Our group showed that a 48-hour acute fast increases net reactive oxygen species (ROS) production in isolated pancreatic islets and alters GSIS [51]. Thus, in the present study, we sought to evaluate the effects of 12 weeks of IF on glucose homeostasis and pancreatic islets isolated from rats.

2. Material and Methods

2.1. Ethical Approval

The Ethical Committee on Animal Research of the Institute of Biomedical Sciences of the University of São Paulo (CEUA) and Brazilian Society of Science in Laboratory Animals (SBCAL) approved the experimental protocols for this study, including the use of 24-hour fasted rats. The approved protocol number is 157/2014/CEUA.

2.2. Animals

Three-week-old female Wistar rats were transferred from the breeding facility to the experimental facility and remained 1 week for acclimatization before IF is initiated. The animals were housed in cages with three animals each in a room with constant temperature of 23 ± 2 °C on a standard 12-hour light/dark cycle. After acclimatization, the animals were randomized into two groups: control (CT) and intermittent fasting (IF) for 12 weeks. During the treatment, the CT group had free access to the standard rodent chow (Nuvilab, Sao Paulo, SP, Brazil – macro-nutrients in supplemental Figure S1) and the IF group was submitted to periods of 24 hours of total food deprivation interspersed with periods of 24 hours of ad libitum access to the same standard rodent chow. Food was withdrawn or made available to the IF group at noon (Figure 1). Both groups had free access to water for the entire period. During the full period, body weight and food intake were recorded. During the last week of treatment, we performed an oral glucose tolerance test (oGTT), an intraperitoneal insulin tolerance test (iITT), and dual x-ray absorptiometry. After 12 weeks of treatment, the animals were killed, and the pancreas, blood, liver, adipose tissue, and muscle were collected for different analyses. All tissue weights collected were normalized by the body weight of the animals. Euthanasia was always performed after the 24-hour period of free access to food, so that both groups were fed, thus eliminating the need to subject the control animals to acute fasting and excluding its effects. The exception occurred in liver collection, in order to analyze glycogen content after acute fasting, so, in this way, liver collection was also performed after a 24-hour fast of both groups in two different cohorts (i.e., both groups with free access to food or both groups in 24-hour fasting).

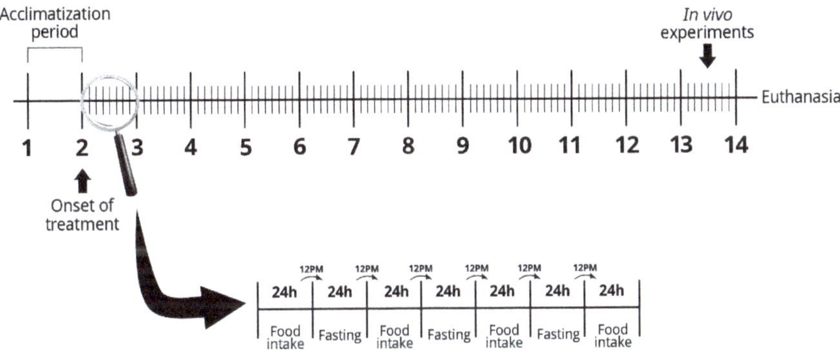

Figure 1. Scheme of the intermittent fasting protocol used.

2.3. Indexes Calculation

The Lee index was calculated using the body weight (grams) and naso-anal length (centimeters) of the animals. The Lee index and adipose tissue mass correlate and can be used as a simple measure of obesity in rats [52].

$$\text{Lee index} = \frac{\sqrt[3]{\text{body weight}}}{\text{naso\_anal length}} \tag{1}$$

The Homeostasis Model Assessment (HOMA) index is based on fasting plasma insulin (ng/ml) and glucose concentrations (mg/dL) [53]. Its purpose is to determine insulin resistance (HOMA-IR) and the functional capacity of pancreatic β-cells (HOMA-BETA).

$$\text{HOMA\_IR} = \frac{(\text{Fasting plasma insulin} \times \text{Fasting blood glucose})}{22.5} \tag{2}$$

$$\text{HOMA\_BETA} = \frac{(\text{Fasting plasma insulin} \times 20)}{(\text{Fasting blood glucose} - 3.5)} \tag{3}$$

2.4. Dual Energy x-Ray Absorptiometry (DEXA)

The animals were anesthetized with Sodium Tiopental (Cristália, Itapira, SP, Brazil) at a dose of 0.2 U/g body weight and placed in a dual energy in vivo radiological absorptiometry device (Biocompare In-Vivo Imaging System FX PRO, San Francisco, CA, USA) for the quantification of abdominal adipose tissue reserves. The images captured had the regions of interest (ROI) delimited and their pixels were quantified for analysis using ImageJ software.

2.5. Glycogen Measurement

The liver was collected after the euthanasia of the animals at two different cohorts: after the 24-hour fasting period or after the 24-hour ad libitum food ingestion period, i.e., either the CT and IF groups with free access to food or the CT and IF groups in 24-hour fasting. In total, 250 g of each sample was weighed and placed in a 15-mL conical tube. Then, 1 mL of 30% potassium hydroxide (Sigma-Aldrich, St Louis, MO, USA) was added to the tissues and boiled for 1 hour. After that, 100 µL of saturated sodium sulfate solution (Sigma-Aldrich, St Louis, MO, USA) and 3.5 mL of 70% ethanol (Synth, Diadema, SP, Brazil) were added to the mixture and the samples were boiled for 15 minutes and centrifuged at 3000 RPM for 7 minutes. The supernatant was discarded and the ethanol wash was repeated. The samples were resuspended in 1 mL of hot water and then 50 µL of the sample and 200 µL of 0.2% antrona concentrated sulfuric acid solution (Sigma-Aldrich, St Louis, MO, USA) were placed in a 96-well plate. Next, absorbance at a wavelength of 650 nm was recorded by a plate reader (Biotek

Sinergy H1 - Winooski, VT, USA). Glycogen concentrations were calculated using the absorbance values, and the liver mass (grams) of tissue was used to normalize glycogen concentrations.

2.6. Oral Glucose Tolerance Test

In the last week of treatment (12th week), both groups were submitted to a 12-hour overnight fast. Then, blood was collected by tail snip for glucose measurement (time 0) on a blood glucose monitor (FreeStyle Potium Neo, Witney, Oxon, Uk). Later, a glucose solution of 2g/kg per body weight was administered to the animals through gavage, and blood glucose concentrations were measured after 5, 10, 15, 30, 60, 90, and 120 min.

2.7. Intraperitoneal Insulin Tolerance Test

In the last week of treatment (12th week), in a cohort different from that in which we performed oGTT, animals were submitted to a 4-hour food restriction. Then, blood was collected by puncture at the animal caudal end for glucose measurement (time 0) on a blood glucose monitor (FreeStyle Potium Neo, Witney, Oxon, UK). After that, a solution of regular human insulin (Humulin, Indianapolis, IN, USA) at a dose of 0.75 U / kg per body weight was administered to the animals by intraperitoneal injection, and blood glucose concentrations were measured after 5, 10, 15, 30, 60, 90, and 120 min.

2.8. Blood Analysis

Blood was collected for insulin concentrations measurement following an Elisa kit protocol (Milipore, Billerica, MA, USA). Hemoglobin A1c (HbA1c) was also assessed by following the protocol of an automated chemistry analyzer by immunoturbidimetry (Labtest, Lagoa Santa, MG, Brazil).

2.9. Isolation of Pancreatic Islets

The isolation of pancreatic islets was carried out by the method of exocrine pancreas digestion using collagenase V [54], in which 20 mL of collagenase solution (0.68 mg/mL—Sigma-Aldrich, St. Louis, MO, USA) was injected through the bile duct. The dissected pancreas was placed in a bath at 37 °C for 25 min and shaken by hand for exocrine pancreas digestion. The sample was washed to remove exocrine tissue, and pancreatic islets were collected using a micropipette and a stereoscope.

2.10. Static Insulin Secretion

Pancreatic islets isolated from Wistar rats were placed in a microtube containing Krebs-Henseleit buffer with 0.1% albumin and 5.6 mM glucose for 30 min. Next, the islets were incubated at 37 °C in a microtube containing Krebs-Henseleit with 2.8, 5.6, 8.3, 11.1, or 16.7 mM glucose. Then, the supernatant and sonicated islets were stored at −20 °C for further measurements. The amounts of secreted and intracellular insulin content were determined by radioimmunoassay (RIA) [55].

2.11. Pancreas Histological Assessment

The pancreas was dissected, collected, and fixed with 40 mL of 10% formalin. Then, the pancreases were paraffin-embedded and sectioned at 4μm using a semi-automated microtome (RM2155 Leica Micro-systems, Wetzlar, Hessen, Germany). Afterward, the tissue sections were mounted on glass slides. The sections were stained with hematoxylin and eosin (H&E). All slides were examined using light microscopy with a camera attached (Nikon Eclipse TS100, Sao Paulo, SP, Brazil) under a magnification of X200. The slides were scanned and all pancreatic islets found were photographed. The images were calibrated and analyzed with the Aperio ImageScope software (Leica Micro-systems, Wetzlar, Hessen, Germany), then the length and width of each islet were measured. Next, the area of each pancreatic islets present on each slide was quantified by multiplying length and width.

2.12. Cell Viability

For the analysis of cell viability, groups of 20 pancreatic islets were dissociated, and the cells were incubated with ViaCount reagent (Millipore, Billerica, MA, USA) for 5 minutes at room temperature. The samples were placed into a 96-well plate for flow cytometer reading (Guava easyCyte ™ 8Ht Sampling - Millipore, Billerica, MA, USA), and cell viability, cell apoptosis, and cell death were quantified by counting 1000 events.

2.13. Measurement of Net ROS Production

Previously isolated islets were pre-incubated with Krebs-Henseleit buffer containing 5.6 mM glucose. Next, the samples were maintained for 1 hour in Krebs-Henseleit buffer containing 2.8 or 16.7 mM glucose. Then, the samples were incubated for 20 min with redox sensitive probe 50 μM dihydroethidium (DHE– Life Technologies, Eugene, Oregon, EUA) or 15 μM MitoSOX Red reagent (Life Technologies, Eugene, Oregon, USA—5 μM). After that, 300 μL of trypsin (Gibco, Grand Island, NY, USA) was added for 2 min to disperse the cells. Afterward, for trypsin inactivation, 600 μL of RPMI-1640 culture medium with 5% fetal bovine serum was added (Life Technologies, Itapevi, SP, Brazil). The islets were recollected and homogenized in 200 μL of RPMI-1640 culture medium for cell dispersion. The samples were placed in a 96-well plate and analyzed by flow cytometry (Guava EasyCyte 8HT- Millipore, Billerica, MA, USA).

2.14. Measurement of Net Hydrogen Peroxide Production

Groups of 120 pancreatic islets were placed in a microtube and pre-incubated for 30 min at 37 °C with in Krebs-Henseleit buffer with 5.6 mM glucose and 0.1% albumin. After this period, the islets were incubated for 1 hour at 37 °C with 2.8 or 16.7 mM glucose. Then, the samples were sonicated and incubated for 30 minutes with 50 μM Amplex Red probe (Life Technologies, Eugene, OR, USA), a specific marker for hydrogen peroxide (H_2O_2). The samples were placed into a 96-well plate for absorbance reading at 560 nm in a microplate reader (Biotek, Winooski, VT, USA).

2.15. Western Blot Analysis

To evaluate protein kinase B phosphorylation (p-AKT), samples of liver, retroperitoneal white adipose tissue (WAT), and extensor digitorum longus muscle were removed before and after an intravenous administration of 10 U regular insulin (Eli Lilly and Company, Indianapolis, IN, USA). To evaluate the expression of mitochondrial superoxide dismutase (SOD2) and glutathione peroxidase 1 (GPX1), groups of 300 pancreatic islets were collected after euthanasia of the animals. The tissues were collected in radioimmunoprecipitation assay buffer (Thermo Scientific, Saint Louis, MI, USA) containing protease and phosphatase inhibitors. The samples were sonicated, laemmli buffer was added, and the samples were boiled for 5 min. After that, polyacrylamide gel electrophoresis was performed followed by transfer to a nitrocellulose membrane. After transfer, the membranes were blocked with 5% bovine albumin solution for one hour at room temperature and incubated overnight at 4 °C with primary antibody (Millipore, Darmstadt, Hesse, Germany). They were incubated with the secondary antibody (Bio-Rad Laboratories Inc., Hercules, CA, USA) for 1 hour at room temperature. Finally, the membranes were developed using the enhanced chemiluminescent (ECL) reagent, and the images were captured by the Image Quant LAS4000 apparatus (GE Healthcare, Uppsala Sweden). The quantitative analysis of the bands was made by densitometry with the program of the Image Quant apparatus. Protein expression was normalized by the expression of the constitutive α-tubulin protein as the control protein or through the stained membranes with Ponceau-S to control protein loading.

2.16. Statistical Analysis

The results are expressed as the mean ± standard error of the mean (SEM). GraphPad Prism 5 software was used for analysis. Differences between multiple conditions were determined by One-way

ANOVA followed by Bartlett's test for equal variances or two-way ANOVA followed by Sidak's multiple comparisons test, as specified in figure legends. In experiments with only two conditions, the differences were determined by Student's t-test. Comparisons were considered significantly different for $p < 0.05$.

3. Results

3.1. Body Weight Gain and Development

Thirty-day-old Wistar rats were randomly divided into two groups: control (CT) and submitted to intermittent fasting (IF) for 12 weeks. Lower weight gain was recorded in the IF group already after the second week of dietary intervention (Figure 2A—week 3). These changes were maintained throughout the whole treatment period; the area under the curve of the treated animals was 20.3% lower than the control animals (Figure 2A). At the end of the treatment, the tibia length and naso-anal length were significantly decreased in the IF group (Figure 2B,C), and this led to an increased Lee index (Figure 2D).

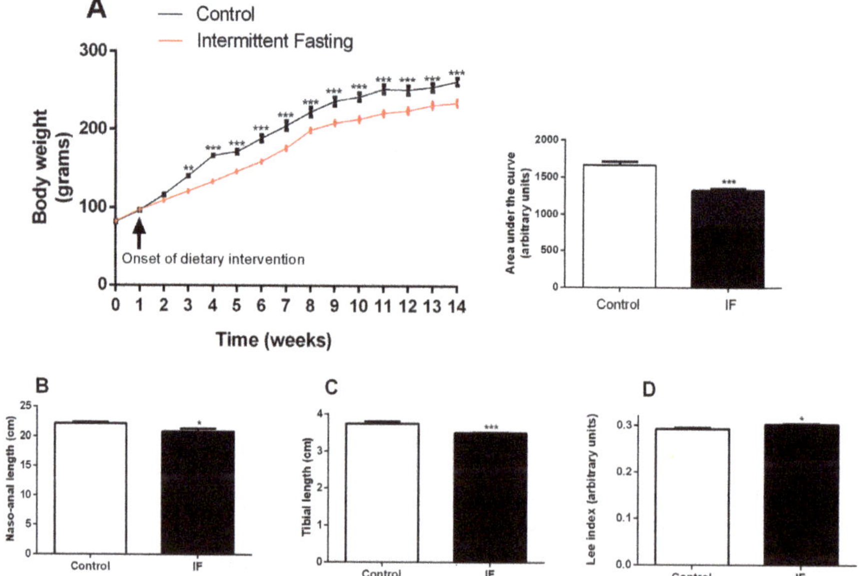

Figure 2. (A) Weekly body weight, (**B**) naso-anal length, (**C**) tibia length, and (**D**) the Lee index of Wistar rats submitted to intermittent fasting (IF) for 12 weeks. The results are presented as the means ± standard error of the mean (SEM) with 10 different animals for each group. * $p <0.05$, ** $p <0.005$, and *** $p <0.0005$ compared to the control of the same period, as indicated by two-way ANOVA followed by Sidak's multiple comparisons test (**A**) or Student's t-test (**B–D**).

3.2. Food Intake and Stomach Disturbances

The IF group consumes 35% less chow compared to the control group if the average total intake is considered, i.e., fasting days (zero consumption) plus feeding days (gorging behavior). However, if we consider only the mean ingestion of ad libitum ingestion days, the consumption in relation to the control is 31% higher, indicating chow overconsumption (Figure 3A). Figure 3B,C show that this hyperphagia caused a large increase in stomach length (by 47.95%) and weight (by 171.66%). Even after emptying stomach contents, we observed increased stomach weight by 12.55% (Figure 3D).

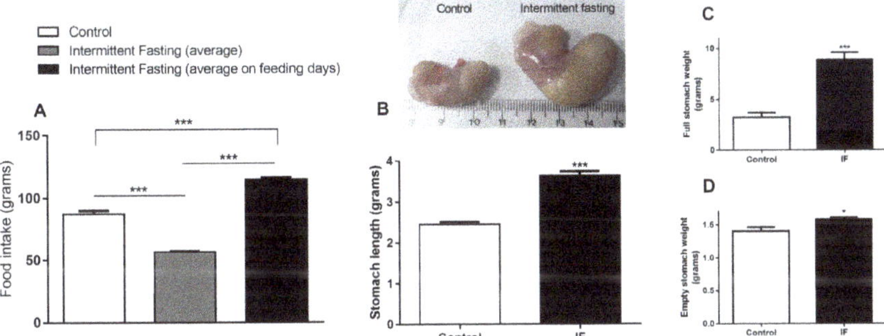

Figure 3. (**A**) Food intake, (**B**) stomach length, and (**C**) full and (**D**) empty stomach weight of Wistar rats submitted to IF for 12 weeks. The results are presented as the means ± standard error of the mean (SEM) with 10 different animals for each group. * p <0.05 and *** p <0.0005 compared to the control of the same period, as indicated by one-way ANOVA followed by Bartlett's test for equal variances (A) or Student's t-test.

3.3. Body Composition

In vivo dual energy x-ray absorptiometry showed increased abdominal adiposity, as can be seen in Figure 4A. In addition, the weights of adipose tissues (Figure 4B–D) and dry muscles (Figure 4E–G) reveals changes in body composition with fat mass gain and muscle loss in the IF group.

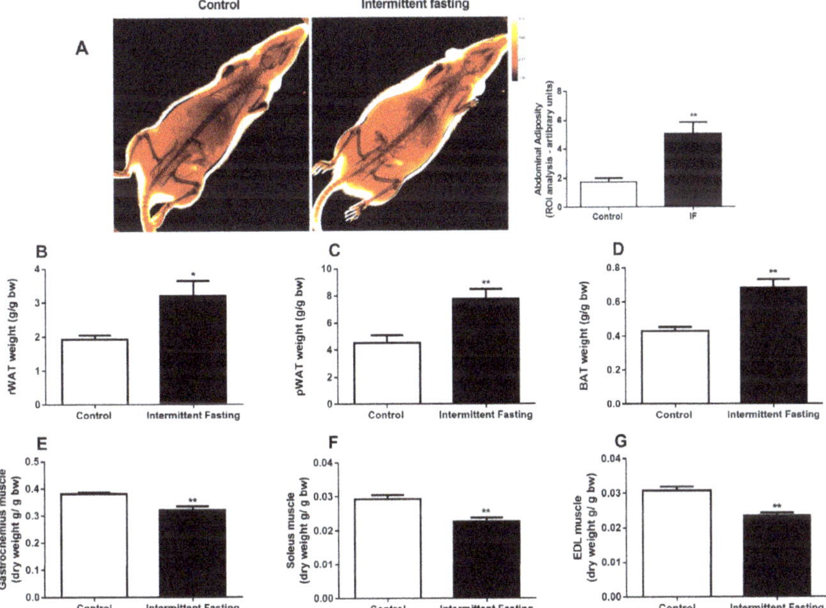

Figure 4. (**A**) Dual energy x-ray absorptiometry (DEXA), (**B**) retroperitoneal, (**C**) perigonadal, and (**D**) brown adipose tissue weight. (**E**) Dry gastrocnemius weight, (**F**) Soleus, and (**G**) Extensor digitorum longus (EDL) muscle of Wistar rats submitted to IF for 12 weeks. The results are presented as the means ± standard error of the mean (SEM) with 10 different animals for each group. * p <0.05 and ** p <0.005 compared to the control of the same period, as indicated by Student's t-test.

3.4. Liver Alterations

IF reduced liver weight in the fed state by 13.8% (Figure 5A) and after fasting by 35.68% (Figure 5B) when compared to the control in a similar state, whereas the reduction in liver weight may be correlated with reduced glycogen stores. We analyzed glycogen content in both states. In the fed state, a 47.68% reduction in glycogen (Figure 5C) was observed, and fasting led to a 98.33% liver glycogen decrease in the IF group (Figure 5D).

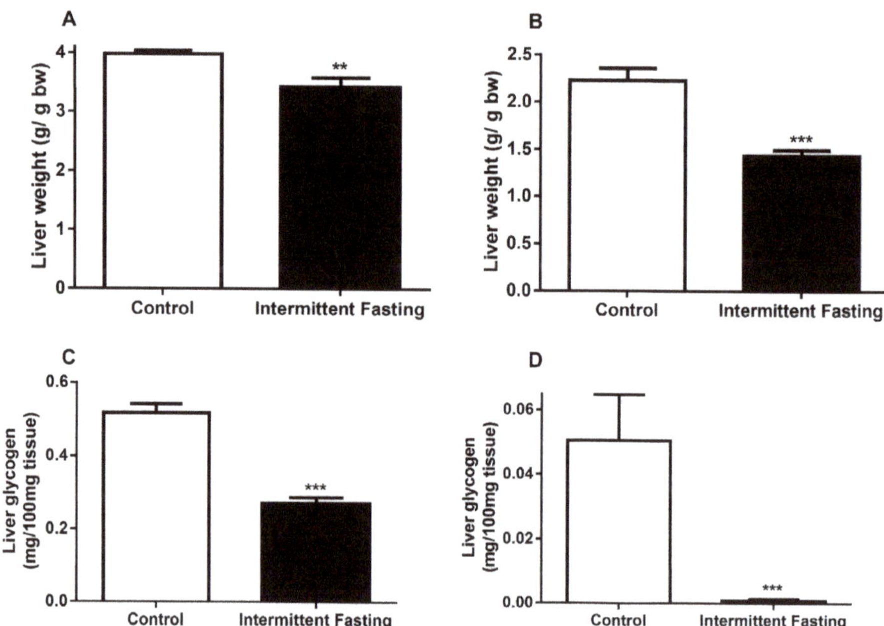

Figure 5. (**A**) Liver weight before and (**B**) after 24h of fasting, (**C**) liver glycogen content before and **D**) after 24h of fasting of Wistar rats submitted to IF for 12 weeks. The results are presented as the means ± standard error of the mean (SEM) with 10 different animals for each group. ** p <0.005 and *** p <0.0005 compared to the control of the same period, as indicated by Student's t-test.

3.5. Glucose Homeostasis

An oral glucose tolerance test (oGTT) performed at the end of treatment showed no differences between the control and IF group (Figure 6A). The area under the curve was significantly lower in this group with a reduction of 6.8% (Figure 6B). However, hemoglobin A1c levels after IF treatment do not differ significantly compared to the control group (Figure 6C).

At the end of the treatment, an intraperitoneal insulin tolerance test (iITT) was also performed. Although the area under the curve was 27% lower in the IF group (Figure 7B), the groups did not show significant differences in blood glucose values at any of the times studied (Figure 7A) and the glucose decay constant (Kitt) did not differ between groups (Figure 7C).

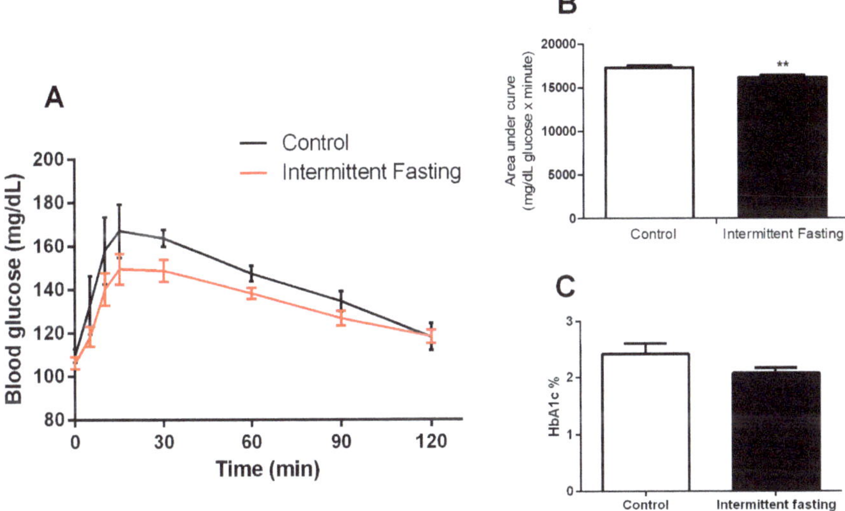

Figure 6. (**A**) Oral glucose tolerance test (oGTT), (**B**) area under curve, and (**C**) hemoglobin A1c (HbA1c) of Wistar rats submitted to IF for 12 weeks. The results are presented as the means ± standard error of the mean (SEM) with 5 different animals for each group. ** p <0.005 compared to the control, as indicated by Student's t-test (B and C), two-way ANOVA followed by Sidak's multiple comparisons test (A).

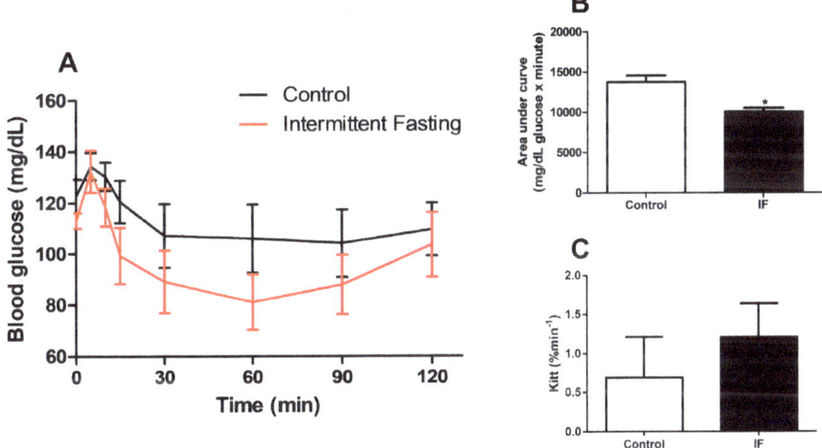

Figure 7. (**A**) Intraperitoneal insulin tolerance test (iITT), (**B**) area under curve and (**C**) glucose decay constant rate during insulin tolerance test (kITT) of Wistar rats submitted to IF for 12 weeks. The results are presented as means ± standard error of the mean (SEM) with 5 different animals for each group. * p <0.05 compared to the control, as indicated by Student's t-test (B and C), two-way ANOVA followed by Sidak's multiple comparisons test (A).

3.6. Insulin Concentrations

Plasma insulin concentrations were measured in the fasted state and after 15, 30, 90, and 120 min of oral glucose administration. Intermittent fasting for 12 weeks greatly increased basal plasma insulin concentrations (by 4-fold). Insulin increase was also observed in the IF group by 155%, 271%, 259.7%, and 304% after 15, 30, 90, and 120 min of oral glucose stimulation (Figure 8A), respectively.

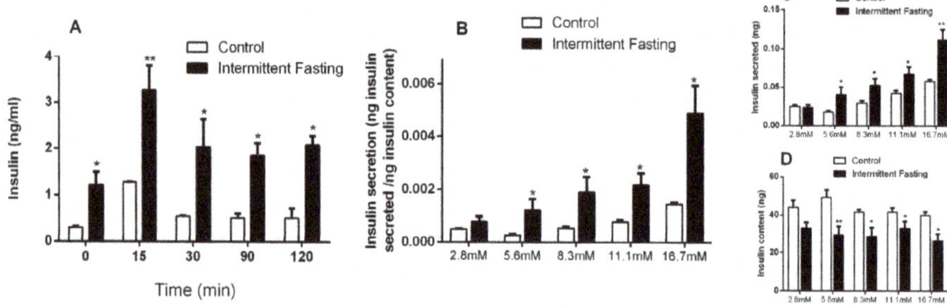

Figure 8. (**A**) Insulin concentrations obtained during oral glucose tolerance tests (oGTT) in Wistar rats submitted to IF for 12 weeks. (**B**) Content-corrected insulin secretion, (**C**) insulin secretion, and (**D**) insulin content values in pancreatic islets isolated from Wistar rats submitted to IF for 12 weeks, after one hour of incubation in the presence of 2.8, 5.6, 8.3, 11.1, and 16.7 mM glucose. The results are presented as the means ± standard error of the mean (SEM) with 5 different animals for each group. * $p <0.05$ and ** $p <0.005$ compared to the control at the same time, as indicated by two-way ANOVA followed by Sidak's multiple comparisons test.

A glucose-stimulated insulin secretion (GSIS) assay was performed with isolated pancreatic islets after glucose stimulation at 2.8, 5.6, 8.3, 11.1, and 16.7 mM levels. Content-corrected GSIS and secreted insulin (Figure 8B,C) were significantly higher after intermittent fasting at all glucose concentrations, except in presence of 2.8 mM glucose. At the same time, the insulin content remaining in the pancreatic islets of the IF group was lower at all glucose concentrations, with the exception of low-level glucose (Figure 8D), which corroborates the higher values of insulin found.

3.7. Homeostasis Model Assessment (HOMA) Indexes

From the values of fasting blood glucose and fasting plasma insulin, both obtained during oGTT performed in the last week of dietary intervention (time 0 – before glucose administration), we calculated HOMA-IR and HOMA-β indexes, which are mathematical models used to evaluate insulin resistance. Intermittent fasting greatly increased values in both models by 3.6-fold (HOMA-IR) and by 4.7-fold (HOMA-β), as can be observed in Figure 9A,B.

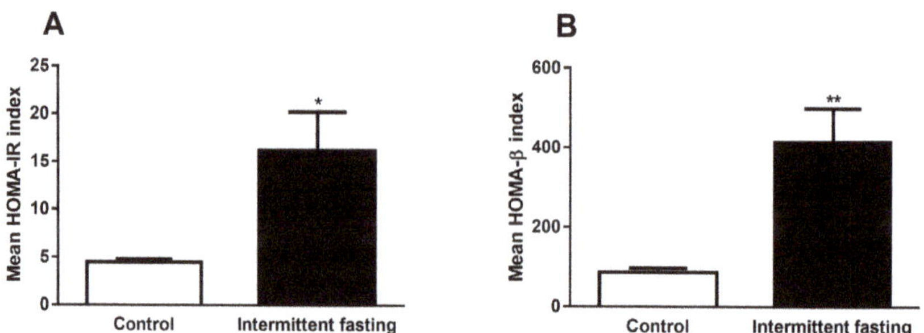

Figure 9. (**A**) Homeostasis model assessment of insulin resistance (HOMA-IR) and (**B**) homeostasis model assessment of β-cell function (HOMA-β) indexes of Wistar rats submitted to IF for 12 weeks. The results are presented as the mean ± standard error of the means (SEM) with 5 different animals for each group. * $p <0.05$ and ** $p <0.005$ compared to the control, as indicated by Student's *t*-test.

3.8. AKT Phosphorylation

Figure 10 shows that 12 weeks of intermittent fasting were able to alter AKT phosphorylation in muscle, liver, and white adipose tissue (WAT). In the control group, an increase in AKT phosphorylation after intravenous insulin stimulation was observed, while the IF group did not show significant differences in AKT phosphorylation, indicating impairment in insulin action (Figure 10A–C).

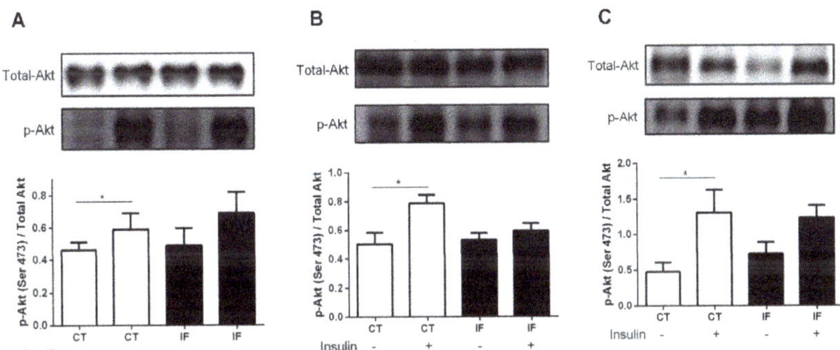

Figure 10. (**A**) Protein kinase B phosphorylation (p-AKT) expression of extensor digitorum longus (EDL) muscle, (**B**) liver, and (**C**) retroperitoneal white adipose tissue from Wistar rats submitted to IF for 12 weeks. The results are presented as the means ± standard error of the mean (SEM) with 5 different preparations for each group. * p <0.05 compared to the respective control, as indicated by Student's t-test.

3.9. Pancreatic Islet Area and Viability of Cells from Pancreatic Islets

Histological analyzes in pancreas revealed a 28.6% decrease in the islet size of the animals submitted to IF treatment (Figure 11A,C). The weight of the pancreas also showed a significant reduction in the IF group of about 26% (Figure 11B).

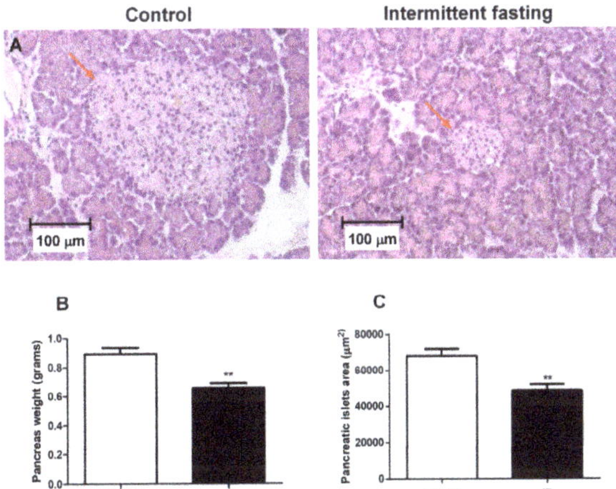

Figure 11. (**A**) Pancreatic islet of the longitudinal cut of pancreas caudal portion, (**B**) pancreas weight related to body weight, and (**C**) pancreatic islets area from Wistar rats submitted to IF for 12 weeks (Hematoxylin-eosin stained tissue (H&E) and 100x magnification). The results were presented as the mean ± standard error of the mean (SEM) with 5 different animals for each group. ** p <0.005 compared to the control, as indicated by Student's *t*-test.

The viability of the dispersed cells from pancreatic islets was assessed and no significant difference in the number of viable cells (Figure 12A) and number of dead cells was observed (Figure 12C). However, we observed a significant increase of 27.8% in the number of apoptotic cells (Figure 11B).

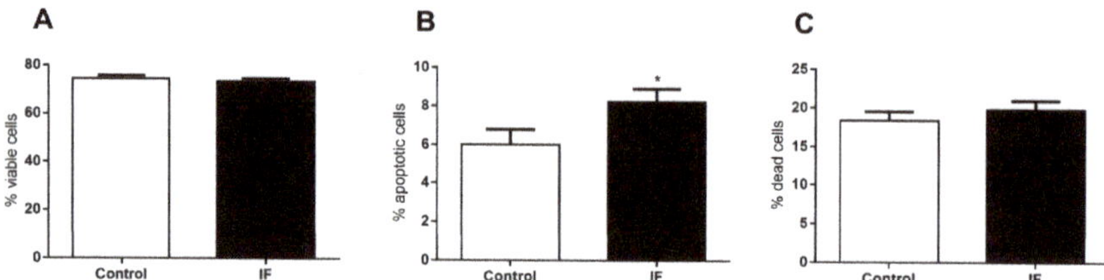

Figure 12. (**A**) Viability, (**B**) the percentage of apoptotic cells, and (**C**) the percentage of dead cells of dispersed cells from pancreatic islet isolated from Wistar rats submitted to IF for 12 weeks. The results are presented as the mean ± standard error of the mean (SEM) with 5 different cellular preparations for each group. * p <0.05 compared to the control, as indicated by Student's *t*-test.

3.10. ROS Production of Dispersed Cells from Pancreatic Islets

We observed an 80.2% increase in the net fluorescence induced by reactive oxygen species (ROS) in the presence of glucose 2.8 mM after incubation with the redox-sensitive probe DHE (Dihydroethidium). No significant difference was found between the groups at 16.7 mM glucose level (Figure 13A).

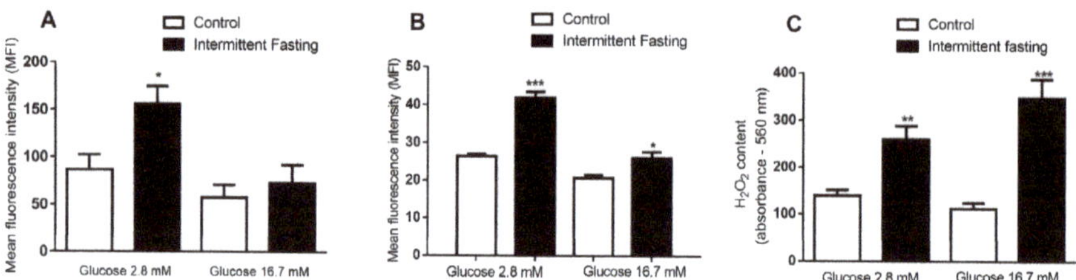

Figure 13. (**A**) Mean fluorescence intensity (MFI) emitted by a dihydroethidium (DHE) probe, (**B**) MFI emitted by a MitoSox Red probe, and (**C**) the absorbance emitted by the Amplex Red probe of dispersed cells from a pancreatic islet isolated from Wistar rats submitted to IF for 12 weeks after one hour of incubation in the presence of 2.8 and 16.7 mM glucose. The results are presented as the means ± standard error of the mean (EPM) with 6 different cellular preparations for each group. * p <0.05, ** p <0.005, and *** p <0.0005, compared to the respective control, as indicated by Student's t-test.

Net mitochondrial ROS production was also measured in the presence of 2.8 and 16.7 mM glucose after incubation with the MitoSox Red probe (Figure 13B). After 12 weeks of intermittent fasting, a significant increase in fluorescence associated to ROS production was revealed (59.2% in low glucose and 26% in high glucose concentration).

Finally, we measured the concentrations of hydrogen peroxide (H_2O_2) in the presence of 2.8 and 16.7 mM of glucose after incubation with the Amplex Red probe (Figure 13C). IF caused a significant increase in H_2O_2 detection in the presence of both glucose concentrations compared to the values of the control group. At low glucose levels, an increment of 85% was observed, while, at at 16.7 mM glucose, an increase of 211% was measured.

3.11. Antioxidant Systems

Lastly, we analyzed two antioxidant enzyme expressions: mitochondrial superoxide dismutase (SOD2), also known as manganese-dependent superoxide dismutase (MnSOD), and glutathione peroxidase 1 (GPX1). The enzymes were measured by electrophoresis gel and calculated in relation to α-tubulin protein expression. Intermittent fasting resulted in a significant increase (by 70.86%) in SOD2 expression (Figure 14A). However, after intermittent fasting, GPX1 expression was not different from the values found in the control group.

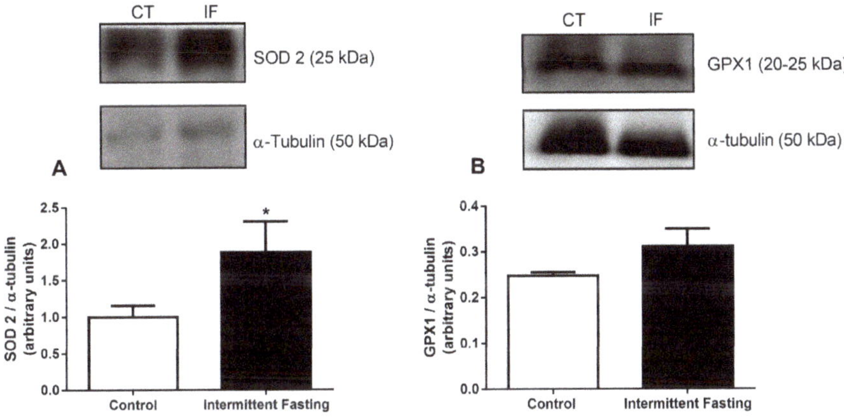

Figure 14. (**A**) Superoxide dismutase 2 (SOD2) and (**B**) glutathione peroxidase 1 (GPX1) expression of pancreatic islets isolated from Wistar rats submitted to IF for 12 weeks. The results are presented as the means ± standard error of the mean (SEM) with 5 different cellular preparations for each group. * $p < 0.05$ compared to the control, as indicated by Student's *t*-test.

4. Discussion

Alternate-day intermittent fasting (IF) is a relatively new dietary approach that has been promoted to help weight management. As a result, the number of people adhering to this new diet grows every day. We are aware that children are not usually placed on IF diets. However, our society is constantly changing, and with increasing levels of obesity and overweight in children, it is possible that this may be a dietary approach recommended by nutrition professionals in the future. Hence, we need to study possible consequences of this practice in young organisms, as the adverse effects of alternate-day IF have not been fully elucidated.

Compared with daily calorie restriction, intermittent fasting in obese volunteers did not produce better adherence, weight loss, weight maintenance, or improvement in risk indicators for cardiovascular disease. Rather, the dropout rate and hunger in the IF fasting group was higher than that in the daily calorie restriction group, and the authors concluded that IF may be less sustainable in the long-term for obese individuals [56,57].

In this study, long-term IF was successful in reducing body weight gain (Figure 2A) in female Wistar rats, and this may be due to lower average food intake (by 35%) (Figure 3A). Weight loss as a consequence of IF has been reported in previous studies in both animal [58,59] and human subjects [60–62]. However, Sakamoto and Grunewald showed that IF in 4-week-old rats markedly reduced the growth rate and the animals had smaller livers, kidneys, hearts, tibias and tibialis anterior muscles [63]. In this study, the treatment was also initiated in young animals (4-weeks old) and, as can be seen in Figure 2B (reduced naso-anal length) and 2C (reduced tibial length), may affect growth since the animal is developing [64].

As previously reported, IF in slim and healthy people for 2 weeks [65] or 1 month [66] showed a significant decrease in resting energy expenditure. Hence, the animals could gain weight in relation to control animals if the food intake in both groups were the same, assuming that IF promotes lower energy expenditure. Additionally, a hyperphagic behavior on free feeding days (Figure 3A) was observed, which caused a large increase in stomach dimensions (3B-D). There are records of IF-induced increased expression of the agouti-related peptide (AGRP), neuropeptide Y (NPY) and orexin due to intermittent fasting in rodents, even when the stomach was full [58,67,68]. These orexigenic neurotransmitters are involved in appetite modulation and metabolic regulation [69], and their increasing concentrations in plasma could explain the increase in food intake. There are also several human studies showing increased subjective appetite sensation as a result of fasting cycles [70–72].

It has been previously reported that IF promotes a greater deposition of triacylglycerides in the white adipose tissue by increasing the expression of genes involved in lipid storage, such as fatty-specific protein 27 (FSP27) [73]. In 1928, Lee described a rapid method to quantify obesity from body weight and naso-anal length values. The result defines the nutritional status, called the Lee index, which correlates positively with adipose tissue mass. [74,75]. In this study, the values obtained from this index reveal a slight, but significant, increase in the IF group values compared to the control group (Figure 2D), which does not mean that the IF animals are obese, but corroborates the largest adipose reserve found (Figure 4A–D).

No significant differences were found between groups neither in blood glucose concentrations after oral glucose administration (Figure 6A) nor in HbA1c (Figure 6C). These results contradict previously published studies, which show a reduction in blood glucose concentrations. [25,76,77]. Nevertheless, in this case, the lower glycaemia can be a consequence of large amounts of circulating insulin concentrations (Figure 8) since insulin is the hypoglycemic hormone that allows glucose to enter in insulin-sensitive target cells. The decrease in liver weight observed may be related to lower glycogen reserves (Figure 5). The significant drop in hepatic glycogen and muscle glycogen concentrations (Supplemental Figure S2) may result in increased fatigue and impaired maintenance of normoglycemia between feeding periods [78].

Our data show changes in body composition with decreased muscle mass (Figure 4E–G) and increased fat mass (Figure 4B–D), in contrast to the literature data, which generally show reduced body fat and maintenance of lean mass. [28,62,79–81]. However, a decrease in lean mass after only 1 month of IF has been reported in adolescent girls [42]. An increase in body mass index was found with IF only in women, with a reduction in this parameter in men [82]. Another study found that improvements in body composition are less pronounced in younger women [62]. In female mice, IF did not cause weight loss [83]. It is possible that intermittent fasting causes more adverse effects on the body composition of females, and this could be accentuated in younger and developing organisms.

Glucose and insulin plasma concentrations under fasting conditions in both healthy and type 2 diabetes mellitus (T2DM) subjects are expressed at characteristic concentrations for each individual's nutritional status. Basal insulin concentrations are a consequence of fasting glucose concentrations, the secretory capacity of pancreatic β-cells, and the rate of pulsatile insulin secretion. Previous work shows that IF increases both basal and glucose-stimulated insulin secretion in mice with diet-induced obesity [83]. This increase in insulin concentrations (Figure 8A), also found in in vitro pancreatic islets (Figure 8B–D), suggests resistance to this hormone, as evidenced by a significant increase in the homeostasis model assessment (HOMA) index (Figure 9), and impaired AKT phosphorylation in muscle, liver, and adipose tissue (Figure 10).

HOMA is a mathematical calculation based on fasting insulinemia and glycaemia, proposed by David Matheus [53] as a simple and fast way to determine insulin resistance (HOMA-IR) and the functional capacity of pancreatic β-cells (HOMA-BETA). Similar data were found in a study performed with male rats submitted to 5 weeks of IF and likewise in volunteers during 1 month of fasting, both showing increased plasma insulin and HOMA indexes [38,84].

One hypothesis for this insulin resistance induced by fasting is increased secretion of ghrelin. The secretion of this gastric hormone is elevated in calorie restricted mice, rats, and humans [85]. Ghrelin is produced and secreted predominantly in the oxyntic mucosa of stomach (approximately 60–70% of circulating ghrelin), and low plasma ghrelin concentrations are associated with elevated fasting insulin concentrations and insulin resistance, suggesting both physiological and pathophysiological roles in glucose metabolism [86,87]. Barazzoni et al. observed that, in rats, sustained ghrelin administration reduced hepatic AKT phosphorylation. The deregulation of AKT phosphorylation has been suggested to occur under insulin resistance conditions and diabetes [88].

Although insulin tolerance tests showed no differences in glycaemia values at any time studied (Figure 7A), nor in glucose decay constant values (Figure 7C), impaired AKT phosphorylation in muscle, liver, and WAT indicates impaired insulin action (Figure 10). Intermittent fasting for 8 months in rats caused impaired glucose clearance, as reported by Cerqueira et al. [89]. In addition, intermittent food restriction in female rats for 6 weeks induced glucose intolerance and detrimental hypothalamic alterations coupled with compulsive eating behavior [90]. Thus, frequent feeding and fasting cycles may be harmful and associated with insulin resistance, increasing risks of T2DM. A hyperinsulinemic-euglycemic clamp was not conducted in this work and should be included in future studies.

There are several studies that suggest that high insulin concentrations are associated with typical pathologies of the metabolic syndrome, such as insulin resistance [91–97], obesity [98–103], hypertension, cardiovascular diseases [104–108], atherosclerosis [109], and hepatic steatosis [110,111]. In general, the development of T2DM begins when the metabolic demand for insulin is greater, due to peripheral insulin resistance. This insulin resistance generally precedes the development of hyperglycaemia. In other words, there is a period of normoglycaemia, where pancreatic β-cells compensate insulin resistance by increasing insulin secretion [112–115]. However, this compensatory hyperinsulinemia can cause damage to the secretory cell in the long run, leading to β-cell apoptosis and the development of T2DM [116–118].

Corroborating with this idea, the decreased pancreas weight and pancreatic islet area (Figure 11), as well as increased apoptotic cells in pancreatic islets (Figure 12), reinforce the hypothesis that IF may be detrimental to the endocrine pancreas and represents an unhealthy long-term diet. Although we must be cautious when comparing the effects to humans, this study in an animal model is an important tool to evaluate the potential impacts of this diet in a standardized manner, with minimal artifact interference. Moreover, animal studies allow us to evaluate the effects directly in important organs, such as the pancreas, given that data in humans is obviously sparse or non-existent as to whether long-term intermittent fasting diets affect pancreatic islets.

Lastly, 8 weeks of IF caused no difference in net ROS production between groups at any of the glucose concentrations studied (Supplemental Figure S3). However, 12 weeks of IF increased total (Figure 13A) and mitochondrial (Figure 13B) net ROS production in the pancreas, besides increasing hydrogen peroxide levels (Figure 13C), together showing that the effects are accentuated by a longer time of treatment and lead to oxidative imbalance, as previously shown by Cerqueira et al. [89] in other tissues. Intermittent fasting in mice showed that, on ad libitum feeding days, when the animal ingests a large amount of food, ROS production is enhanced due to increased mitochondrial respiration [21].

These ROS increments caused the enhancement antioxidant defenses, as evidenced by increased SOD2 (Figure 14A), that catalyzes the dismutation of superoxide into O_2 and H_2O_2, which are less damaging molecules. This cellular response promoted by 12 weeks of IF diet may be important to mitigate oxidative stress, since increased ROS production has been reported to cause a loss of function and even apoptosis in several cell types [119–126], including pancreatic islets [127–133].

The insulin receptor is essential for pancreatic β-cell function and survival. The insulin signaling networks contain many proteins that are established regulators of apoptosis and proliferation. Insulin directly prevents apoptosis in human and mouse islets, as well as in β-cell lines [134,135]. Cerqueira et al. showed that long-term intermittent feeding leads to redox imbalance and peripheral insulin receptor nitration [89]. Thus, increased oxidative stress induced by 12 weeks IF may be accompanied by the oxidation of the insulin receptor in pancreatic β-cells and the impairment of cell survival. However, a more detailed study of oxidative stress induced by fasting on the protective effects of autocrine insulin signaling is needed.

Several studies show that the effects of IF depend on several factors, such as the period of day in which fasting is practiced. Time restricted feeding (TRF) is a type of IF that involves eating all nutrients within a few hours every day, usually up to 12 hours. Studies suggest that, depending on the time of the eating window, TRF leads to opposite effects. Restricting food intake to the resting phase worsened fasting and postprandial glucose concentrations, blood pressure, and lipid concentrations in humans [32,136], and induced leptin resistance that contributes to the development of obesity and metabolic disorders in mice [137]. However, restricting food intake to the active phase improved insulin sensitivity, blood pressure, glucose tolerance, and oxidative stress [8,138,139].

Corroborating these results, there are several long-term studies in humans showing detrimental effects of skipping breakfast, which is a type of IF by prolonging overnight fasting to the active phase. Breakfast skipping is associated with a significantly increased risk of overweight and obesity, poorer glycemic control, insulin resistance, and an increased risk of T2DM [140–145]. Considering that the present study evaluated juvenile rats and that they present a different endocrinology response to adults, our results only apply to juveniles. In developing organisms, growth hormone may play an important role in inducing insulin resistance under fasting stress conditions, which may be significant for the defense against hypoglycemia [146]. Besides, further long-term research is necessary to investigate which IF protocol is more suitable to reduce these side effects and improve health, before IF can be considered a good alternative for weight management in young individuals.

5. Conclusions

Intermittent fasting is effective for weight loss, but the long-term safety has been questioned. Considering increasing levels of obesity and overweight in children, we studied possible consequences of this practice in young animals. We have shown here that 12 weeks of alternate-day intermittent fasting in young female Wistar rats causes several changes that can be detrimental in the long run to young individuals, including the elevation of pancreatic islet cells apoptosis and ROS production. We also found a reduction in pancreatic islet mass, a large increase in insulin secretion, and signs of insulin resistance by reduced AKT phosphorylation in muscle and adipose tissue, which may be a risk factor for T2DM. Besides that, we noticed a remodeling of body composition, with increased body fat and decreased muscle mass. Taken together, these findings suggest that caution may be warranted when unrestrictedly recommending intermittent fasting to young individuals, especially for people with compromised glucose metabolism. This study was conducted in healthy juvenile rats and these findings may not translate into adult humans. Further studies are required.

Supplementary Materials: The following are available online at http://www.mdpi.com/2072-6643/12/4/1029/s1. Figure S1 Macro-nutrients of the standard rodent chow (Nuvilab, Sao Paulo, SP, Brazil), Figure S2 Gastrocnemius glycogen content, Figure S3 Mean fluorescence intensity (MFI) emitted by dihydroethidium (DHE) probe of dispersed cells from pancreatic islet isolated from Wistar rats submitted to IF for 8 weeks after one hour of incubation in the presence of 2.8 and 16.7 mM glucose.

Author Contributions: Conceptualization, A.C.M., P.R.; methodology, A.C.M.; formal analysis, A.C.M.; investigation, A.C.M., E.A.V.-B., A.C.P.-C., J.S.M.L., C.F.L., P.R.; data curation, A.C.M., E.A.V.-B., A.R.C., A.C.P.-C., C.F.L.; writing—original draft preparation, A.C.M.; writing—review and editing, A.C.M. and H.E.; visualization, H.E.; supervision, A.R.C.; project administration, A.C.M.; funding acquisition, A.C.M. and A.R.C. All authors have read and agreed to the published version of the manuscript.

Funding: This research was funded by National Council for Scientific and Technological Development (CNPq), grant number 166017/2014-0.

Acknowledgments: The authors would like to thank Marlene Santos da Rocha for the excellent technical assistance, Alicia Juliana Kowaltowski for revising the English writing, and Altay Alves Lino de Souza for revising statistical analyses.

Conflicts of Interest: The authors declare that there is no conflict of interest.

References

1. Tuomilehto, J.; Lindström, J.; Eriksson, J.G.; Valle, T.T.; Hämäläinen, H.; Ilanne-Parikka, P.; Keinänen-Kiukaanniemi, S.; Laakso, M.; Louheranta, A.; Rastas, M.; et al. Prevention of type 2 diabetes mellitus by changes in lifestyle among subjects with impaired glucose tolerance. *N. Engl. J. Med.* **2001**, *344*, 1343–1350. [CrossRef]
2. Thent, Z.C.; Das, S.; Henry, L.J. Role of exercise in the management of diabetes mellitus: the global scenario. *PLoS ONE* **2013**, *8*, e80436. [CrossRef]
3. Pilon, B. *Eat Stop Eat*; Medwin House Inc.: Ontario, Canada, 2007.
4. Mosley, M.; Spencer, M. *A Dieta dos 2 Dias—The Fast Diet*; Sextante: Belo Horizonte, Brazil, 2013.
5. Harris, L.; Hamilton, S.; Azevedo, L.B.; Olajide, J.; De Brún, C.; Waller, G.; Whittaker, V.; Sharp, T.; Lean, M.; Hankey, C.; et al. Intermittent fasting interventions for treatment of overweight and obesity in adults: A systematic review and meta-analysis. *JBI Database Syst. Rev. Implement. Rep.* **2018**, *16*, 507–547. [CrossRef] [PubMed]
6. Harvie, M.; Wright, C.; Pegington, M.; McMullan, D.; Mitchell, E.; Martin, B.; Cutler, R.G.; Evans, G.; Whiteside, S.; Maudsley, S.; et al. The effect of intermittent energy and carbohydrate restriction v. daily energy restriction on weight loss and metabolic disease risk markers in overweight women. *Br. J. Nutr.* **2013**, *110*, 1534–1547. [CrossRef] [PubMed]
7. Gabel, K.; Hoddy, K.K.; Haggerty, N.; Song, J.; Kroeger, C.M.; Trepanowski, J.F.; Panda, S.; Varady, K.A. Effects of 8-hour time restricted feeding on body weight and metabolic disease risk factors in obese adults: A pilot study. *Nutr. Healthy Aging* **2018**, *4*, 345–353. [CrossRef] [PubMed]
8. Sutton, E.F.; Beyl, R.; Early, K.S.; Cefalu, W.T.; Ravussin, E.; Peterson, C.M. Early Time-Restricted Feeding Improves Insulin Sensitivity, Blood Pressure, and Oxidative Stress Even without Weight Loss in Men with Prediabetes. *Cell Metab.* **2018**, *27*, 1212–1221.E3. [CrossRef] [PubMed]
9. Harvie, M.N.; Pegington, M.; Mattson, M.P.; Frystyk, J.; Dillon, B.; Evans, G.; Cuzick, J.; Jebb, S.A.; Martin, B.; Cutler, R.G.; et al. The effects of intermittent or continuous energy restriction on weight loss and metabolic disease risk markers: A randomized trial in young overweight women. *Int. J. Obes.* **2011**, *35*, 714–727. [CrossRef] [PubMed]
10. Corley, B.T.; Carroll, R.W.; Hall, R.M.; Weatherall, M.; Parry-Strong, A.; Krebs, J.D. Intermittent fasting in Type 2 diabetes mellitus and the risk of hypoglycaemia: A randomized controlled trial. *Diabet. Med.* **2018**. [CrossRef]
11. Marosi, K.; Moehl, K.; Navas-Enamorado, I.; Mitchell, S.J.; Zhang, Y.; Lehrmann, E.; Aon, M.A.; Cortassa, S.; Becker, K.G.; Mattson, M.P. Metabolic and molecular framework for the enhancement of endurance by intermittent food deprivation. *FASEB J.* **2018**. [CrossRef]
12. Almeneessier, A.S.; Alzoghaibi, M.; BaHammam, A.A.; Ibrahim, M.G.; Olaish, A.H.; Nashwan, S.Z.; BaHammam, A.S. The effects of diurnal intermittent fasting on the wake-promoting neurotransmitter orexin-A. *Ann. Thorac. Med.* **2018**, *13*, 48–54. [CrossRef]
13. Goodrick, C.L.; Ingram, D.K.; Reynolds, M.A.; Freeman, J.R.; Cider, N. Effects of intermittent feeding upon body weight and lifespan in inbred mice: Interaction of genotype and age. *Mech. Ageing Dev.* **1990**, *55*, 69–87. [CrossRef]
14. Heilbronn, L.K.; Smith, S.R.; Martin, C.K.; Anton, S.D.; Ravussin, E. Alternate-day fasting in nonobese subjects: Effects on body weight, body composition, and energy metabolism. *Am. J. Clin. Nutr.* **2005**, *81*, 69–73. [CrossRef] [PubMed]
15. Johnson, J.B.; Laub, D.R.; John, S. The effect on health of alternate day calorie restriction: eating less and more than needed on alternate days prolongs life. *Med. Hypotheses* **2006**, *67*, 209–211. [CrossRef] [PubMed]
16. Caviezel, F.; Margonato, A.; Slaviero, G.; Bonetti, F.; Vicedomini, G.; Cattaneo, A.G.; Pozza, G. Early improvement of left ventricular function during caloric restriction in obesity. *Int. J. Obes.* **1986**, *10*, 421–426. [PubMed]
17. Bruce-Keller, A.J.; Umberger, G.; McFall, R.; Mattson, M.P. Food restriction reduces brain damage and improves behavioral outcome following excitotoxic and metabolic insults. *Ann. Neurol.* **1999**, *45*, 8–15. [CrossRef]
18. Duan, W.; Guo, Z.; Mattson, M.P. Brain-derived neurotrophic factor mediates an excitoprotective effect of dietary restriction in mice. *J. Neurochem.* **2001**, *76*, 619–626. [CrossRef]

19. Kim, J.; Kang, S.W.; Mallilankaraman, K.; Baik, S.H.; Lim, J.C.; Balaganapathy, P.; She, D.T.; Lok, K.Z.; Fann, D.Y.; Thambiayah, U.; et al. Transcriptome Analysis Reveals Intermittent Fasting-Induced Genetic Changes in Ischemic Stroke. *Hum. Mol. Genet.* **2018**. [CrossRef]

20. Wan, R.; Camandola, S.; Mattson, M.P. Intermittent fasting and dietary supplementation with 2-deoxy-D-glucose improve functional and metabolic cardiovascular risk factors in rats. *FASEB J.* **2003**, *17*, 1133–1134. [CrossRef]

21. Anson, R.M.; Guo, Z.; de Cabo, R.; Iyun, T.; Rios, M.; Hagepanos, A.; Ingram, D.K.; Lane, M.A.; Mattson, M.P. Intermittent fasting dissociates beneficial effects of dietary restriction on glucose metabolism and neuronal resistance to injury from calorie intake. *Proc. Natl. Acad. Sci. USA* **2003**, *100*, 6216–6220. [CrossRef]

22. Bouguerra, R.; Jabrane, J.; Maâtki, C.; Ben Salem, L.; Hamzaoui, J.; El Kadhi, A.; Ben Rayana, C.; Ben Slama, C. Ramadan fasting in type 2 diabetes mellitus. *Ann. Endocrinol. (Paris)* **2006**, *67*, 54–59. [CrossRef]

23. Khaled, B.M.; Bendahmane, M.; Belbraouet, S. Ramadan fasting induces modifications of certain serum components in obese women with type 2 diabetes. *Saudi Med. J.* **2006**, *27*, 23–26. [PubMed]

24. Shariatpanahi, Z.V.; Shariatpanahi, M.V.; Shahbazi, S.; Hossaini, A.; Abadi, A. Effect of Ramadan fasting on some indices of insulin resistance and components of the metabolic syndrome in healthy male adults. *Br. J. Nutr.* **2008**, *100*, 147–151. [CrossRef] [PubMed]

25. Belkacemi, L.; Selselet-Attou, G.; Bulur, N.; Louchami, K.; Sener, A.; Malaisse, W.J. Intermittent fasting modulation of the diabetic syndrome in sand rats. III. Post-mortem investigations. *Int. J. Mol. Med.* **2011**, *27*, 95–102. [CrossRef] [PubMed]

26. Teng, N.I.; Shahar, S.; Rajab, N.F.; Manaf, Z.A.; Johari, M.H.; Ngah, W.Z. Improvement of metabolic parameters in healthy older adult men following a fasting calorie restriction intervention. *Aging Male* **2013**, *16*, 177–183. [CrossRef] [PubMed]

27. Gnanou, J.V.; Caszo, B.A.; Khalil, K.M.; Abdullah, S.L.; Knight, V.F.; Bidin, M.Z. Effects of Ramadan fasting on glucose homeostasis and adiponectin levels in healthy adult males. *J. Diabetes Metab. Disord.* **2015**, *14*, 55. [CrossRef] [PubMed]

28. Tinsley, G.M.; La Bounty, P.M. Effects of intermittent fasting on body composition and clinical health markers in humans. *Nutr. Rev.* **2015**, *73*, 661–674. [CrossRef] [PubMed]

29. Castrogiovanni, P.; Li Volti, G.; Sanfilippo, C.; Tibullo, D.; Galvano, F.; Vecchio, M.; Avola, R.; Barbagallo, I.; Malaguarnera, L.; Castorina, S.; et al. Fasting and Fast Food Diet Play an Opposite Role in Mice Brain Aging. *Mol. Neurobiol.* **2018**. [CrossRef]

30. Hu, Y.; Zhang, M.; Chen, Y.; Yang, Y.; Zhang, J.J. Postoperative intermittent fasting prevents hippocampal oxidative stress and memory deficits in a rat model of chronic cerebral hypoperfusion. *Eur. J. Nutr.* **2018**. [CrossRef]

31. Chentli, F.; Azzoug, S.; Amani, M.l.A.; Elgradechi, A. Diabetes mellitus and Ramadan in Algeria. *Indian J. Endocrinol. Metab.* **2013**, *17*, S295–S298. [CrossRef]

32. Stote, K.S.; Baer, D.J.; Spears, K.; Paul, D.R.; Harris, G.K.; Rumpler, W.V.; Strycula, P.; Najjar, S.S.; Ferrucci, L.; Ingram, D.K.; et al. A controlled trial of reduced meal frequency without caloric restriction in healthy, normal-weight, middle-aged adults. *Am. J. Clin. Nutr.* **2007**, *85*, 981–988. [CrossRef]

33. al-Hadramy, M.S.; Zawawi, T.H.; Abdelwahab, S.M. Altered cortisol levels in relation to Ramadan. *Eur. J. Clin. Nutr.* **1988**, *42*, 359–362. [PubMed]

34. Bogdan, A.; Bouchareb, B.; Touitou, Y. Ramadan fasting alters endocrine and neuroendocrine circadian patterns. Meal-time as a synchronizer in humans? *Life Sci.* **2001**, *68*, 1607–1615. [CrossRef]

35. Ben Salem, L.; Bchir, S.; Bouguerra, R.; Ben Slama, C. Cortisol rhythm during the month of Ramadan. *East Mediterr. Health J.* **2003**, *9*, 1093–1098. [PubMed]

36. Roky, R.; Houti, I.; Moussamih, S.; Qotbi, S.; Aadil, N. Physiological and chronobiological changes during Ramadan intermittent fasting. *Ann. Nutr. Metab.* **2004**, *48*, 296–303. [CrossRef]

37. Haouari, M.; Haouari-Oukerro, F.; Sfaxi, A.; Ben Rayana, M.C.; Kâabachi, N.; Mbazâa, A. How Ramadan fasting affects caloric consumption, body weight, and circadian evolution of cortisol serum levels in young, healthy male volunteers. *Horm. Metab. Res.* **2008**, *40*, 575–577. [CrossRef]

38. Bahijri, S.; Borai, A.; Ajabnoor, G.; Abdul Khaliq, A.; AlQassas, I.; Al-Shehri, D.; Chrousos, G. Relative metabolic stability, but disrupted circadian cortisol secretion during the fasting month of Ramadan. *PLoS ONE* **2013**, *8*, e60917. [CrossRef]

39. Guerrero-Morilla, R.; Ramírez-Rodrigo, J.; Ruiz-Villaverde, G.; Sánchez-Caravaca, M.A.; Pérez-Moreno, B.A.; Villaverde-Gutiérrez, C. Endocrine-metabolic adjustments during Ramadan fasting in young athletes. *Arch. Latinoam. Nutr.* **2013**, *63*, 14–20.

40. Ajabnoor, G.M.; Bahijri, S.; Borai, A.; Abdulkhaliq, A.A.; Al-Aama, J.Y.; Chrousos, G.P. Health impact of fasting in Saudi Arabia during Ramadan: association with disturbed circadian rhythm and metabolic and sleeping patterns. *PLoS ONE* **2014**, *9*, e96500. [CrossRef]

41. Heilbronn, L.K.; Civitarese, A.E.; Bogacka, I.; Smith, S.R.; Hulver, M.; Ravussin, E. Glucose tolerance and skeletal muscle gene expression in response to alternate day fasting. *Obes. Res.* **2005**, *13*, 574–581. [CrossRef]

42. Reiches, M.W.; Moore, S.E.; Prentice, A.M.; Ellison, P.T. Endocrine responses, weight change, and energy sparing mechanisms during Ramadan among Gambian adolescent women. *Am. J. Hum. Biol.* **2014**, *26*, 395–400. [CrossRef]

43. Munsters, M.J.; Saris, W.H. Effects of meal frequency on metabolic profiles and substrate partitioning in lean healthy males. *PLoS ONE* **2012**, *7*, e38632. [CrossRef] [PubMed]

44. Bosboom, R.S.; Zweens, J.; Bouman, P.R. Effects of feeding and fasting on the insulin secretory response to glucose and sulfonylureas in intact rats and isolated perfused rat pancreas. *Diabetologia* **1973**, *9*, 243–250. [CrossRef] [PubMed]

45. Svenningsen, A.; Bonnevie-Nielsen, V. Effects of fasting on beta-cell function, body fat, islet volume, and total pancreatic insulin content. *Metabolism* **1984**, *33*, 612–616. [CrossRef]

46. Gasa, R.; Sener, A.; Malaisse, W.J.; Gomis, R. Apparent starvation-induced repression of pancreatic islet glucokinase. *Biochem. Mol. Med.* **1995**, *56*, 99–103. [CrossRef]

47. Bone, A.J.; Howell, S.L. Alterations in regulation of insulin biosynthesis in pregnancy and starvation studied in isolated rat islets of langerhans. *Biochem. J.* **1977**, *166*, 501–507. [CrossRef]

48. Hedeskov, C.J.; Capito, K. The effect of starvation on insulin secretion and glucose metabolism in mouse pancreatic islets. *Biochem. J.* **1974**, *140*, 423–433. [CrossRef]

49. Tadayyon, M.; Bonney, R.C.; Green, I.C. Starvation decreases insulin secretion, prostaglandin E2 production and phospholipase A2 activity in rat pancreatic islets. *J. Endocrinol.* **1990**, *124*, 455–461. [CrossRef]

50. Hedeskov, C.J.; Capito, K. The pentose cycle and insulin release in isolated mouse pancreatic islets during starvation. *Biochem. J.* **1975**, *152*, 571–576. [CrossRef]

51. Munhoz, A.C.; Riva, P.; Simões, D.; Curi, R.; Carpinelli, A.R. Control of Insulin Secretion by Production of Reactive Oxygen Species: Study Performed in Pancreatic Islets from Fed and 48-Hour Fasted Wistar Rats. *PLoS ONE* **2016**, *11*, e0158166. [CrossRef]

52. Bernardis, L.L.; Patterson, B.D. Correlation between "Lee index" and carcass fat content in weanling and adult female rats with hypothalamic lesions. *J. Endocrinol.* **1968**, *40*, 527–528. [CrossRef]

53. Matthews, D.R.; Hosker, J.P.; Rudenski, A.S.; Naylor, B.A.; Treacher, D.F.; Turner, R.C. Homeostasis model assessment: Insulin resistance and beta-cell function from fasting plasma glucose and insulin concentrations in man. *Diabetologia* **1985**, *28*, 412–419. [CrossRef] [PubMed]

54. Lacy, P.E.; Kostianovsky, M. Method for the isolation of intact islets of Langerhans from the rat pancreas. *Diabetes* **1967**, *16*, 35–39. [CrossRef] [PubMed]

55. Scott, A.M.; Atwater, I.; Rojas, E. A method for the simultaneous measurement of insulin release and B cell membrane potential in single mouse islets of Langerhans. *Diabetologia* **1981**, *21*, 470–475. [CrossRef] [PubMed]

56. Trepanowski, J.F.; Kroeger, C.M.; Barnosky, A.; Klempel, M.C.; Bhutani, S.; Hoddy, K.K.; Gabel, K.; Freels, S.; Rigdon, J.; Rood, J.; et al. Effect of Alternate-Day Fasting on Weight Loss, Weight Maintenance, and Cardioprotection Among Metabolically Healthy Obese Adults: A Randomized Clinical Trial. *JAMA Intern. Med.* **2017**, *177*, 930–938. [CrossRef]

57. Sundfør, T.M.; Svendsen, M.; Tonstad, S. Effect of intermittent versus continuous energy restriction on weight loss, maintenance and cardiometabolic risk: A randomized 1-year trial. *Nutr. Metab. Cardiovasc. Dis.* **2018**, *28*, 698–706. [CrossRef]

58. Chausse, B.; Solon, C.; Caldeira da Silva, C.C.; Masselli Dos Reis, I.G.; Manchado-Gobatto, F.B.; Gobatto, C.A.; Velloso, L.A.; Kowaltowski, A.J. Intermittent fasting induces hypothalamic modifications resulting in low feeding efficiency, low body mass and overeating. *Endocrinology* **2014**, *155*, 2456–2466. [CrossRef]

59. Chausse, B.; Vieira-Lara, M.A.; Sanchez, A.B.; Medeiros, M.H.; Kowaltowski, A.J. Intermittent fasting results in tissue-specific changes in bioenergetics and redox state. *PLoS ONE* **2015**, *10*, e0120413. [CrossRef]

60. Varady, K.A.; Bhutani, S.; Klempel, M.C.; Kroeger, C.M.; Trepanowski, J.F.; Haus, J.M.; Hoddy, K.K.; Calvo, Y. Alternate day fasting for weight loss in normal weight and overweight subjects: A randomized controlled trial. *Nutr. J.* **2013**, *12*, 146. [CrossRef]

61. Coutinho, S.R.; Halset, E.H.; Gåsbakk, S.; Rehfeld, J.F.; Kulseng, B.; Truby, H.; Martins, C. Compensatory mechanisms activated with intermittent energy restriction: A randomized control trial. *Clin. Nutr.* **2017**. [CrossRef]

62. López-Bueno, M.; González-Jiménez, E.; Navarro-Prado, S.; Montero-Alonso, M.A.; Schmidt-RioValle, J. Influence of age and religious fasting on the body composition of Muslim women living in a westernized context. *Nutr. Hosp.* **2014**, *31*, 1067–1073. [CrossRef]

63. Sakamoto, K.; Grunewald, K.K. Beneficial effects of exercise on growth of rats during intermittent fasting. *J. Nutr.* **1987**, *117*, 390–395. [CrossRef] [PubMed]

64. Sengupta, P. The Laboratory Rat: Relating Its Age with Human's. *Int. J. Prev. Med.* **2013**, *4*, 624–630. [PubMed]

65. Soeters, M.R.; Lammers, N.M.; Dubbelhuis, P.F.; Ackermans, M.; Jonkers-Schuitema, C.F.; Fliers, E.; Sauerwein, H.P.; Aerts, J.M.; Serlie, M.J. Intermittent fasting does not affect whole-body glucose, lipid, or protein metabolism. *Am. J. Clin. Nutr.* **2009**, *90*, 1244–1251. [CrossRef] [PubMed]

66. el Ati, J.; Beji, C.; Danguir, J. Increased fat oxidation during Ramadan fasting in healthy women: An adaptive mechanism for body-weight maintenance. *Am. J. Clin. Nutr.* **1995**, *62*, 302–307. [CrossRef] [PubMed]

67. Gotthardt, J.D.; Verpeut, J.L.; Yeomans, B.L.; Yang, J.A.; Yasrebi, A.; Roepke, T.A.; Bello, N.T. Intermittent Fasting Promotes Fat Loss with Lean Mass Retention, Increased Hypothalamic Norepinephrine Content, and Increased Neuropeptide Y Gene Expression in Diet-Induced Obese Male Mice. *Endocrinology* **2016**, *157*, 679–691. [CrossRef] [PubMed]

68. Zhang, L.N.; Mitchell, S.E.; Hambly, C.; Morgan, D.G.; Clapham, J.C.; Speakman, J.R. Physiological and behavioral responses to intermittent starvation in C57BL/6J mice. *Physiol. Behav.* **2012**, *105*, 376–387. [CrossRef]

69. Kalra, S.P.; Kalra, P.S. Neuropeptide Y: A physiological orexigen modulated by the feedback action of ghrelin and leptin. *Endocrine* **2003**, *22*, 49–56. [CrossRef]

70. Finch, G.M.; Day, J.E.; Razak; Welch, D.A.; Rogers, P.J. Appetite changes under free-living conditions during Ramadan fasting. *Appetite* **1998**, *31*, 159–170. [CrossRef]

71. Thivel, D.; Finlayson, G.; Miguet, M.; Pereira, B.; Duclos, M.; Boirie, Y.; Doucet, E.; Blundell, J.E.; Metz, L. Energy depletion by 24-h fast leads to compensatory appetite responses compared with matched energy depletion by exercise in healthy young males. *Br. J. Nutr.* **2018**, 1–10. [CrossRef]

72. Clayton, D.J.; Burrell, K.; Mynott, G.; Creese, M.; Skidmore, N.; Stensel, D.J.; James, L.J. Effect of 24-h severe energy restriction on appetite regulation and ad libitum energy intake in lean men and women. *Am. J. Clin. Nutr.* **2016**, *104*, 1545–1553. [CrossRef]

73. Karbowska, J.; Kochan, Z. Intermittent fasting up-regulates Fsp27/Cidec gene expression in white adipose tissue. *Nutrition* **2012**, *28*, 294–299. [CrossRef] [PubMed]

74. Malafaia, A.B.; Nassif, P.A.; Ribas, C.A.; Ariede, B.L.; Sue, K.N.; Cruz, M.A. Obesity induction with high fat sucrose in rats. *Arq. Bras. Cir. Dig.* **2013**, *26*, 17–21. [CrossRef] [PubMed]

75. Lasheen, N.N. Pancreatic functions in high salt fed female rats. *Physiol. Rep.* **2015**, *3*. [CrossRef] [PubMed]

76. Carter, S.; Clifton, P.M.; Keogh, J.B. The effects of intermittent compared to continuous energy restriction on glycaemic control in type 2 diabetes; a pragmatic pilot trial. *Diabetes Res. Clin. Pract.* **2016**, *122*, 106–112. [CrossRef] [PubMed]

77. Joslin, P.M.N.; Bell, R.K.; Swoap, S.J. Obese mice on a high-fat alternate-day fasting regimen lose weight and improve glucose tolerance. *J. Anim. Physiol. Anim. Nutr.* **2017**, *101*, 1036–1045. [CrossRef] [PubMed]

78. Costill, D.L.; Hargreaves, M. Carbohydrate nutrition and fatigue. *Sports Med.* **1992**, *13*, 86–92. [CrossRef] [PubMed]

79. Alsubheen, S.A.; Ismail, M.; Baker, A.; Blair, J.; Adebayo, A.; Kelly, L.; Chandurkar, V.; Cheema, S.; Joanisse, D.R.; Basset, F.A. The effects of diurnal Ramadan fasting on energy expenditure and substrate oxidation in healthy men. *Br. J. Nutr.* **2017**, *118*, 1023–1030. [CrossRef]

80. Moro, T.; Tinsley, G.; Bianco, A.; Marcolin, G.; Pacelli, Q.F.; Battaglia, G.; Palma, A.; Gentil, P.; Neri, M.; Paoli, A. Effects of eight weeks of time-restricted feeding (16/8) on basal metabolism, maximal strength, body composition, inflammation, and cardiovascular risk factors in resistance-trained males. *J. Transl. Med.* **2016**, *14*, 290. [CrossRef]

81. Norouzy, A.; Salehi, M.; Philippou, E.; Arabi, H.; Shiva, F.; Mehrnoosh, S.; Mohajeri, S.M.; Mohajeri, S.A.; Motaghedi Larijani, A.; Nematy, M. Effect of fasting in Ramadan on body composition and nutritional intake: A prospective study. *J. Hum. Nutr. Diet* **2013**, 97–104. [CrossRef]

82. Yarahmadi, S.; Larijani, B.; Bastanhagh, M.H.; Pajouhi, M.; Baradar Jalili, R.; Zahedi, F.; Zendehdel, K.; Akrami, S.M. Metabolic and clinical effects of Ramadan fasting in patients with type II diabetes. *J. Coll. Physicians Surg. Pak.* **2003**, *13*, 329–332.

83. Liu, H.; Javaheri, A.; Godar, R.J.; Murphy, J.; Ma, X.; Rohatgi, N.; Mahadevan, J.; Hyrc, K.; Saftig, P.; Marshall, C.; et al. Intermittent fasting preserves beta-cell mass in obesity-induced diabetes via the autophagy-lysosome pathway. *Autophagy* **2017**, *13*, 1952–1968. [CrossRef] [PubMed]

84. Park, S.; Yoo, K.M.; Hyun, J.S.; Kang, S. Intermittent fasting reduces body fat but exacerbates hepatic insulin resistance in young rats regardless of high protein and fat diets. *J. Nutr. Biochem.* **2017**, *40*, 14–22. [CrossRef] [PubMed]

85. Zhao, T.J.; Liang, G.; Li, R.L.; Xie, X.; Sleeman, M.W.; Murphy, A.J.; Valenzuela, D.M.; Yancopoulos, G.D.; Goldstein, J.L.; Brown, M.S. Ghrelin O-acyltransferase (GOAT) is essential for growth hormone-mediated survival of calorie-restricted mice. *Proc. Natl. Acad. Sci. USA* **2010**, *107*, 7467–7472. [CrossRef] [PubMed]

86. Sangiao-Alvarellos, S.; Cordido, F. Effect of ghrelin on glucose-insulin homeostasis: Therapeutic implications. *Int. J. Pept.* **2010**, *2010*. [CrossRef] [PubMed]

87. Antunes, L.A.C.; Jornada, M.N.; Elkfury, J.L.; Foletto, K.C.; Bertoluci, M.C. Fasting ghrelin but not PYY(3-36) is associated with insulin-resistance independently of body weight in Wistar rats. *Arq. Bras. Endocrinol. Metabol.* **2014**, *58*, 377–381. [CrossRef] [PubMed]

88. Barazzoni, R.; Zanetti, M.; Cattin, M.R.; Visintin, L.; Vinci, P.; Cattin, L.; Stebel, M.; Guarnieri, G. Ghrelin enhances in vivo skeletal muscle but not liver AKT signaling in rats. *Obesity* **2007**, *15*, 2614–2623. [CrossRef] [PubMed]

89. Cerqueira, F.M.; da Cunha, F.M.; Caldeira da Silva, C.C.; Chausse, B.; Romano, R.L.; Garcia, C.C.; Colepicolo, P.; Medeiros, M.H.; Kowaltowski, A.J. Long-term intermittent feeding, but not caloric restriction, leads to redox imbalance, insulin receptor nitration, and glucose intolerance. *Free Radic. Biol. Med.* **2011**, *51*, 1454–1460. [CrossRef]

90. Rosas Fernández, M.A.; Concha Vilca, C.M.; Batista, L.O.; Ramos, V.W.; Cinelli, L.P.; Tibau de Albuquerque, K. Intermittent food restriction in female rats induces SREBP high expression in hypothalamus and immediately postfasting hyperphagia. *Nutrition* **2017**, *48*, 122–126. [CrossRef]

91. Marangou, A.G.; Weber, K.M.; Boston, R.C.; Aitken, P.M.; Heggie, J.C.; Kirsner, R.L.; Best, J.D.; Alford, F.P. Metabolic consequences of prolonged hyperinsulinemia in humans. Evidence for induction of insulin insensitivity. *Diabetes* **1986**, *35*, 1383–1389. [CrossRef]

92. Del Prato, S.; Leonetti, F.; Simonson, D.C.; Sheehan, P.; Matsuda, M.; DeFronzo, R.A. Effect of sustained physiologic hyperinsulinaemia and hyperglycaemia on insulin secretion and insulin sensitivity in man. *Diabetologia* **1994**, *37*, 1025–1035. [CrossRef]

93. Koopmans, S.J.; Ohman, L.; Haywood, J.R.; Mandarino, L.J.; DeFronzo, R.A. Seven days of euglycemic hyperinsulinemia induces insulin resistance for glucose metabolism but not hypertension, elevated catecholamine levels, or increased sodium retention in conscious normal rats. *Diabetes* **1997**, *46*, 1572–1578. [CrossRef] [PubMed]

94. Bertacca, A.; Ciccarone, A.; Cecchetti, P.; Vianello, B.; Laurenza, I.; Maffei, M.; Chiellini, C.; Del Prato, S.; Benzi, L. Continually high insulin levels impair Akt phosphorylation and glucose transport in human myoblasts. *Metabolism* **2005**, *54*, 1687–1693. [CrossRef]

95. Bertacca, A.; Ciccarone, A.; Cecchetti, P.; Vianello, B.; Laurenza, I.; Del Prato, S.; Benzi, L. High insulin levels impair intracellular receptor trafficking in human cultured myoblasts. *Diabetes Res. Clin. Pract.* **2007**, *78*, 316–323. [CrossRef] [PubMed]

96. Kim, B.; McLean, L.L.; Philip, S.S.; Feldman, E.L. Hyperinsulinemia induces insulin resistance in dorsal root ganglion neurons. *Endocrinology* **2011**, *152*, 3638–3647. [CrossRef] [PubMed]

97. Catalano, K.J.; Maddux, B.A.; Szary, J.; Youngren, J.F.; Goldfine, I.D.; Schaufele, F. Insulin resistance induced by hyperinsulinemia coincides with a persistent alteration at the insulin receptor tyrosine kinase domain. *PLoS ONE* **2014**, *9*, e108693. [CrossRef] [PubMed]

98. Genuth, S.M.; Przybylski, R.J.; Rosenberg, D.M. Insulin resistance in genetically obese, hyperglycemic mice. *Endocrinology* **1971**, *88*, 1230–1238. [CrossRef]

99. Odeleye, O.E.; de Courten, M.; Pettitt, D.J.; Ravussin, E. Fasting hyperinsulinemia is a predictor of increased body weight gain and obesity in Pima Indian children. *Diabetes* **1997**, *46*, 1341–1345. [CrossRef]

100. Sigal, R.J.; El-Hashimy, M.; Martin, B.C.; Soeldner, J.S.; Krolewski, A.S.; Warram, J.H. Acute postchallenge hyperinsulinemia predicts weight gain: A prospective study. *Diabetes* **1997**, *46*, 1025–1029. [CrossRef]

101. Ishikawa, M.; Pruneda, M.L.; Adams-Huet, B.; Raskin, P. Obesity-independent hyperinsulinemia in nondiabetic first-degree relatives of individuals with type 2 diabetes. *Diabetes* **1998**, *47*, 788–792. [CrossRef]

102. Mehran, A.E.; Templeman, N.M.; Brigidi, G.S.; Lim, G.E.; Chu, K.Y.; Hu, X.; Botezelli, J.D.; Asadi, A.; Hoffman, B.G.; Kieffer, T.J.; et al. Hyperinsulinemia drives diet-induced obesity independently of brain insulin production. *Cell Metab.* **2012**, *16*, 723–737. [CrossRef]

103. Templeman, N.M.; Clee, S.M.; Johnson, J.D. Suppression of hyperinsulinaemia in growing female mice provides long-term protection against obesity. *Diabetologia* **2015**, *58*, 2392–2402. [CrossRef] [PubMed]

104. Reaven, G.M.; Hoffman, B.B. A role for insulin in the aetiology and course of hypertension? *Lancet* **1987**, *2*, 435–437. [CrossRef]

105. Takatori, S.; Zamami, Y.; Mio, M.; Kurosaki, Y.; Kawasaki, H. Chronic hyperinsulinemia enhances adrenergic vasoconstriction and decreases calcitonin gene-related peptide-containing nerve-mediated vasodilation in pithed rats. *Hypertens. Res.* **2006**, *29*, 361–368. [CrossRef] [PubMed]

106. Zamami, Y.; Takatori, S.; Hobara, N.; Yabumae, N.; Tangsucharit, P.; Jin, X.; Hashikawa, N.; Kitamura, Y.; Sasaki, K.; Kawasaki, H. Hyperinsulinemia induces hypertension associated with neurogenic vascular dysfunction resulting from abnormal perivascular innervations in rat mesenteric resistance arteries. *Hypertens. Res.* **2011**, *34*, 1190–1196. [CrossRef] [PubMed]

107. Nakamura, A.; Monma, Y.; Kajitani, S.; Noda, K.; Nakajima, S.; Endo, H.; Takahashi, T.; Nozaki, E. Effect of glycemic state on postprandial hyperlipidemia and hyperinsulinemia in patients with coronary artery disease. *Heart Vessels* **2015**. [CrossRef] [PubMed]

108. Zidi, W.; Allal-Elasmi, M.; Zayani, Y.; Zaroui, A.; Guizani, I.; Feki, M.; Mourali, M.S.; Mechmeche, R.; Kaabachi, N. Metabolic Syndrome, Independent Predictor for Coronary Artery Disease. *Clin. Lab.* **2015**, *61*, 1545–1552. [CrossRef] [PubMed]

109. Madonna, R.; De Caterina, R. Prolonged exposure to high insulin impairs the endothelial PI3-kinase/Akt/nitric oxide signalling. *Thromb. Haemost.* **2009**, *101*, 345–350. [CrossRef]

110. Steneberg, P.; Rubins, N.; Bartoov-Shifman, R.; Walker, M.D.; Edlund, H. The FFA receptor GPR40 links hyperinsulinemia, hepatic steatosis, and impaired glucose homeostasis in mouse. *Cell Metab.* **2005**, *1*, 245–258. [CrossRef]

111. Bril, F.; Lomonaco, R.; Orsak, B.; Ortiz-Lopez, C.; Webb, A.; Tio, F.; Hecht, J.; Cusi, K. Relationship between disease severity, hyperinsulinemia, and impaired insulin clearance in patients with nonalcoholic steatohepatitis. *Hepatology* **2014**, *59*, 2178–2187. [CrossRef]

112. Westermark, P.; Wilander, E. The influence of amyloid deposits on the islet volume in maturity onset diabetes mellitus. *Diabetologia* **1978**, *15*, 417–421. [CrossRef]

113. Bonner-Weir, S. Regulation of pancreatic beta-cell mass in vivo. *Recent Prog. Horm. Res.* **1994**, *49*, 91–104. [PubMed]

114. Butler, A.E.; Janson, J.; Bonner-Weir, S.; Ritzel, R.; Rizza, R.A.; Butler, P.C. Beta-cell deficit and increased beta-cell apoptosis in humans with type 2 diabetes. *Diabetes* **2003**, *52*, 102–110. [CrossRef] [PubMed]

115. Kasuga, M. Insulin resistance and pancreatic beta cell failure. *J. Clin. Investig.* **2006**, *116*, 1756–1760. [CrossRef] [PubMed]

116. Király, M.A.; Bates, H.E.; Kaniuk, N.A.; Yue, J.T.; Brumell, J.H.; Matthews, S.G.; Riddell, M.C.; Vranic, M. Swim training prevents hyperglycemia in ZDF rats: Mechanisms involved in the partial maintenance of beta-cell function. *Am. J. Physiol. Endocrinol. Metab.* **2008**, *294*, E271–E283. [CrossRef] [PubMed]

117. Karaca, M.; Magnan, C.; Kargar, C. Functional pancreatic beta-cell mass: Involvement in type 2 diabetes and therapeutic intervention. *Diabetes Metab.* **2009**, *35*, 77–84. [CrossRef] [PubMed]

118. Cerf, M.E. Beta cell dysfunction and insulin resistance. *Front. Endocrinol.* **2013**, *4*, 37. [CrossRef]

119. Sies, H. Oxidative stress: From basic research to clinical application. *Am. J. Med.* **1991**, *91*, 31S–38S. [CrossRef]
120. Halliwell, B.; Cross, C.E. Oxygen-derived species: Their relation to human disease and environmental stress. *Environ. Health Perspect.* **1994**, *102*, 5–12.
121. Vincent, A.M.; Brownlee, M.; Russell, J.W. Oxidative stress and programmed cell death in diabetic neuropathy. *Ann. N. Y. Acad. Sci.* **2002**, *959*, 368–383. [CrossRef]
122. Loh, K.P.; Huang, S.H.; De Silva, R.; Tan, B.K.; Zhu, Y.Z. Oxidative stress: Apoptosis in neuronal injury. *Curr. Alzheimer Res.* **2006**, *3*, 327–337. [CrossRef]
123. Ryter, S.W.; Kim, H.P.; Hoetzel, A.; Park, J.W.; Nakahira, K.; Wang, X.; Choi, A.M. Mechanisms of cell death in oxidative stress. *Antioxid. Redox Signal.* **2007**, *9*, 49–89. [CrossRef] [PubMed]
124. Choi, K.; Kim, J.; Kim, G.W.; Choi, C. Oxidative stress-induced necrotic cell death via mitochondira-dependent burst of reactive oxygen species. *Curr. Neurovasc. Res.* **2009**, *6*, 213–222. [CrossRef] [PubMed]
125. Navarro-Yepes, J.; Burns, M.; Anandhan, A.; Khalimonchuk, O.; del Razo, L.M.; Quintanilla-Vega, B.; Pappa, A.; Panayiotidis, M.I.; Franco, R. Oxidative stress, redox signaling, and autophagy: Cell death versus survival. *Antioxid. Redox Signal.* **2014**, *21*, 66–85. [CrossRef] [PubMed]
126. Filomeni, G.; De Zio, D.; Cecconi, F. Oxidative stress and autophagy: The clash between damage and metabolic needs. *Cell Death Differ.* **2015**, *22*, 377–388. [CrossRef] [PubMed]
127. Zraika, S.; Hull, R.L.; Udayasankar, J.; Aston-Mourney, K.; Subramanian, S.L.; Kisilevsky, R.; Szarek, W.A.; Kahn, S.E. Oxidative stress is induced by islet amyloid formation and time-dependently mediates amyloid-induced beta cell apoptosis. *Diabetologia* **2009**, *52*, 626–635. [CrossRef]
128. Lee, B.W.; Chae, H.Y.; Kwon, S.J.; Park, S.Y.; Ihm, J.; Ihm, S.H. RAGE ligands induce apoptotic cell death of pancreatic β-cells via oxidative stress. *Int. J. Mol. Med.* **2010**, *26*, 813–818.
129. Lu, T.H.; Su, C.C.; Chen, Y.W.; Yang, C.Y.; Wu, C.C.; Hung, D.Z.; Chen, C.H.; Cheng, P.W.; Liu, S.H.; Huang, C.F. Arsenic induces pancreatic β-cell apoptosis via the oxidative stress-regulated mitochondria-dependent and endoplasmic reticulum stress-triggered signaling pathways. *Toxicol. Lett.* **2011**, *201*, 15–26. [CrossRef]
130. Chen, K.L.; Liu, S.H.; Su, C.C.; Yen, C.C.; Yang, C.Y.; Lee, K.I.; Tang, F.C.; Chen, Y.W.; Lu, T.H.; Su, Y.C.; et al. Mercuric compounds induce pancreatic islets dysfunction and apoptosis in vivo. *Int. J. Mol. Sci.* **2012**, *13*, 12349–12366. [CrossRef]
131. Du, S.C.; Ge, Q.M.; Lin, N.; Dong, Y.; Su, Q. ROS-mediated lipopolysaccharide-induced apoptosis in INS-1 cells by modulation of Bcl-2 and Bax. *Cell Mol. Biol.* **2012**, *58*, OL1654-9.
132. Chang, K.C.; Hsu, C.C.; Liu, S.H.; Su, C.C.; Yen, C.C.; Lee, M.J.; Chen, K.L.; Ho, T.J.; Hung, D.Z.; Wu, C.C.; et al. Cadmium induces apoptosis in pancreatic β-cells through a mitochondria-dependent pathway: The role of oxidative stress-mediated c-Jun N-terminal kinase activation. *PLoS ONE* **2013**, *8*, e54374. [CrossRef]
133. Tachibana, K.; Sakurai, K.; Yokoh, H.; Ishibashi, T.; Ishikawa, K.; Shirasawa, T.; Yokote, K. Mutation in insulin receptor attenuates oxidative stress and apoptosis in pancreatic beta-cells induced by nutrition excess: Reduced insulin signaling and ROS. *Horm. Metab. Res.* **2015**, *47*, 176–183. [CrossRef] [PubMed]
134. Johnson, J.D.; Bernal-Mizrachi, E.; Alejandro, E.U.; Han, Z.; Kalynyak, T.B.; Li, H.; Beith, J.L.; Gross, J.; Warnock, G.L.; Townsend, R.R.; et al. Insulin protects islets from apoptosis via Pdx1 and specific changes in the human islet proteome. *Proc. Natl. Acad. Sci. USA* **2006**, *103*, 19575–19580. [CrossRef] [PubMed]
135. Beith, J.L.; Alejandro, E.U.; Johnson, J.D. Insulin stimulates primary beta-cell proliferation via Raf-1 kinase. *Endocrinology* **2008**, *149*, 2251–2260. [CrossRef] [PubMed]
136. Carlson, O.; Martin, B.; Stote, K.S.; Golden, E.; Maudsley, S.; Najjar, S.S.; Ferrucci, L.; Ingram, D.K.; Longo, D.L.; Rumpler, W.V.; et al. Impact of reduced meal frequency without caloric restriction on glucose regulation in healthy, normal-weight middle-aged men and women. *Metabolism* **2007**, *56*, 1729–1734. [CrossRef]
137. Oishi, K.; Hashimoto, C. Short-term time-restricted feeding during the resting phase is sufficient to induce leptin resistance that contributes to development of obesity and metabolic disorders in mice. *Chronobiol. Int.* **2018**, 1–19. [CrossRef]
138. Cote, I.; Toklu, H.Z.; Green, S.M.; Morgan, D.; Carter, C.S.; Tumer, N.; Scarpace, P.J. Limiting feeding to the active phase reduces blood pressure without the necessity of caloric reduction or fat mass loss. *Am. J. Physiol. Regul. Integr. Comp. Physiol.* **2018**. [CrossRef]
139. Woodie, L.N.; Luo, Y.; Wayne, M.J.; Graff, E.C.; Ahmed, B.; O'Neill, A.M.; Greene, M.W. Restricted feeding for 9h in the active period partially abrogates the detrimental metabolic effects of a Western diet with liquid sugar consumption in mice. *Metabolism* **2018**, *82*, 1–13. [CrossRef]

140. Bi, H.; Gan, Y.; Yang, C.; Chen, Y.; Tong, X.; Lu, Z. Breakfast skipping and the risk of type 2 diabetes: A meta-analysis of observational studies. *Public Health Nutr.* **2015**, *18*, 3013–3019. [CrossRef]

141. Reutrakul, S.; Hood, M.M.; Crowley, S.J.; Morgan, M.K.; Teodori, M.; Knutson, K.L. The relationship between breakfast skipping, chronotype, and glycemic control in type 2 diabetes. *Chronobiol. Int.* **2014**, *31*, 64–71. [CrossRef]

142. Azami, Y.; Funakoshi, M.; Matsumoto, H.; Ikota, A.; Ito, K.; Okimoto, H.; Shimizu, N.; Tsujimura, F.; Fukuda, H.; Miyagi, C.; et al. Long working hours and skipping breakfast concomitant with late evening meals are associated with suboptimal glycemic control among young male Japanese patients with type 2 diabetes. *J. Diabetes Investig.* **2018**. [CrossRef]

143. Watanabe, Y.; Saito, I.; Henmi, I.; Yoshimura, K.; Maruyama, K.; Yamauchi, K.; Matsuo, T.; Kato, T.; Tanigawa, T.; Kishida, T.; et al. Skipping Breakfast is Correlated with Obesity. *J. Rural Med.* **2014**, *9*, 51–58. [CrossRef] [PubMed]

144. Sakurai, M.; Yoshita, K.; Nakamura, K.; Miura, K.; Takamura, T.; Nagasawa, S.Y.; Morikawa, Y.; Kido, T.; Naruse, Y.; Nogawa, K.; et al. Skipping breakfast and 5-year changes in body mass index and waist circumference in Japanese men and women. *Obes. Sci. Pract.* **2017**, *3*, 162–170. [CrossRef] [PubMed]

145. Uemura, M.; Yatsuya, H.; Hilawe, E.H.; Li, Y.; Wang, C.; Chiang, C.; Otsuka, R.; Toyoshima, H.; Tamakoshi, K.; Aoyama, A. Breakfast Skipping is Positively Associated with Incidence of Type 2 Diabetes Mellitus: Evidence From the Aichi Workers' Cohort Study. *J. Epidemiol.* **2015**, *25*, 351–358. [CrossRef] [PubMed]

146. Møller, N.; Jørgensen, J.O. Effects of growth hormone on glucose, lipid, and protein metabolism in human subjects. *Endocr. Rev.* **2009**, *30*, 152–177. [CrossRef] [PubMed]

© 2020 by the authors. Licensee MDPI, Basel, Switzerland. This article is an open access article distributed under the terms and conditions of the Creative Commons Attribution (CC BY) license (http://creativecommons.org/licenses/by/4.0/).

MDPI

St. Alban-Anlage 66

4052 Basel

Switzerland

Tel. +41 61 683 77 34

Fax +41 61 302 89 18

www.mdpi.com

Nutrients Editorial Office

E-mail: nutrients@mdpi.com

www.mdpi.com/journal/nutrients

www.ingramcontent.com/pod-product-compliance
Lightning Source LLC
LaVergne TN
LVHW070921170726
843515LV00005B/840